THE DAWN OF COGNITIVE SCIENCE

SYNTHESE LIBRARY

STUDIES IN EPISTEMOLOGY,

LOGIC, METHODOLOGY, AND PHILOSOPHY OF SCIENCE

VOLUME 295

THE DAWN OF COGNITIVE SCIENCE

Early European Contributors

Edited by

LILIANA ALBERTAZZI
University of Trento, Italy and
Central European Foundation,
Bolzano, Italy

KLUWER ACADEMIC PUBLISHERS
DORDRECHT / BOSTON / LONDON

A C.I.P. Catalogue record for this book is available from the Library of Congress.

ISBN 0-7923-6799-5

Published by Kluwer Academic Publishers,
P.O. Box 17, 3300 AA Dordrecht, The Netherlands.

Sold and distributed in North, Central and South America
by Kluwer Academic Publishers,
101 Philip Drive, Norwell, MA 02061, U.S.A.

In all other countries, sold and distributed
by Kluwer Academic Publishers,
P.O. Box 322, 3300 AH Dordrecht, The Netherlands.

Printed on acid-free paper

Printed in the Netherlands.

TABLE OF CONTENTS

LILIANA ALBERTAZZI / Introduction: Back to the Origins 1

LILIANA ALBERTAZZI / The Primitives of Presentation. Wholes, Parts and Psychophysics 29

MARTIN KUSCH / The Politics of Thought: A Social History of the Debate between Wundt and the Würzburg School 61

RICCARDO LUCCIO / Representation in Psychophysics 89

ROBIN D. ROLLINGER / Lotze on the Sensory Representation of Space 103

JAN SEBESTIK / Ernst Mach's Evolutionary Theory of Representation 123

GIANNI ZANARINI / Hermann von Helmholtz and Ernst Mach on Musical Consonance 135

SALVO D'AGOSTINO / The *Bild* Conception of Physical Theories from Helmholtz to Hertz 151

KARL SCHUHMANN / Representation in Early Husserl 167

JENS CAVALLIN / Contents, Psycho-physical Products and Representations 185

ELISABETH R. VALENTINE / G.F. Stout's Philosophical Psychology 209

THEO HERRMANN and STEFFI KATZ / Otto Selz and the Würzburg School 225

SERENA CATTARUZZA / The Intrumental Model of Language in Karl Bühler 237

ALFRED ZIMMER / The concept of Perceptual 'Field' and the Revolution in Cognition Caused by Köhler's *physische Gestalten* 251

DAVID WOODRUFF SMITH / Consciousness and Actuality in Whiteheadean Ontology 269

WOLFANG WILDGEN / Kurt Lewin and the Rise of the 'Cognitive Sciences' in Germany: Cassirer, Bühler, Reichenbach 299

RUGGERO PIERANTONI / Children's Drawings as Sensible Probes into the Realm of Representations 333

BACK TO THE ORIGINS

Start again from zero. Picasso

1. EUROPEAN TRADITION AND THE SCIENCE OF 'MIND'

In recent decades European thought has been subjected to at least two forms of negative or restrictive assessment of its tradition, or of a part of it. This criticism has derived largely from a fundamental misunderstanding, and often from ignorance of the sources. The misunderstanding concerns the purely analytic invention of a 'Continental' philosophy (usually made to coincide with modern 'German' philosophy represented by Kant, idealism, Heidegger and derivations of his thought) which has been accused of an *irrationalism* and an *obscurity*, although these charges have not always well argued with regard to individual authors and themes. This invention, which has often displayed markedly ideological features, has counterposed 'Continental' philosophy with a prevalently Anglo-American school of thought which, since the specific interpretation given to Frege by Carnap and Russell, has developed a strongly logical-linguistic body of thought in the twentieth century. Dummett's thesis that analytic thought has a partly 'Austrian' origin in not only Frege but also Meinong is no exception, given the interpretation and analytic developments of Meinong's thought.[1]

The analytic trend in its various forms has characterised numerous disciplinary areas, thereby introducing the need for a *formal* (or better *formalised*) *theory of reference*. Both a consequence and a distinctive feature of this trend has been the removal of *cognitive* considerations (in particular those to do with *perceptive* processes) from every field of inquiry, not only in logic, epistemology and ontology but also, and paradoxically, in theory of the mind, where the only the merely inferential approach has long prevailed. The analytic paradigm has therefore long been paramount, at least in philosophy, and only recently have problems raised by research in artificial intelligence and the cognitive sciences – in their investigations of analogue reasoning, common sense, the understanding of form, and natural language recognition – redirected attention to the importance of a 'model of the mind' which is not restricted to logical-inferential analysis alone, and to the need for a sophisticated *theory of perception*.[2]

L. Albertazzi (ed.), The Dawn of Cognitive Science. Early European Contributors, 1–27.

A second criticism of the Western tradition has recently been raised by certain sectors of artificial intelligence research, which have emphasised its eminently rationalist, Cartesian and mentalist character (the only one to have generally aroused a certain interest among analytic thinkers). Moreover, some elements of 'mentalist' thought drawn from Western rationalism – especially from Descartes (and which we may somewhat brutally summarise as the question of the *homunculus* in the mind)[3] – have demonstrated their usefulness in some sectors of computational analysis: for example in investigation of the cognitive processes involved in problem-solving,[4] in the development of 'models of mind,'[5] in the parallel development of a modular conception of the mind,[6] and in the logico-formal analysis of natural language[7] (see also Wildgen, this volume). The few references made to Gestalt psychology by some of these authors, Newell and Simon for example, mainly concern the mental processes involved in problem solving or memory retrieval, inasmuch as they are based on translations of Gestaltist texts on those themes published in English in the 1940's and 1950's.[8]

However, these 'mentalist' elements – which bore fruit in the first and second generations of artificial intelligence – have rendered the Western tradition (unlike its Eastern counterpart) entirely extraneous to, and unusable by, any further *ecological* development of artificial intelligence.[9] The Western tradition has little to offer to those sectors of research, from robotics to analysis of vision, from analysis of common sense[10] to the cognitive semantics of natural language,[11] which assume an *embodied mind*[12] and an *enactive* viewpoint in perception[13] antecedent to every form of abstraction and idealisation of individually analysed cognitive processes.

But is the European tradition of modern thought really as obscure and/or mentalistic as its critics claim?

2. Central-European tradition

For a scholar of 'Continental' extraction, the question seems to have been framed in highly naïve and substantially erroneous terms. S/he is aware of the enormous richness and theoretical complexity of the European tradition, and specifically as regards certain theories (philosophical and epistemological) which arose between the end of the nineteenth century and the beginning of the twentieth before the so-called 'linguistic turn,' and which in certain respects were connected with it. I refer in particular to Brentanism and its theoretical and experimental development in Gestalt psychology and experimental phenomenology.[14]

These theories, though developed in a variety of disciplinary areas, have a number of features in common. For example, they provide a *phenomenological description* of the events or facts under observation (the so called 'immediately given') which makes them immune to the danger of psychophysical and/or neurophysiological reductionism, and they display a markedly *dynamic approach* – whether they investigate the genesis of objects from the cognitive point of view, or whether they concern themselves with physical and perceptive phenomena connected with movement – as well as a pronounced sensitivity towards problems of *interaction* and *dependence*.

In many respects, these theories conducted pioneering analysis of problems that are currently the focus of debate in the cognitive sciences and artificial intelligence: from the problem of reference to that of representation, to problems of 'model and categorial classification', of ontology and the cognitive analysis of natural language. In several cases, moreover, these theories seem to display clearer awareness of the complexity of problems and of the interrelations among different areas of inquiry. And they also seem to offer elements for the solution of problems currently under investigation. As a consequence, when analysing the contemporary literature, the 'Continental' scholar often has a curious feeling of *déjà vu*: that of 'having already met the question,' as if in many cases contemporary research is rediscovering phenomena and issues already widely known and thoroughly discussed at the beginning of the twentieth century (see Schuhmann, this volume). The examples abound and they are just as frequent in the literature on the auditory stream (temporal apprehension) as they are in the literature on visual perception (point of view, three-dimensional construction of objects), on intentionality, on imagery, on categorial classification (prototypicity and typicity) or on semantics (conceptual spaces).

A typical example of the situation that arose between 1870 and 1930 – although traces of it are obviously apparent in the whole of European thought – is provided by a group of theories developed within a particular *cultural* setting and which we may broadly term *Central-European*. These theories are generally unknown to Anglo-American scholars (sometimes to European ones as well) for various reasons, most notably the fate of numerous European intellectuals during and after the Second World War,[15] the lack of translations from German and, once again, the decades-long dominance of an alternative scientific paradigm.

At first sight, the distinctive feature of these theories is some sort of "complementarity of scientific spirit and metaphysical tension, of empirical analysis and research into the absolute," or to cite a character in Zagajewski's novels, "people whose lives are determined by metaphysics."[16] This feature is displayed not only by pre-eminently 'philosophical' and, so to speak, 'Austrian' theories, like those of Brentano and Meinong, who pursued a programme of experimental metaphysics,[17] but also and particularly by Polish thought and its ramifications in Britain:[18] in fact, a non-independent piece of *Mitteleuropa* was transplanted to Britain via Lotze, and then invaded Munich (through Stumpf and subsequently the realist trend in phenomenology) and even the stronghold of German thought represented by Berlin (through Stumpf, Koffka and Lewin), thence to return to Leipzig and Göttingen (through Lotze) and Frankfurt (through Wertheimer, Koffka and Köhler).[19] Through a series of shifts and influences, Florence – the most Central-European city of the Mediterranean region – also joined this set of concentric circles.[20]

Moreover, the majority of scholars in the aforementioned universities worked on the boundaries among various disciplines, addressing common themes from different points of view (that of consonance or melody, for example) and opting for the conception of a *hierarchy of intellectual constructions* in conceptualisation.

This type of inquiry, which usually drew on laboratory experiments for its development, also in philosophy, displayed a number of *variations* in its base assumptions. By way of simplification, it was more 'cognitivist' among the scholars of the Graz school of Meinong and more 'realist' in various branches of Husserlian phenomenology; and the same difference can be discerned in experimental psychology within *Gestaltpsychologie* and its various ramifications (see Albertazzi, this volume).[21] In no sense, however, was it either a psychology *à la* Dilthey (that is, a *Geisteswissenschaftspsychologie* opposed the experimental psychology of Ebbinghaus and rooted in German idealism) or a neo-Kantian philosophy *à la* Rickert or Windelband.[22] A history of chairs of philosophy and psychology in German universities between the end of the nineteenth century and the beginning of the twentieth exemplifies the so-called 'thought psychology controversy' among very different schools of thought (from the point of view of the sociology of science see Kusch, this volume).[23]

These authors, I would stress again, addressed issues still central to current debate in the cognitive sciences: from the perception of form to the nature of images and their modelling in different formats of representation, as well as the problem of measuring psychological experiences. As I have already pointed out, the problem at the core of the whole of contemporary research is that of *psychophysics*, or the so called mind-body problem. The hypothesis which formed the basis for the Weber-Fechner law – the existence of an interdependence between an increment in the physical stimulus and variation in the corresponding sensations, verified experimentally – was one of the turning-points in the emancipation of psychology of science. Fechner, moreover, was well aware that the law failed to explain certain fundamental distinctions, like the difference between 'perception' and 'perception of difference,' between 'psychophysical activity,' 'sensation' and 'conscious psychic phenomenon,' or between 'quantitative difference' and 'qualitative difference.'[24]

Furthermore, the fact that Heinrich Weber's paradigm of *equating sensory perception* with the *functioning of measurement devices in physics* has come to predominate is one of the reasons for the general invisibility of the Central-European intellectual milieu in modern science (on the problem of representation in psychophysics, see in particular Luccio, this volume). And yet it was a milieu that produced a series of scientific results of great importance: for example, to cite just one area of research, acoustic and visual perception (see Zanarini, on Helmholtz and Mach, this volume).

These various theories – which concerned not only philosophy but also mathematics, psychology, physics, logic, aesthetics, ethics, and even literature – display a set of distinctive features, most notably their already-mentioned *phenomenological* and *descriptive* approach, also in experimental practice, and the use of an *exact* method in metaphysics similar to that of the natural sciences. There is, in fact, a close connection of style and topic among theories developed by the philosophers, psychologists and physicists of the time (on this subject in particular see D'Agostino, Schuhmann, and Zanarini, this volume).

More generally, all these theories conduct analysis of the *foundational concepts* or *primitives* of various disciplines. Indeed, whether their focus is the concept of *number*, *local signs*, *intentional reference*, *existential judgements*, *Bild* (a Kantian concept which emphasises its character as a mental construct by which reference is made to empirical contents), *direct factors*, *objectual traits*, *particulars*, *forces* and *Gestalten* in the perceptual field, the general pattern that emerges is that of a group of scientists whose work centred on the concept of *representation* and who sought to draw a *boundary* between the internal and external psychophysics inherited from Fechner and mediated by the philosophy of Lotze.[25] More generally, these theories shared a monistic assumption,[26] given that independently of their specific fields of inquiry, they all assumed the *inner nature of representation* (see Albertazzi, this volume). The basis for this assumption was Johannes Müller's theory of specific energies, which for different reasons and in different ways exerted just as much influence on physicists (Helmholtz, for example) as on philosophers (Lotze, for instance, or Brentano via Trendelenburg).

A *parallel* reading of these texts and the cognitive sciences literature has convinced me that resumption of some of the theories developed – in the exact sciences as well as in philosophy – at the beginning of this century might be useful not only in philosophy but also in various areas of contemporary cognitive science. I cite as examples the theory of vision and the temporal structure of spatial representation; the analysis of the structure and formation of mental images and experimental studies of memory; the ecology of perception, phenomenologically-based structuralism and naïve physics; and finally cognitive semantics as a natural approach to semantics on a representational basis. The ideation of this Reading derives from this conviction. Obviously, it cannot offer a detailed survey of all the works and authors mentioned, and it must concentrate on the most important of them in the various fields covered, paying particular attention to the topic of *representation*. Nevertheless, the volume intends to provide first insights into certain theoretical aspects of the contemporary development of the cognitive sciences. As a collateral effect, the book seeks to eliminate certain prejudices that, in the twentieth century, long surrounded Kantism, Herbartism and even some aspects of psychologism, exemplifying their presence and fecundity (especially as regards the concept of presentation, *Vorstellung*) in both the works considered and in contemporary debate.[27]

3. The non-independent pieces of Central Europe

An idea of the complexity of the frame of reference is provided by the fact that between 1870 and 1930 some of the most brilliant, though *less-known*, theories produced by European thinkers were produced. They were preceded a few decades earlier by two works, by Bolzano and by Lotze, published respectively in 1835 and 1856, which acted as a sort of prelude. Beginning with Brentano, the impact of these two texts was varyingly reflected in the theories in question and traversed the boundaries of Europe to influence the work of Bradley, Stout, Moore, Russell, James and Santayana, amongst others.[28] Echoes of the theory, although at the moment no close dependence can be shown, are also to be found in Whitehead's ontology of processes (as Woodruff Smith, this volume, shows).

Lotze was widely read in Great Britain, where his thought was generally treated in anti-idealist terms (on Lotze see Rollinger, this volume). Evidence of this interest is provided by the works of Ward, Hicks, McKenzie, Stout, Moore, Caird and especially Green, who was much taken with Lotze, drawing on both his psychophysical and pantheistic theories mediated by Fischer.[29] The majority of these writings were published in the issues of the *Proceedings of the Aristotelian Society* and *Mind* during those years. Paradoxically, Lotze's ideas were also incorporated into British idealism, as developed by Bradley in particular, and traces of his thought are apparent in the early writings of Russell, who was greatly influenced by Bradley's *Logic*.

Lotze also had an influence on psychological inquiry. Indeed, elements of his theories are to be found in the Külpe school (Mayer, Orth, Ach), in the Leipzig school (Cornelius, Stern) and in the Graz school (Meinong, Ehrenfels, Witasek, Höfler and Benussi), thereby linking with Brentanism.

As for Brentano, the influence of his thought was enormous, and, through his pupils it gave rise to phenomenology, Prague lingustics, Meinong's theory of objects, and the Lvov-Warsaw logical-philosophical school. Through Meinong it was also associated in various ways with the Vienna Circle (his name appears in the *Manifesto* of 1929), with Cambridge philosophy (Moore and Russell in particular), with Alexander, with American realism; and its influence is still apparent in contemporary analytic philosophy (Chisholm).[30]

However, the influence of Brentano's thought extended beyond the bounds of philosophy to play an important role in the psychology of his period and the years that followed. Highly regarded by James, and one of the first critics of Fechnerian psychophysics on the concept of *order* and of the *qualitative* trait of phenomenic experience, Brentano engendered several currents of psychological thought, outstanding among which were the schools of Graz and Berlin (he was therefore one of the origins of *Gestaltpsychologie*, see Albertazzi, and Zanarini, this volume),[31] while, after various derivations, traces of his ideas can be detected in the Florence school (De Sarlo, Bonaventura), in the Würzburg school (in particular Otto Selz, on whom, in particular, see Hermann and Katz, this volume),[32] in the Vienna school (Bühler, on whom in particular see Cattaruzza, and Wildgen, this volume, and Brunswik), in the Leipzig school of the *Ganzeitspsychologie* (Krueger, Sander),[33] the Rostock school (Jaensch), the Marburg school (Katz), the Hamburg school (Stern, Werner, Muchow), and subsequently in the schools of Louvain (Michotte, Fraisse), Frankfurt (Metzger, Rausch) and New York (Ash, Wallach, Arhneim). The tradition of his thought still continues in Italy and the schools of Padua (from Benussi, Musatti, Metelli, to Zanforlin, Vicario and Massironi) Trieste (Kanizsa, Bozzi, Luccio, Gerbino), and in Germany, at Bremen (Stadler, Kruse) and Regensburg (Zimmer).

In their turn, all Brentano's pupils were influenced by Lotze. In the course of time awareness of the enormous influence of Lotze and Brentano at the beginning of the century has faded.[34] Consequently, in order to fix the frame of reference in terms of chronology and content, a table can be constructed of some of the works that at the turn of the century were most representative of this twofold line of thought.

As a point of departure one may take Franz Brentano's *Psychology from an Empirical Standpoint*, for its original recasting of the problem of *reference on an intentional basis*, and then Twardowski's work on the *object* and *content* of representations, a turning point in Brentanism, and which was highly influential on British philosophers as well (on Twardowski see Cavallin, this volume).[35] Another seminal text was Stout's *Analytic Philosophy*, which anticipated some ideas of Gestalt psychology (documented also by his correspondence with Koffka) and introduced descriptive psychology to Great

Britain, although his volume of 1931, *Mind and Matter*, took a further step towards a realist theory (these last aspects, in particular, are emphasised by Valentine, this volume).[36] One should also consider Lipps' *Raumaesthetik*, which expounded *intuitive mechanics*, the experimental analysis of *empathy*, and an excellent piece of theory in support, for example, of current cognitive semantics;[37] Stumpf's *Tonpsychologie*, especially as regards its treatment of the phenomenon of *fusion*, and his theory of *wholes*, which in turn influenced Twardowski and Stout (on some aspects of the question see Albertazzi, and Zanarini, this volume). There followed a seminal discussion of von Ehrenfels' article on figural qualities, as a preliminary theory of the *emergence of qualitative forms* from perceptual continua, which constituted a sort of preview of modern phenophysics;[38] Husserl's *Logical Investigations*, in particular the third and fifth, for the mereological theory of the *whole of consciousness* – a text with huge influence on his contemporaries; Meinong's *Untersuchungen zur Gegenstandstheorie und Psychologie* and its analysis of the relationship between the *theory of objects* and *psychological foundation*; Witasek's book on the *forms of completion* of the act of presentation – a text which influenced the Graz school's psychological research in its entirety; Wertheimer's work on the *temporal structure* of phenomenal configurations; and Benussi's *experimental analysis of consciousness*, which provides a sort of 'parallel' to Husserl's investigations.[39]

Nor should we overlook Katz and his study of the *modes of appearance* of colours, as well as his book of 1925 on the "inexhaustible richness of the touchable world," nor Köhler for his courageous *isomorphist manifesto* (on Köhler see Zimmer, this volume), Koffka on the *different types* of presentation in defence of the Würzburg school, Lewin, who resumed Lipps' endeavour to apply *topological* concepts to *psychological forces*, and finally Brunswik for his analysis of *phenomenal objects*.

I note in passing that not only Husserl's *Logical Investigations* but also the works of Köhler, Koffka and Brunswik were either directly dedicated to Stumpf or acknowledged the influence of his theory of *functions*. In fact, Wertheimer, Koffka, Köhler, von Allesch, Lewin and Gelb were among Stumpf's close collaborators. Brunswik also acknowledges his debt to Brentano, whose position is *constructivist in flavour*. One simultaneously finds partly critical statements concerning the actualism of the Berlin school, in favour of a *Denkpsychologie* (Psychology of thought) which emphasises the role of the *functions* or processes of thought, a feature which became even more evident in the *Ganzheitspsychologie* (Psychology of totality) of Leipzig (on the political aspects of the controversies among the schools of psychology of the time see again the contribution by Kusch, this volume).

In my opinion, Bühler's *Sprachtheorie* of 1934 is a transitional text from the intentionalistic to the logical approach, to the concept of *representation* (on Bühler see Cattaruzza, and Wildgen, this volume).[40] Bühler's work, in fact, was a theoretical shift, as also evidenced by his 1913 book, *Die Gestaltwahrnehmungen*, and which, in other respects, is also recognisable in Köhler's *On Physical Forms in Equilibrium and Stationary State*, published in 1920 (on this latter see again Zimmer, this volume).[41]

The change of perspective was announced in further studies by Schlick, Carnap and then Neurath, among others, which appeared on *Erkenntnis* in the 1930's and, for all their differences, sowed the seeds of what became a general cleavage. In fact, the previous interweaving of metaphysics and psychology characteristic of the decades between 1870 and 1930, the boom years of the formal revolution, was considered 'off limits.'[42] As for Reichenbach (on whom see Wildgen's essay) and Felix Kaufmann, they sat on the fence, so to speak: for example, the Berlin *Society for Scientific Philosophy* had several contacts with the Stumpfians, while Kaufmann's work reflects the influence of Husserlian phenomenology.[43] One of the last works produced in the previous theoretical climate was Boring's *History of Psychology*, published in 1929. After 1930, the scene was dominated by analytic philosophy and by behaviourism, which in many respects are two sides of the same coin.

I am entirely aware that my claim for the relevance of Central-European theories to current developments in science – both from a theoretical (broadly Brentanist) and experimental (Gestaltist) point of view – may be attributed to a 'Continental' bias. And yet, when reviewing the history of cognitivism, Gardner himself asks whether we have truly moved forward from Gestalt psychology and the Würzburg school, or whether in fact we are merely rediscovering what they already knew.[44] Support for my thesis is provided by Murray, for example, who as a historian of psychology has found several reasons for emphasising the importance of Gestalt analysis, both as an anticipation and as a reference for the development of the contemporary cognitive sciences (on this see Albertazzi, Wildgen, and Zimmer, this volume).[45] But what is it that make these theories so interesting to contemporary science? What are the old and new issues on the agenda?

4. New and Old Topics

To illustrate my thesis, I shall consider a series of issues characteristic of both contemporary debate and the analyses of those thinkers whom we may effectively describe as the early European contributors to modern cognitive science.

As my own contribution to this book shows, one of the starting-points of mainstream contemporary cognitive science is the assumption that knowledge essentially consists of the manipulation of *inner representations*, variously defined as neurophysiological states, mental images or symbolic codifications. The main question is this: Are those different approaches to the topic of representation antagonistic, or may we instead speak of *levels* of representation, and in particular of levels of *actual* representations?

Among the problems immediately arising from the analysis of representation are the following: (i) are there *primitive aspects* of representing connected to a type of *concrete* representation which differentiates its format with respect to abstract, mnestic or symbolic representations? (ii) is there a *continuum* between the two (or more) levels and/or it is possible to draw a demarcation line, at least *descriptively*, between the different aspects of the representation? In other words, is it possible to identify aspects of the *structure* and of the perceptual *content* that differ from substantially *symbolic* and *definitional* aspects of the representation? When addressing the problem of representation one is compelled to take account of the multigranular processes and hierarchies of knowledge representation, as researchers currently working in artificial intelligence know very well.

Another question raised by the assumption of the inner nature of representation is the difference between the component of the *object*, of the *content* and of the *act* of a representation: a question of considerable importance, given that it drove the development of the school of Brentano.

In fact, the inner nature of the representation engenders a multitude of inner objects of various kinds and diverse modalities of ontological existence, in which the distinction between 'object' and 'content' is often blurred.

In what way are *acts* components of the representative structure, what kind of spatio-temporal events are they and what kind of 'energy,' if any, do they express? The importance of this component of presentation, which relates to the temporality and the modality of the appearance of objects, was clear to the scientists working at the turn of the century. Several hypotheses were erected upon it (from Brentano to Köhler to Husserl), but it has been almost entirely eliminated from contemporary scientific research.

A particular importance is due to those aspects of knowledge representation more closely tied to phenomena of *assimilation* and of *perceptual organization*, as evident, for example, in phenomena of 'amodal' perception (in Kanizsa's sense) which shed light on the characters of emergence at different levels (within the *actual* representation itself) of multiscaled levels of cognitive integration. Stout's concept of 'presentation' as *perceptual appearance* should be understood in precisely this sense (see Valentine, this volume).

In cognitive science, sensorial presentations and mental presentations are often viewed as mutually exclusive, and they are also given separate treatment by neurophysiologists and by cognitive scientists, whereas it seems that some phenomena (like anomalous surfaces or various types of imagery) comprise both aspects in a continuous and functional manner.[46]

This question is discussed by Pierantoni's essay in this volume, which analyses the development of drawing and geometry as a manifestation in pure form of the *schematic level* around which meaning is structured. As an 'empirical' proof of the validity of the *theories of inner representation* developed between the end of the nineteenth century and the beginning of the twentieth – although their origin were much earlier, as he underlines – Pierantoni provides the reader with material and reflection in favour of the latter hypothesis. The example of Nadia, an autistic child who seemed 'to have direct access to the retinal image', as she later put it, is a very strong argument in support of the priority of a inner psychophysics in Brentanian terms and a plea for 'creativity' itself, intended in a Whiteheadian sense (see for the hypothesis Albertazzi, this volume).

Another important aspect of current research on information processing is the issue of imagery, for decades ignored by behaviourist research. The revival of interest in this theme has been due to certain problems arising in computer science[47] and psycholinguistics[48] and to the birth of the cognitive sciences,[49] in particular research in the theory of vision.[50]

Since the pioneering research of Shepard and Cooper, whose first studies were published in the 1960's and 1970's,[51] the theme has undergone major development from the point of view of experimental analysis, which has concerned itself mainly with *psychological* and *neurophysiological* research. Developed in particular has been analysis of the quantity of information processed, flanked by research in the theory of vision.[52] There has been less interest, however, in what we may call the *conceptual* level of *phenomenological description* and *classification* of images as mental contents, the laws of dependence between contents and objects, among objects, contents and their parts, or their role in a theory of intentionality: questions which have been largely left in the background. The focus has been instead on finding answers to such questions as the type of representation constituted by images, their role in the dynamics of information processing, and the level of processing at which they make their appearance.

Staring from the assumption that representation is internal, it is obviously of great importance to determine the form taken by imagery phenomena with respect to their physical and/or neurophysiological correlates, and the manner in which they differ, for example, from more properly perceptive phenomena. Does imagery constitute a *common background* for every *type* of representation of objects (concrete, perceptive, actual, but also mnestic, ideal, symbolic, etc.), or does it only concern the configuration and representation of substantially *absent* objects (mnestic, anomalous like Kanizsa's triangle, fictitious like the golden mountain, impossible like the round square, merely epistemological, etc.)? In other words, is *every act of presentation* at least in part imaginative or is it not? When we speak of images, are we referring to mental contents understood as pictorial representations, to forms of linguistic symbolisation, or to something else? And in the construction of an image, of any form of image, what role is played by, for example, the *anticipation of the objectual form* of the content?

From a theoretical point of view, debate in the cognitive sciences has been grounded on two broad interpretations. On the one hand, there are those who have theorised the nature of imagery as essentially *depictive*, based on an internal location. In this case, the image is represented by a set of points encoding the location in the two-dimensional space of the contour points of an object which is imagined. The key role in this account is performed by the *similarity* between the depicted object and the image of that object.[53] On the other hand, there are those who sustain a propositional interpretation of imagery couched in terms of 'mental sentences' based on abstract symbols and therefore substantially part of discourse.[54] It should be noted that the former interpretation does not necessarily exclude the latter, but rather argues that the pictorial aspect of the image is essential, or in other words, rules out that all the inner representations of imagery are propositional.[55] Further support for the pictorial view has been provided by recent findings in neurophysiological research which demonstrate that imagery relies on topographically organised regions of the cortex.[56]

Another important research finding is that visual *mental imagery* and visual *perception* are very similar in processing information, and in particular that imagery plays a role in the earliest stages of information processing at the perceptive level. Imagery, therefore, should not be considered to be a second or third stage, so to speak, of information processing in accordance with a rigid bottom-up construction, but rather as being present from the beginning with top-down inputs, thereby performing a crucial role.

Once again, the Continental scholar finds these findings very familiar. An assumption of this kind, in fact, traverses large part of Western reflection on the matter, and in various forms it is prefigured in Aristotle (*phantasia*) and in Kant (*Einbildungskraft*). In this case too, a historical approach, although it does not magically produce a solution, at least sharpens the focus on the problem and averts pointless research, suggesting that exploration of areas at the moment of difficult access would be more fruitful. As far as imagery in particular is concerned, the morphological and phenomenological description of the nature, type and interconnections among mental images has long been amply developed in the Western tradition.

At the turn of the century, these themes were the focus of the theory of intentional reference developed by Brentano, Stumpf and Twardowski.[57] More specifically, they were an integral part of Husserl's theory of intentionality. As I have pointed out on several occasions, those authors who subscribed to a theory of inner representations on realistic bases – so called immanentist realism – described the morphology (and genesis) of the various types of mental image as an inner variant of the distinction among *act*, *content* and intentional *object* and their modes of cross-reference.

These authors also offered a description of the phenomenological level which, as already pointed out, has been left in the background by research. To them we owe the distinction between at least *two* fundamental levels of imagery: (i) *intuitions*, which are actual presentations tied to perception and having *direct reference* to the object ('table,' for example) through the content; (ii) *presentations of fantasy* (*Phantasievorstellungen*) which, to use a Gestaltist expression, are directed at 'absent' objects ('Othello,' for example). The latter are weaker presentations and they are characterised by *direct reference to the content*, which is their true 'object' (*Ding*). In the case of presentations of fantasy, *the reference to the object is indirect*, so that one may speak of 'two' objects of reference, one primary (the content) and one secondary (the object).

The distinction and the functional connection between these two types of presentation, as well as analysis of the laws of dependence between object and content in the various forms of imagery, are distinctive of a large part of Husserl's theories (see Schuhmann, this volume). Characteristically, much of Husserl's thought sustains the pictorial theory of imagery from a phenomenological point of view. When Kosslyn, for example, distinguishes between the generation and transformation of images in the case of familiar objects or in that of unknown objects (in the former case, the image appears to be *distinct* from the object, in the latter the image seems instead *to spring from the object*), he is only proposing a variation on the law of dependence between object and content of imagery in the various types of presentation analysed by Husserl in the Third Logical Investigation, or by Twardowski in his book of 1894 on the content and object of presentations.[58]

Another important aspect of the functional structure of imagery highlighted by phenomenology is the *objectual component* of the mental image relative to the *form of the specific act*, for example the act of seeing, touching, rotating, separating objects, etc., and then imagining, remembering them, etc. This is the Husserlian concept of signifying intention (*Bedeutungsintention*) always present in any type of presentation, so that one may define imagining as *something undifferentiated* (very similar to the imageless component of thought of the Würzburgians),[59] which receives its *form of specification* only from the content (functionally understood as *filling intention*, *Erfüllungsintention*).

Husserl's distinction therefore once again raises the problem of the inner presentation of knowledge in radical form: if imagery is a constitutive part of every presentative and representative form, from the point of view of phenomenological description and classification the boundaries between intuitive presentations and presentations of fantasy become, if not indistinct, at least not discontinuous. In other words, Husserl's theory suggests the existence of a continuum in representation, in which elementary components of imagery or top-down processes are active from the first presentative stage onwards, as well as those more closely connected to the sensory contents triggered by stimuli. Husserl's theory, in short, argues that visual mental imagery and visual perception share a common mechanism and that, to use Kosslyn's expression, imagery is not merely a 'parasite' on perception.

The previous point immediately introduces another issue at the centre of contemporary debate: the hypothetical or direct nature of perception, which Gregory and Gibson have argued for respectively.[60] The issue concerns in particular the epistemological status of sensation and the presence of *unconscious presentations* (be they mnestic traces, hypotheses, etc.) in the initial stages of information processing; a theme that is central to research in the theory of vision, for example.[61]

Many of the authors discussed in the previous section assumed that *sensation* is at least partly the result of cognitive integrations. This was the assumption of Brentano, of the Graz school, of the Würzburg school, and also of Helmholtz, for whom sensations are signs which simply inform us that we are dealing with the same or with different objects, but nothing else (see Zanarini, this volume). In the light of these presuppositions, the problem of how to construct an epistemological *model* in the various areas of research becomes of great importance, as end-of-century debate in physics or psychology has demonstrated. That 'sensation' at least partially is 'reflexive,' as Husserl thought, and that it is endowed with cognitive elements, was demonstrated by analysis of phenomena such as consonance or melody, or optical-geometric illusions conducted at the turn of the century by both philosophers and scientists.[62]

As I have already pointed out, an extremely important feature of all these analyses is that they interweave science, ontology and phenomenological description. Examples are provided by Helmholtz and Mach, according to whom the identification of complex sounds (i.e., a chord) as *unitary objects* makes them irreducible to mere sensory phenomena. Which also entails consideration of the cognitive integrations and intellectual constructs that intervene as soon as they appear, turning them into what in Meinongian terminology are called 'higher order objects.' In this case too, therefore, a decisive role is played by phenomenological description in scientific reflection: for example, Mach's and Stumpf's idea that consonance is closely connected to high level properties of harmony, or that according to Ehrenfels, Meinong and Husserl, a melody is a construct comprising higher-order figural qualities (see Zanarini, and Albertazzi, this volume). Moreover, the *evidence of sensation* has been a theme with ontological implications in the European tradition since the time of Wolff.

A careful re-reading of these authors also throws light on the complex theoretical core of Gestalt, which despite a set of generally accepted assumptions (viz., the apprehension of form is a specific psychic process not reducible to others; Gestalten are wholes characterised by the non-summativeness of their parts and transposition of structure; apprehension of the whole determines the order and the connection of the parts, and certain fundamental laws of the organisation of the percept) comprises a variety of positions anticipated by those who may be regarded as its predecessors, like Brentano, Mach or Stout. However, the peculiarity of Gestalt theory did not prevent the onset of various problems centred on the problem of the relation among the parts of a whole (Husserl) and the nature of this relation as 'objective modification' (Stout), as 'perceptive contents' (von Ehrenfels), as 'unitary forms' (Stumpf, Koffka), or as 'components productive' of cognitive completion in the presentation of the whole (Meinong and Benussi versus Köhler).[63]

These issues also relate closely to the hypothesis of the presence of *phases* in actual representation, a question which still today divides scholars, in particular those who support Gregory and those who support Gibson. In various respects, however, this question, too, was addressed by experimentalists like Benussi and by descriptive psychologists like Stout, who joined the debate begun by Helmholtz on unconscious perceptions to offer innovative solutions (see Albertazzi, and Valentine, this volume). In particular, one finds in these authors the first clear confutation of the bundle or mosaic hypothesis and of associationism, theses which later became integral to Gestalt theory[64] (see Herrmann and Katz, and Valentine, this volume).

Another important aspect developed by European scholars between the end of the nineteenth century and the beginning of the twentieth concerns the *theory of relations*. In those years, substantially three different points of view were taken up on the mechanisms that compose the relations between objects and individuals.

The first point of view was sustained by *logical atomism* (Russell and the Wittgenstein of the *Tractatus*), and it envisaged a mechanism which composed relations *from the bottom up*, or in other words, from the 'atomistic' components. It was the point of view that predominated in the first years of the century and for many decades thereafter.

The second point of view was the *holistic* position taken up at the end of the nineteenth century by Hegel, certain exponents of British idealism and (between 1890 and 1920 circa) by Driesch and Uexküll in experimental biology. It conceived a mechanism which composed relations *from the top down*, or in other words not from the atomistic components but from the *totality* (*Totalität, Ganzheit*), a theory which found its main supporters in the *Ganzheitpsychologie* of Leipzig.[65]

Again at the turn of the century, a third conception of relations, entirely original and anti-atomistic, developed which viewed them in *molar* terms, being opposed to both elementarism and structuralism.[66] The molar conception of relations (in certain respects already present in James, who talked of a 'feeling' of relations in the sense of figural qualities)[67] was developed by Brentanism as a whole, and also by its exponents in Britain, notably Stout. In this case, assumed as paramount were the inner relations among the components of the *whole* (*Ganze*) which constitute a variant on the laws of ontological dependence. A classic formulation of the theory was propounded by Husserl, in particular in his Third Logical Investigation with its distinction between *part* and *piece* (or between non-independent part and independent part of the whole).[68] A variation on the theory was offered by Ehrenfels in his essay on the figural qualities and by Meinong in his essay on the higher-order objects already mentioned, which emphasised the presence of *components* in the apprehension of the whole, albeit not in the elementarist sense.[69] In psychology, the molar point of view merged naturally with the conceptual core of Gestalt.[70]
As these various theories show, the problem of the structure of the relations, of its components, mechanisms of apprehension and composition, was closely connected with other problems, and therefore analysed with them – the problems of truth conditions, of the nature of meaning and language, of abstraction – which subsequently characterised a large part of twentieth-century thought, though in an entirely different manner. Independently of individual positions, the essence of the problem was perhaps best grasped by Wertheimer in his book of 1920 and a series of essays which in many respects reflect the theories of Husserl. In his Second Logical Investigation, when analysing the relationship between the experienced sense of a statement and its correlative state of affairs, Husserl had emphasised that the *objects* we deal with, and as they 'phenomenologically appear,' *are not reducible to the sensorial data*.[71] The Würzburg school had conducted significant analysis of the connections between mere 'sensations' and contents, images, acts of thought and feeling. In discussion of some problems connected with the

teaching of mathematics, Wertheimer showed that the components of a whole (for example the lines of a rectangle or of another geometric figure) had a role and a function such as to determine the inner relation the operations and the final outcome in a *functional hierarchy*.[72] Indeed, we perceive the *shape* of an object relative to (and simultaneously with) with its position in space, its direction, distance, velocity, etc., and these latter *relational qualities* are one condition for the (phenomenal) identity itself of objects (Musatti 1926, vii).

On another occasion, Wertheimer started from the conventional principles that 'it is propositions that are true or false,' and that the presupposition of formal logic is that all propositions are distinct, to argue instead that states of affairs are *not independent of each other in the real world*, and that consequently a truth theory should take account of every statement and its corresponding object as *parts* of the *related wholes*. He emphasised that it may be the case that something can be true in the *piecemeal* sense and false as a *part* in its whole. It was therefore extremely important, in the theory of knowledge as well as in ontology and logic, to distinguish the object as a *piece* (*a*), the object as a *part* of its whole (*a* in the series *abc*), the object as part of another whole (*a* in the series *amn*):[73] in fact, even in linguistic statements, as in the epistemological sphere, there may be phenomena analogous to the perceptive ones of masking analysed by Gottschadt and Köhler,[74] or ones of optical deformation. In short, truth does not consist solely in the correspondence between a proposition and an object, 'its' object taken in isolation: if the object turns out to be a *part in a pattern* of a definite situation, the truth of a proposition is such when it corresponds not only with the part, but with the *role* that it plays in the whole.[75]

From a more general point of view, these observations by Wertheimer are a critique against the ontological existence of 'atomic facts' in isolation from each other; a conception which in logical terms has been expressed both by traditional logic and logistics, neither of which, Wertheimer points out, distinguished between content as 'piece' and content as 'part,' and both of which had a substantially *summative* view of relations. Moreover, a distinctive feature of traditional logic was its emphasis on the correctness of every step of the proof, not on finding solutions. More than a formal logic, Wertheimer argued, what we need is a logic of 'objects' and of their laws of dependence, to which Gestalt might be able to furnish a number of fundamental concepts.[76]

The *cognitive* counterpart to the 'objectual' aspect of logical relations is exemplified by the Graz school's analysis of the foundation of higher-order objects, as well as by numerous Gestaltist studies, starting from Rubin's classic analysis of the *fundamental relation* of *figure/ground*.[77] But one may also cite Köhler's study of the ability of animals to recognise relations of *clarity* among different objects,[78] Katz's analysis of the relationship between colour and area of the visual field,[79] those by Cornelius, Krueger and especially Köhler on colour tone,[80] and Asch's research into the *base conditions* for associative processes, which were usually restricted to processes of association among units.[81] Asch's experiments extended the concept of associative conditions to particular types of perceptive relations, like the *constitutive* relation and those of *figure/ground* and *inclusion* which demonstrate that association does not have the primary function of joining unrelated contents of mental life as discrete, irreducible sense data – the hypothesis accredited by British empiricism – but is rather some sort of *after-effect* of the laws of organisation in immediate experience. The same conclusion was reached, for example, by Wever's tachistoscopic research on the nature of attentive processes, which showed the prime importance of the concepts of 'clearness' and 'attentiveness' on the basis of their connection with the fund-amental relation of figure/ground.[82]

In this case too, therefore, psychological and cognitive research at the beginning of the twentieth century has an ontological and semantic counterpart, in the sense that analysis of the relations of perceptive organisation shed light on the psychological constitution of *objects*, and on the formation of the *meaning relation* as a specific case of molar 'conceptual unit' formation.

5. Conclusions

This brief introductory discussion should have been sufficient to illustrate the conceptual richness and the thematic complexity of theories developed between 1870 and 1930. It should also have clarified the purpose of this book, which is to demonstrate that, in both analysis and theory, positions were developed in that period which anticipated not only much of the *content* of the contemporary cognitive sciences but also many of the *modes* of research typical of the late twentieth-century mainstream: content and modes to be found, for example, among scholars who might be considered to be 'descriptive psychologists,' like Gibson, Neisser[83] and Johnson-Laird, or those more specifically engaged in experimental work, like Marr, Fodor, Kosslyn or Phylyshyn.

The overall intention of this book is to suggest that at the beginning of the twentieth century, in a variety of disciplinary areas, a scientific paradigm arose which fostered, guided and often created research through an endeavour to develop *conceptual categories* which subsequently disappeared from science and its vocabulary for decades. From the 1930's onwards, in fact, partly as a result of the failure to justify one of the crucial components of that paradigm (the concept of *evidence*),[84] an entirely different approach came to predominate in many areas of scientific inquiry. On reading the literature, one is struck by how brusquely the logico-formal paradigm took the place of the cognitive paradigm. Scientists no longer talked of 'objects' and/or 'things' but of 'individuals;' 'qualities,' especially figural ones, were supplanted by 'predicates' even in the philosophy of mind; the complexity of the 'relations' that had characterised debate among the various branches of Brentanism and Gestalt was restricted to inclusion, negation and conjunction; 'dynamic and functional invariants' became 'logical constants;' and so on.

A Continental scholar is amused by the coincidence that the revival of a conceptual apparatus based essentially on categories of the mind and psyche, and on an immanentist ontology of the kind discussed in European debate at the end of the last century, has been substantially due to the entry of robots into the machine civilisation.

ACKNOWLEDGEMENTS

The idea of producing this book arose from a series of meetings between philosophers and scientists on various issues at the focus of current debate. The excellent results achieved by these meetings, and in particular during the conference on *The Origins of Cognitive Science. Theories of Representation 1870-1930* (Bolzano, 12-13 December 1997) gave rise to this volume. The project is part of the research programme pursued by the Central European Institute of Bolzano. I wish to thank the President of the Institute once again for the support given to the initiative.

The Editor

NOTES

[1] Dummett 1988. On the revival of Meinong's theories in formal semantics see, among others, Routley 1980; Lambert 1983; Zalta 1983. A corollary to the idea of a Continental philosophy is the so-called Neurath-Haller thesis of the existence of an Austrian as opposed to German philosophy. Cf. Haller 1979, ch. 1; Grassl and Smith 1986; Smith 1994, ch. 1, § 3. For more historical discussion of the sources and the framework cf. Albertazzi 1996a.

[2] Cf. Albertazzi, forthcoming a.

[3] Cf. Donald 1991. A criticism of the little man in the head is to be found in Gibson 1979.

[4] Cf. Newell, Shaw and Simon 1958.

[5] Cf. Johnson-Laird 1883; Jackendoff 1992.

[6] Fodor 1983.

[7] Cf. Chomsky 1957; Montague 1974.

[8] Duncker 1935 was translated in 1945; Wertheimer 1920 in 1938; Katona's book appeared in 1940.

[9] On the ecological theory of perception see Gibson 1979; Turvey et al. 1981.

[10] Cf. Davis 1990.
[11] Cf. Lakoff 1987; Langacker 1987; Talmy 1988.
[12] Cf. Varela, Thompson and Rosch 1991; Maturana and Varela 1987.
[13] Cf. Varela 1989.
[14] On the concept of 'experimental phenomenology' see Thinès 1977. On the topic cf. Albertazzi 1998a, b, and c.
[15] Cf. Fleming and Bailyn 1969; Mandler and Mandler 1969.
[16] Cf. Zagajewski 1986.
[17] Cf. Albertazzi 1998c.
[18] Cf. Coniglione, Poli and Wolenski 1993.
[19] In fact, in1844 Lotze was appointed to the chair vacated by Herbart at Göttingen.
[20] Cf. Albertazzi 1997a.
[21] Cf. also Albertazzi 1996a.
[22] Dilthey 1974; Windelband 1909.
[23] Cf. also Kusch 1995; Ash 1995.
[24] Meinong's school closely studied this latter difference. Cf. Meinong 1896.
[25] Lotze had studied at Leipzig under E.H. Weber and was a close friend of Fechner, whose theory of universal animation (*Allbeseelung*) he endorsed, with some modifications. On Fechner cf. Heidelberger 1994. For the concept of inner psychophysics see Fechner 1860 and Metzger 1986, IV.
[26] Also Schrödinger 1956 identified the sentient, perceptual and thinking self with the scientific image of the world. On the concept of *Bild* and mental representation see D'Agostino 1994.
[27] This opinion is also shared by Boring 1929, Mandler 1964, Robinson 1976 and Schulz 1981. Of contrary opinion are Smith 1988 and Ash 1995.
[28] Cf. Dappiano 1995; Santayana 1971, 111. Lotze's great works were translated into English by B. Bosanquet and T.H. Green. The editor of *Mind*, G.C. Robertson was, like James, an admirer of Lotze and disseminated his thought. Moreover, J. Royce and C. Cantoni studied under Lotze at Göttingen. Almost all the Brentanists acknowledged his influence. Brentano cites him in his *Psychologie* of 1874 and owed his appointment in Vienna to him. Stumpf, whose dissertation had been supervised by Lotze, dedicated his *Über den psychologischen Ursprung der Raumvorstellung* of 1873 to him; Husserl cites Lotze with admiration in the first volume of *Logische Untersuchungen*.
[29] Cf. Hicks 1904/5; Stout 1896; Moore 1903; Russell 1906; Bradley 1883.
[30] For a survey of the school of Brentano cf. Albertazzi, Libardi and Poli 1996. For an introduction to Brentano's thought cf. Albertazzi 1999.
[31] Cf. Smith 1988.
[32] In particular, on the relation between Selz's ideas and the computational approach see Newell et al. 1958, and Simon 1981.
[33] There were close contacts between Berlin and Würzburg, as testified by the presence of Wertheimer and Koffka at both universities.
[34] Cf. Albertazzi and Poli 1993; Poli 1998.
[35] Twardowski's volume of 1894 was reviewed in *Mind*, probably by Stout, and called 'an excellent piece of psychological analysis.'
[36] On the influence of Meinong's thought in Anglo-Saxon countries see Findlay 1963, 9–19; van der Schaar 1996.
[37] Lipps's theory anticipated many experimental observations of the masking of surfaces, so-called amodal shrinking, the shapes of figures, and the shifting of modified images onto others. Cf. Albertazzi 1998d, and forthcoming c.

[38] Cf. Thom 1972; Petitot 1985.
[39] Witasek 1910; Meinong 1904; Wertheimer 1912; Benussi 1912. See also Albertazzi 1996a, and b.
[40] Cf. also Albertazzi 1995.
[41] Köhler's book was the first application of non-linear field theory to perception and order formation. I owe this observation to Alf Zimmer.
[42] Cf. Poli 1995, Introduction.
[43] On Kaufmann cf. Albertazzi 1997b.
[44] Gardner 1985.
[45] Murray 1995, 2 ff.; Albertazzi 1998a, b, and c. For a broad sociological and conceptual reconstruction of the original framework of Gestalt psychology see again Ash 1995.
[46] On the concept of 'amodal' see Albertazzi forthcoming d.
[47] Cf. again Newell, Simon and Shaw 1958.
[48] Cf. Paivio 1971; Chomsky1957.
[49] Cf. Neisser 1976; Gardner 1985.
[50] Cf. Marr 1982; Biederman 1985.
[51] See, for example, Cooper and Shepard 1973; Shepard and Cooper 1982.
[52] Cf. Biederman 1985; Ullmann 1984; Pomerantz 1983.
[53] Cf. Kosslyn 1980.
[54] Cf. Pylyshyn 1973. There is also an intermediate proposal, as in Anderson 1978.
[55] Cf. for example Anderson and Bower 1973; Kosslyn and Pomerantz 1977; Kosslyn 1980; Pomerantz 1983; Paivio 1971; Pylyshyn 1973. For an up-to-date summary of the imagery debate and an overview of the current state of research see Kosslyn 1994. See also Barlow, Blackmore and Weston-Smith 1990.
[56] Cf. Kosslyn 1994, 12 ff.
[57] Cf. also Zeglen 1994.
[58] Cf. Twardowski 1894.
[59] Cf. Bühler 1907.
[60] Cf. Albertazzi 1998b. As to the similarity between the ecological theory of direct perception and Gestalt ideas, note that Gibson attended Smith College at Winsconsin University, when Koffka was lecturing.
[61] Cf. Marr 1982.
[62] Cf. Brentano 1907; Lipps 1897; Benussi 1914.
[63] Cf. Benussi 1912; Koffka-Kenkel 1913; Koffka 1915. On this issue at the centre of the polemic between Benussi and Koffka see Albertazzi forthcoming d.
[64] Cf. Wertheimer 1920.
[65] Cf. Krueger 1926.
[66] Cf. respectively Wundt 1896 and Tichener 1910.
[67] Cf. James 1890, 224.
[68] Cf. Husserl 1900/1.
[69] Cf. Ehrenfels 1890; Meinong 1899. Cf. also Gelb 1910; Höfler 1910.
[70] Cf. Albertazzi 1996a, and 1998c.
[71] Cf. Husserl 1900/1, Second Logical Investigation.
[72] Cf. Wertheimer 1920, § 39. Cf. also Katona 1940.
[73] Wertheimer 1934 (1961), 21.
[74] Gottschadt 1926, and 1928; Köhler 1969.
[75] Idem.
[76] For an intermediate position between phenomenology and Gestalt see Brunswig 1910.

[77] Cf. Rubin 1921.
[78] Cf. Köhler 1915.
[79] Cf. Hering 1905, sec. 24; Mach 1865; Katz 1906 and 1911.
[80] Cf. Köhler 1910.
[81] Cf. Asch 1961.
[82] Cf. Wever 1927a, and b.
[83] Note that Neisser attended Swarthmore College when Köhler was a teacher there.
[84] Cf. Albertazzi 1997b.

References

Albertazzi, L.: 1995, 'Il campo figurale della lingua: esprimere, indicare, rappresentare,' in Negrini, ed., 1995, 9–42.

Albertazzi, L. ed.: 1996a, *The Philosophy of Alexius Meinong*, special issue of *Axiomathes*, **7**.

Albertazzi, L.: 1996b, 'Comet Tails, Fleeting Objects and Temporal Inversions,' in Albertazzi, ed., 1996a, 111–136.

Albertazzi, L.: 1997a, 'Science and Avant-garde in Early Nineteenth Century Florence,' in Poli, ed., 1997, 177–212.

Albertazzi, L.: 1997b, 'The 'Open Texture' of Concepts – Kaufmann and the Brentanists,' in Stadler, ed., 1997, 47–71.

Albertazzi, L. ed.: 1998a, *Shapes of Forms: From Phenomenology to Gestalt Psychology to Ontology and Mathematics*. Dordrecht, Kluwer.

Albertazzi, L. ed.: 1998b, 'Form Aesthetics,' in Albertazzi, ed., 1998a, 1–17.

Albertazzi, L.: 1998c, 'Form Metaphysics,' in Albertazzi, ed., 1998a, 257–305.

Albertazzi, L.: 1998d, 'The Aesthetics of Particulars: A Case of Intuitive Mechanics,' in Poli and Scaramuzza, eds., 1998, 169–196.

Albertazzi, L.: 1999, *Introduzione a Brentano*. Bari, Laterza.

Albertazzi, L. ed.: Forthcoming a, *Meaning and Cognition: A Multidisciplinary Approach*. Amsterdam, Benjamins Publishing Company.

Albertazzi, L.: Forthcoming b, 'Which Semantics?,' in Albertazzi, ed., forthcoming a.

Albertazzi, L.: Forthcoming c, 'Directions and Perspective Points in Spatial Perception,' in Albertazzi, ed., forthcoming a.

Albertazzi, L.: Forthcoming d, 'Vittorio Benussi,' in L. Albertazzi, D. Jacquette and R. Poli, eds.

Albertazzi, L. and Poli, R. eds.: 1993, *Brentano in Italia*. Milan, Guerini.

Albertazzi, L., Libardi, M. and Poli, R. eds.: 1996. *The School of Franz Brentano*. Dordrecht, Kluwer.

Albertazzi, L., Jacquette, D. and Poli, R. eds.: Forthcoming, *The School of Alexius Meinong*. Aldershot, Ashgate.

Anderson, J.R.: 1978, 'Arguments Concerning Representations for Mental Imagery,' *Psychological Review*, **85**, 249–277.

Anderson, J.R. and Bower, G.H.: 1973, *Human Associative Memory*. New York, Winston.

Ash, M.G.: 1995, *Gestalt Psychology in German Culture, 1890-1967*. Cambridge, Cambridge University Press.

Asch, S.: 1961, 'Perceptual Conditions of Association,' in Henle, ed., 1961, 187–200.

Barlow, H., Blackmore, C. and Weston-Smith, M. eds.: 1990, *Images Understanding*. Cambridge, Cambridge University Press.

Benussi, V.: 1912, 'Stroboskopische Scheinbewegungen und geometrisch-optische Gestalttäuschungen,' *Archiv für die gesamte Psychologie,* **24**, 31–62.
Benussi, V.: 1914, Besprechung zu Koffka-Kenkel, "Beiträge zur Psychologie der Gestalt- und Bewegungserlebnisse I," *Archiv für die gesamte Psychologie,* **32**, 50– 57.
Benussi, V.: 1913, *Psychologie der Zeitauffassung*. Heidelberg, Winter.
Bertotti, B. and Curi, U. eds.: 1994, *Erwin Schrödinger scienziato e filosofo*. Padova, Il Poligrafo.
Biederman, I.: 1985, 'Human Image Understanding: Recent Research and a Theory,' *Computer vision, graphics and image processing,* **32**, 29–73.
Boring, E.G.: 1929 (1950), *A History of Experimental Psychology*. Englewood Cliffs, NJ, Prentice Hall, 2 ed.
Bradley, F.H.: 1883, *Appearance and Reality*. London, Swan Sonnenschein.
Brentano, F.: 1907, *Untersuchungen zur Sinnespychologie*. Leipzig, Duncker & Humblot.
Brunswig, A.: 1910, *Das Vergleichen und die Relationserkenntnis*. Leipzig and Berlin, Teubner.
Bühler, K.: 1907, 'Tatsachen und Probleme zu einer Psychologie der Denkvorgänge. I. Über Gedanken,' *Archiv für die gesamte Psychologie,* **9**, 1907, 297-365.
Chase, W.G., ed.: 1973. *Visual Information Processing*. New York, Academic Press.
Chomski, N.: 1957, *Syntactic Structures*. Paris, Mouton-Gravenhage.
Cooper, L.A. and Shepard R.N.: 1973, 'Chronometric Studies of the Rotation of Mental Images,' in W.G. Chase, ed., 1973, 76–176.
Coniglione, F., Poli, R. and Wolenski, J. eds.: 1993. *Polish Scientific Philosophy: The Lvov-Warsaw School*. Amsterdam, Rodopi.
D'Agostino, S.: 1994, 'Erwin Schrödinger: scienza e Wissenschaft,' in Bertotti and Curi, eds., 1994, 101–110.
Dappiano, L.: 1995, 'Cambridge and the Austrian Connection,' in Poli, ed., 1995, 99–124.
Davis, E.: 1990, *Representations of Commonsense Knowledge*. San Francisco, Morgan Kaufmann.
Dilthey, W.: 1974 (1894), *Ideen über eine beschreibende und zergliedernde Psychologie*. In *Gesammelte Schriften*, ed. by G. Misch, Teubner, Stuttgart, vol. IV, 136–240.
Donald, M.: 1991, *Origins of the Modern Mind*. Cambridge, Mass., Harvard University Press.
Dummett, M.: 1988, *Urprünge der analytischen Philosophie*. Frankfurt a.Main, Suhrkamp.
Duncker, K.: 1935, *Zur Psychologie des produktiven Denkens*. Berlin, Julius Springer. En. tr. 1945 as 'On problem solving,' *Psychological Monographs,* **58**, 113 ff.
Ellis, W.D. ed.: 1938, *A Source Book of Gestalt Psychology*. London, Routledge & Kegan Paul.
Ehrenfels, Ch. von.: 1890, 'Über Gestaltqualitäten,' *Vierteljharschrift für wissenschaftliche Philosophie,* **14**, 242–292. En. tr. in Smith, ed., 1988, 82–117.
Fechner, G.T.: 1860, *Elemente der Psychophysik*. Leipzig, Breitkopf & Härtel.
Findlay, J.N.:1952, 'The Influence of Meinong in Anglo-Saxon Countries,' in *Meinong-Gedenkschrift*, Graz, Styria Steirische Verlaganstalt, 9–20.
Findlay, J.N.: 1963, *Meinong's Theory of Objects and Values*. Oxford, Clarendon.
Fleming, D. and Bailyn, B. eds.: 1969, *The Intellectual Migration. Europe and America, 1930-1960*. Cambridge, Mass., Harvard University Press.
Fodor, J.A.: 1983, *Modularity of Mind: An Essay on Faculty Psychology*. Cambridge, Mass., MIT Press.
Frijda, N.H. and de Groot, A.D. eds.: 1981. *Otto Selz: His Contribution to Psychology*. New York, Mouton.

Gardner, H.: 1985, *The Mind's New Science*. New York, Basic Books.
Gelb, A.: 1910, *Theoretisches über "Gestaltqualitäten"*. Berlin University, Doctoral dissertation.
Gibson, J.J.: 1979, *The Ecological Approach to Visual Perception*. Boston, Houghton-Mifflin.
Gottschadt, K.: 1926/1928, 'Über den Einfluß der Erfahrung auf die Wahrnehmung von Figuren,' *Psychologische Forschung*, **8**, 261–317; **12**, 1–87.
Grassl, W. and Smith, B.: 'A Theory of Austria,' in Niyri, ed., 1986, 11–30.
Katona, G.: 1940, *Organizing and Memorizing: Studies in the Psychology of Learning and Teaching*. New York, Columbia University Press.
Katz, D.: 1906, 'Experimentelle Beiträge zur Psychologie des Vergleichs im Gebiet des Zeitsinns,' *Zeitschrift für Psychologie und Physiologie der Sinnesorgane*, **42**, 302–340; 414–450.
Katz, D.: 1911, 'Die Erscheinungsweisen der Farben und ihre Beeinflussung durch die individuelle Erfahrung,' *Zeitschrift für Psychologie*, **7**, 236-00. Rev. ed. 1930, *Der Aufbau der Farbwelt*. Leipzig, Barth. En. part. tr. 1935, London, Kegan Paul.
Kijania-Placek, K. and Wolenski, J. eds.: 1996, *The Lvov-Warsaw School and Contemporary Philosophy*, Part II, special issue of *Axiomathes*, **7**.
Koffka, K.: 1915, "Beiträge zur Psychologie der Gestalt und Grundlegung der Wahrnehmungspsychologie. Eine Auseinandersetzung mit V. Benussi," *Zeitschrift für Psychologie*, **73**, 11–90. En. tr. in Ellis, ed., 1938, 371–378.
Koffka, K. and Kenkel, F.: 1913, "Beiträge zur Psychologie der Gestalt- und Bewegungserlebnisse I. Untersuchungen über den Zusammenhang zwischen Erscheinungsgrösse und Erscheinungsbewegung bei einigen sogenannten optische Täuschungen," *Zeitschrift für Psychologie*, **67**, 353–449.
Köhler, W.: 1915, 'Nachweis einfacher Strukturfunkionen beim Schimpansen und beim Haushuhn. Über eine neue Methode zur Untersuchung des bunten Farbensystems,' *Abhandlungen der Königlich Preussischen Akademie der Wissenschaften*, Phys.-Math. Klasse, **2**, 1–101.
Köhler, W.: 1920, *Die physische Gestalten in Ruhe und im stationären Zustand*. Braunschweig, Vieweg. En. part. tr. in Ellis, ed., 1938, 17–54.
Köhler, W.: 1969, *The Task of Gestalt Psychology*. Princeton NJ, Princeton University Press.
Kosslyn, S.M.: 1980, *Image and Mind*. Cambridge, Mass., Harvard University Press.
Kosslyn, S.M.: 1994, *Image and Brain*. Cambridge, Mass., MIT Press.
Kosslyn, S.M. and Pomerantz, J.R.: 1977, 'Imagery, Propositions, and the Form of Internal Representations,' *Cognitive Psychology*, **9**, 52–76.
Krueger, F.: 1926, 'Über psychische Ganzheit,' *Neue Psychologische Studien*, **1**, München, Beck, 1–121.
Kusch, M.: 1955, *Psychologism. A Case Study in the Sociology of Philosophical Knowledge*. London, Routledge.
Jackendoff, R.: 1992, *Languages of the Mind: Essays on Mental Representation*. Cambridge, Mass., MIT Press.
James, W.: 1890, *Principles of Psychology*. Boston, Holt & Co. Repr. 1950, New York, Dover Publications.
Johnson-Laird. P.N.: 1983, *Mental Models*. Cambridge, Mass., Harvard University Press.
Haller, R.: 1979, *Studien zur österreichische Philosophie*. Amsterdam, Rodopi.
Heidelberger, M.: 1994, 'The Unity of Nature and Mind: Gustav Theodor Fechner's non Reductive Materialism,' in Poggi and Bossi, eds., 1994, 215–236.

Henle, M. ed.: 1961, *Documents of Gestalt Psychology*. University of California Press, Berkeley.

Hering, E.: 1905, *Outlines of a Theory of Light Sense*, En. tr. by L.M. Hurwich and D. Jameson, Cambridge, Mass., Cambridge University Press.

Hicks, G.D.: 1904/5, 'Idealism and the Problem of Knowledge and Existence,' *Proceedings of the Aristotelian Society*, **5**, 136–178.

Höfler, A.: 1910, 'Gestalt und Beziehung – Gestalt und Anschauung,' *Zeitschrift für Psychologie*, **60**, 161–228.

Husserl, E.: 1900-1, *Logische Untersuchungen*. Halle: Niemeyer. Repr. 1975. Vol. I ed. by E. Holenstein, Husserliana XVII, Den Haag, Nijhoff 1975. Repr. 1984. Vols. I and II ed. by U. Panzer, Husserliana XIX , *ivi*, 1, 2. En. tr. 1970, *Logical Investigations*, ed. by J.N. Findlay, London, Routledge & Kegan Paul.

Lakoff, G.: 1987, *Women, Fire and Dangerous Things*. Chicago, University of Chicago Press.

Lambert, K.: 1983, *Meinong and the Principle of Independence: Its place in Meinong's Theory of Objects and its Significance in Contemporary Philosophical Logic*. Cambridge, Cambridge University Press.

Langacker, R.: 1987, *Foundations of Cognitive Grammar. Theoretical Prerequisites. I*. Stanford, Stanford University Press.

Lipps, T.: 1897, *Raumaesthetik und geometrisch-optische Täuschungen*. Leipzig, Barth.

Mach, E.: 1866, 'Über den physiologischen Effect räumlich vertheilter Lichtreize,' Part II, *Sitzungsbericht der Kaiserlichen Akademie der Wissenschaften, Mathematisch-Naturwissenschaftliche Classe*, **54**, 131–144.

Mandler, G.: 1964, *Thinking: From Association to Gestalt*. New York, John Wiley. Repr. 1982, New York, Greenwood Press.

Mandler, J. and Mandler, G.: 1969, 'The Diaspora of Experimental Psychology,' in Fleming and Bailyn, eds., 1969, 371–419.

Marr, D.: 1982, *Vision: A Computational Investigation into the Human Representation and Processing of Visual Information*. New York, Freeman.

Maturana, H. and Varela, F.J.: 1987, *The Tree of Knowledge: The Biological Roots of Human Understanding*. Boston, New Science Library.

Meinong, A.: 1896, 'Über die Bedeutung des Weber'schen Gesetzes. Beiträge zur Psychologie des Messens,' *Zeitschrift für Psychologie und Physiologie der Sinnesorgane*, **11**, 81–133, 230–285, 353–404. Repr. in Meinong 1971, 215-480.

Meinong, A. ed.: 1904, *Untersuchugen zur Gegenstandstheorie und Psychologie*. Leipzig, Barth 1904.

Meinong, A.: 1971, *Abhandlungen zur Erkenntnistheorie und Gegenstandstheorie*, ed. by R. Haller, Graz, Akademische Druck- u. Verlagsanstalt.

Metzger, W.: 1986, *Gestaltpsychologie*, ed. by M. Stadler and H. Crabus, Frankfurt a.Main, Verlag Waldemar Kramer.

Moore, G.E.: 1903, 'Refutation of Idealism,' *Mind*, **12**, 433–53.

Montague R.: 1974, *Formal Philosophy: Selected Papers of Richard Montague*, ed by R. Thomason, New Haven, Yale University Press.

Murray, D.J.: 1995, *Gestalt Psychology and the Cognitive Revolution*. New York, Harvester and Wheatsheaf.

Musatti, C.L.: 1926, *Analisi del concetto di realtà empirica*, Città di Castello, Il Solco. Repr. 1964, as *Condizioni dell'esperienza e fondazione della psicologia*. Florence, Editrice Universitaria.

Negrini, G., ed.: 1995, *Categorie, oggetti e strutture della conoscenza*, Proceedings of the CNR seminar organised in Rome 1–2 December 1994, Rome.

Newell, A., Shaw, J.C and Simon, H.A.: 1958, 'Elements of a Theory of Human Problem Solving,' *Psychological review*, **65**, 151–66.

Neisser, U.: 1976, *Cognition and Reality*. San Francisco, Freeman.

Nyiri, J.C. ed.: 1986, *Von Bolzano zu Wittgenstein. Zur Tradition der österreichische Philosophie*. Wien, Hölder-Pichler-Tempsky.

Paivio, A.: 1971, *Imagery and Verbal Processes*. New York, Holt, Rinehart & Winston.

Petitot, J.: 1985, *Morphogenèse du sens*. Paris, PUF.

Poggi, S. and Bossi, M. eds.: 1994, *Romanticism in Science*. Dordrecht, Kluwer.

Poli, R. ed.: 1995, *In Itinere. European Cities and the Birth of Modern Scientific Philosophy*. Amsterdam, Rodopi.

Poli, R. ed.: 1998, *The Brentano Puzzle*. Aldershot, Ashgate.

Poli, R. and Scaramuzza, G., eds.: 1998, *Phenomenological Aesthetics*, special issue of *Axiomathes*, **9**.

Pomerantz, J.R.: 1983, 'Global and Local Precedence: Selective Attention in Form and Motion Perception,' *Journal of Experimental Psychology: General*, **112**, 516–540.

Pylyshyn, Z.: 1973, "What's the Mind's Eye Tells the Mind's Brain: A Critique of Mental Imagery," *Psychological Bulletin*, **80**, 1–24.

Robinson, D.: 1976, *An Intellectual History of Psychology*. New York, McMillan.

Routley, R.: 1980, *Exploring Meinong's Jungle and Beyond*. Canberra, Australian National University.

Rubin, E.: 1921, *Visuell wahrgenommene Figuren. Studien in psychologischer Analyse*. Copenhagen, Gyldenalske.

Russell, B. A.W.: 1906, 'The Nature of Truth,' *Mind*, **15**, 528–533.

Santayana, G.: 1971, *Lotze's System of Philosophy*. Bloomington-Indiana, Indiana University Press.

Schrödinger, E.: 1956, *Mind and Matter*. Cambridge, Cambridge University Press.

Schulz, D.: 1981, *A History of Modern Psychology*. New York, Academic Press, 3 ed.

Shepard, R.N. and Cooper, L.A.: 1982, *Mental Images and their Transformations*. Cambridge, Mass., MIT Press.

Simon, H.A.: 1981, 'Otto Selz and Information-Processing Psychology,' in Frijda and de Groot, eds., 1981, 147–63.

Smith, B. ed.: 1988, *Foundations of Gestalt Theory*. Munich, Philosophia.

Smith, B. 1994. *Austrian Philosophy. The Legacy of Franz Brentano*. Chicago and La Salle, Open Court.

Stadler, F. ed.: 1997, *Phänomenologie und logischer Empirismus*. Wien, Springer.

Stout, G.F.: 1896, *Analytic Psychology*. London, Swan, Sonnenschein.

Talmy, L.: 1988, 'Force Dynamics in Language and Thought,' *Cognitive science*, **12**, 49–100.

Thinès, G.: 1977, *Phenomenology and the Science of Behaviour. An Historical and Epistemological Approach*. London, Allen & Unwin.

Thom, R.: 1972, *Stabilité structurelle et morphogénèse. Essai d'une théorie générale des modèles*. Paris, InterEditions.

Tichener, E.B.: 1910, *A Text Book of Psychology*. New York, Macmillan.

Turvey, M.T., Shaw, R.E., Reed, E.S. and Mace, W.M.: 1981, 'Ecological Laws of Perceiving and Acting: In Reply to Fodor and Pylyshyn,' *Cognition*, **9**, 237–304.

Twardowski, K.: 1894, *Zur Lehre vom Inhalt und Gegenstand der Vorstellungen: Eine psychologische Untersuchung*. Wien, Hölder. Repr. 1982, München, Philosophia. En. tr. 1977 *On Content and Object of Presentations*, by R. Grossman, The Hague, Nijhoff.

van der Schaar, M.: 1996, 'From Analytic Psychology to Analytic Philosophy: The Reception of Twardowski's Ideas in Cambridge,' in Kijania-Placek and Wolenski, eds., 1996, 295–350.

Ullmann, S.: 1984, 'Visual Routines,' *Cognition*, **18**, 97–159.

Varela, F.J.: 1989, *Connaitre: Les sciences cognitives, tendances et perspectives*. Paris, Editions du Seuil.

Varela, F. J., Thompson, E. and Rosch, E.: 1991, *The Embodied Mind*. Cambridge, Mass., MIT Press.

Wertheimer, M.: 1912, 'Experimentelle Studien über das Sehen von Bewegung,' *Zeitschrift für Psychologie*, **61**, 161–265.

Wertheimer, M.: 1934, 'On Truth,' *Social Research*, **1**, 135–46. Repr. in Henle, ed., 1961, 19–28.

Wertheimer, M.: 1920, *Über Schlussprozesse im produktiven Denken*. Berlin, Weltkreisverlag. Repr. 1925, in *Drei Abhandlungen zur Gestalttheorie*, Erlangen, Verlag der philosophischen Akademie. En. tr. in Ellis, ed., 1938, 274–82.

Wever, E.G.:1927a, 'Figure and Ground in the Visual Perception of Form,' *American Journal of Psychology*, **38**, 194–226.

Wever, E.G.: 1927b, 'Attention and Clearness in the Perception of Figure and Ground,' *American Journal of Psychology*, **40**, 51–74.

Windelband, W.: 1909, *Die Philosophie in Deutschen Geistesleben des XIX. Jahrhunderts. Fünf Vorlesungen*. Tübingen, J.C.B. Mohr.

Witasek, S.: 1910, *Psychologie des Raumwahrhnemung des Auges*. Heidelberg, Winter.

Wundt, W.: 1896, *Grundiß der Psychologie*. Leipzig, Engelmann. En. tr. 1902, New York, Stechert.

Zagajewski, A.: 1986, *Solidarnosc i samotnosc* (Solidarity and Solitude). Paris, Instytut Literacki.

Zalta, E.: 1983, *Abstract Objects: An Introduction to Axiomatic Metaphysics*. Dordrecht, Reidel.

Zeglen, U. ed.: 1994, *Philosophy of Mind in the Lvov-Warsaw School*, special issue of *Axiomathes*, **6**.

LILIANA ALBERTAZZI

THE PRIMITIVES OF PRESENTATION WHOLES, PARTS AND PSYCHOPHYSICS

Die Monaden das Universum darstellen, indem sie es vorstellen. Leibniz

[...] cum dicitur quod finis prior est in intentione, intentio sumitur pro actu mentis, qui est intendere. Cum autem comparamus intentionem boni et veri, intentio sumitur pro ratione quam significat definitio; unde aequivoce accipitur utrobique. Thomas Aquinas

Das Äussere ist also jeweils, im strengen Sinne dieses Wortes, ein – mehr oder weniger zutreffenden – Bild des inneren. Metzger

1. 'Ein Altes Scheinproblem'

A theoretical issue that has always been central to research in artificial intelligence is the concept of *representation*. For the sake of simplicity, one may say that analysis of the matter has moved in two main directions: the *symbolic direction* summed up by research in knowledge representation, and the *sub-symbolic direction* of connectionist analysis.

In short, the initial approach to the problem of representation framed in mentalistic, inferential and logico-computational terms has been matched by a different one which contests the view of representation as a *separate module* to be codified in merely symbolic and/or linguistic form. For connectionism, in fact, the representation is what emerges from the workings of the neuronal network and it is located in certain *states* of that network.[1]

Of course, there have been a number of attempts to integrate the two points of view, but a distinctive feature of the symbolic approach to the theme of representation is its predominant interest in the *abstractive* and *inferential* cognitive processes that fall under the heading of 'problem solving', while connectionism has been mainly interested in modelling aspects of *perception*.

L. Albertazzi (ed.), The Dawn of Cognitive Science. Early European Contributors, 29–60.

At first sight, therefore, attention has concentrated on different 'formats' of representation. A feature instead shared by the two approaches is their notion of the *inner character* of a representation, whether this is defined as a neurophysiological state, mental image or symbolic codification.

The current situation is characterized by a number of unresolved problems: for instance, the connection, prevalence or priority of the various representative formats, and the presence and/or role in the latter of unconscious, hypothetical or mnestic elements.[2]

In recent decades the theme of representation has also acquired considerable importance in linguistics. Indeed, certain semantic analyses, as opposed to Chomskian generative grammar, and following cognitive research by Neisser, Jackendoff and Johnson-Laird, have emphasised that *natural language*, in all its forms, bears the traces of Gestalt *conceptualization* and, in particular, of *perceptive schemes*.[3]

This area of research has not yet 'met' artificial intelligence research, in the sense that it has not yet encountered the problem of implementing its results in the machines. Nevertheless, it has raised the theoretical problem of the connection between language and representation in a manner different from the logico-symbolic approach, and in many respects it meets current requirements of common-sense reasoning, and also those of certain sectors of naïve physics and connectionism.[4]

In various ways, therefore, from research in artificial intelligence to the cognitive sciences, the debate on representation displays greater complexity than in the 1950s, and it highlights both aspects pertinent to *different formats* of conceptualization (from perceptive to those to do with processes of abstraction and idealization) and, more generally, *ontological* aspects (inner and/or outer) inasmuch as they concern the problem of *direct reference*.[5]

This complexity internally to the problem of representation is entirely familiar to the European tradition, where the concept of representation – insofar as it relates to a general 'theory of knowledge' (*Erkenntnislehre*) – has always been *structurally* flanked by other aspects concerning ontology, logic, ethics and the theory of the emotions. The two main components of Western thought, rationalism and empiricism, can in fact be classified on the basis of the importance attributed to the various components of the theory of representation.

A further problem concerning representation is *psychophysical* in nature, or in modern parlance the 'mind-body problem'. Although the term 'psychophysics' can only be properly applied to inquiry conducted after publication, in 1869, of Fechner's *Elemente der Psychophysik*, it also covers Herbart, J. Müller, Lotze, Mach, Hering and Helmholtz. Varyingly understood as interaction or parallelism, since J. Müller the psychophysical account of the relationship between *interior* and *exterior* in the representation has assumed that the *perceptive determinations* manifest features of their underlying somatic processes.

After Fechner, psychophysics was distinguished by two main parameters: (i) the attempt to make *mathematical measurement* of psychic phenomena, and (ii) the assumption of *physics* as its model, so that the psychophysical correlates reflected the spatial relations in a system of Cartesian coordinates.[6] This '*altes Scheinproblem*', as Köhler called it,[7] was subsequently challenged by Gestalt, with its view of a (*functional*) correspondence between the *spatio-temporal structure* of events and their *modes of behaviour* in the field.[8]

2. A MOST SUBTLE QUESTION

But what is a representation? This simple question immediately raises a series of sub-issues, i.e.:

a. Are there *primitive aspects* of representing connected to a type of *concrete* presentation which differentiate its format with respect to abstract, mnestic or symbolic representations?

b. Does the use of diverse terms like *Vorstellung* and *Darstellung* in German have a specific *reference*, or is it merely a matter of the preferential linguistic *use* of various authors? Examination of the history of the term and its etymology may yield useful information, starting with Kant, who translates '*repraesentatio*' as '*Vorstellung*' and '*exibitio*' (*hypothiposis*) as '*Darstellung*' – which in turn may be either schematic or symbolic, depending on whether or not it rests on an intuition.[9] Moreover, the German prepositions *vor* and *dar* refer to different spatial relationships, from the exterior to the interior and vice versa.[10]

Vorstellung as mental presentation is a term which assumed increasingly psychological connotations from Leibniz to Tetens.[11] As such, *Vorstellung* was in extremely widespread use between 1830 and 1900, and thereafter, but sometimes with very different meanings. Consider Brentano's use of the term in 1874, where presentations or psychic phenomena are not contents or ideas, but *acts*. In fact, for Brentano examples of presentations are *seeing* a colour, *hearing* a sound, or *thinking* of a centaur.[12] These are therefore *concrete* actual presentations, no matter whether they are perceptual or mental, and to use Brentano's own expression, they involve *psychic energy*.

But then consider Helmholtz's use of the term in *Optics*, where he defines presentations as "the *images* of a visual object as retained in the memory, without being accompanied by any sensation," that is, in other words, *mnestic contents.*[13]

In both cases, however, whether *acts* or *contents* are concerned, the *Vorstellen* in the psychological sense denotes the *inner* aspects of the theory of representation. A particular case is Bolzano's theory according to which *real* presentations, as individual substances, have a counterpart in the *non-real* domain as *Vorstellungen an sich*, as complex meanings, independent of any mental act, a theory which subsequently figured in the architecture of Husserlian phenomenology.[14]

The passage from the use of *Vorstellung* to *Darstellung* (that is, from *inner presentation* to *external representation* or *depiction*) is clearly exemplified in Bühler's work, but undergoes a shift of meaning. Instead of being tied directly to that of *direct notification of a psychic content*, Bühler sets the concept of representation, as in the case of paintings or iconic depictions, in a twofold relation with:

(i) the concept of *representance* (*Stellvertretung*), that is, with the function of symbolising objects and states of affairs which in particular characterises the representative function of *language* (*Darstellungsfunktion der Sprache*);[15]

(ii) the concept of *communication* (*Mittheilung*), drawing on a notion developed by Bernhardi, and which was later central to Marty's philosophy of language.[16]

Thus, inserted between the *presented* inner object of direct intentional reference and the *represented* object are a series of 'idealistic,' segnic or propositional passages; an operation which today is one of the central issues of the debate on imagery, from Fodor to Pylyshyn to Kosslyn.[17] Indeed, Fodor's 'language of thought' and the 'computational approach to human information processing' of Pylyshyn can be read as an example of *representation* theory, *linguistic oriented*, and in this sense non-brentanian. Viceversa, Shepard, Cooper, and Kosslyn approach to the topic are.

From this point of view, other sub-issues immediately arise:

c. Is it possible to draw a demarcation line, at least *descriptively*, between the different aspects of the representation? In other words, are there aspects of the *structure* and of the perceptual *content* that differ from substantially *symbolic* and *definitional* aspects of the representation? Even more drastically, is it possible to identify a 'matter' of the representation or 'direct factors' in Fechner' sense which are sense-data as in the later Stout, or ilethic data as in

Husserl's early writings, or even muscular sensations, visual copies of retinal images, and the like?

d. What difference is there between the *object* and the *content* of a re–presentation? This question, complicated by the sometimes synonymous use of *Gegenstand* and *Object* in German, is crucial for a theory of intentional reference. In fact Brentanism developed a jungle of objects (immanent, transcendent, negative, impossible, incomplete, etc.) based, first of all, on the primary distinction between 'having something *as* (*wie*) an object,' and 'having something *for* (*als*) an object,' i.e. on the difference between the 'internal correlate *of* the act' and 'the object *presented by* the act.'[18] In particular, Brentanisms analyses not only *how* meaning is *represented*, once semantically and syntactically processed, but the very *origin* of meaning itself in its direct intentional reference, where the 'directness of the act' does not mean immediately a 'representation of the I.'[19]
Starting from the structure of the intentional presentation, Brentanism developed a number of distinctions concerning contents, and correlates of consciousness, ranging from psychology to ontology and semantics. And indeed, a representation appears as a whole endowed with parts of different types (objects, contents, characteristic notes, aspects, moments of formal connections among the parts, etc.).[20]
A formulation of the problem is to be found in Höfler's logic, when he explains the use of the concept of 'object' in two senses:

> on one hand it means the *independently existing thing* [...] to which our presentation or judgement directs itself, on the other hand the more or less accurate *image* of such a reality, which exists in our mind; this quasi-image or rather *sign* is identical with the *content* [...] which lies *entirely* within the subject, like the presenting or the judging *act* itself.[21]

e. Then, what type of component of the representative structure are *acts*? Indeed, even in Brentanism, which most closely examined the structure of the representation, the nature of acts was almost invariably taken for granted and neglected, and because of the *unitariness* of the actual psychic phenomenon, it was often made to coincide with the *content* of the representation. This amounts to saying that, of the *existential relation of dependence* between act and intentional object, only the second aspect was actually analysed, amongst other things because of the constraints on experimental enquiry, which concentrates on the analysis and measurement of the actual *content* and its extension. In its original version, however, the act of presentation was psychic energy or psychic activity, in a sense similar to that given to the concept by Tetens.[22] Successively, that aspect of the question was almost completely neglected in the analytic interpretation of the Brentano's theory of intentional reference. For example, British philosophers sometimes translated 'act' as 'attitude,' which is a partial misunderstanding of the theory, towards a propositional interpretation of the nature of intentionality.[23]

As regards the psychophysics of the act, both Brentano and the experimentalists attributed *intensity* to the content and not to the act. This, however, is a rather troublesome aspect of psychophysics. Brentano thought that sensory *intensity* was a measure of the *density* of the sensation and therefore concerned the *objective* aspect of the sensation – that is, what he called the *space of sensations* (*Empfindungsraum*). In other words, a sensation is *pointillistically* drawn in this space, although gaps are left between one point and another, and the more intense the sense-datum is, the more dense this *pointillieren* becomes. For example, according to Brentano colours do not have intensity, but are punctualised differently and varyingly dispersed in this space, so that intensity is some sort of *ratio* between fullness and emptiness.

Conversely, Brentano denied that from a *subjective* point of view *thought* could have intensity. However, from the point of view of the intentional presentation, given that what exists is only the actual perceptual *act* of that particular shade of red, or of that particular chord, these are *concrete universals* which as such seemingly possess intensity. The situation becomes even more problematic with regard to emotional presentations – a case study in Brentanism.[24]

In fact, the structure of Brentano's intentional presentation, in his last dvelopment (from 1900 onwards), asserts that the act has something (i.e. itself) as its object (*zum Objekt*) *directly*, of which however it is conscious (*als Objekt*) *indirectly* through a content. Consequently, we are *primarily* oriented to the inner object, of which state one can identify only the relationship between the variation in the stimulus and arithmetic variation, as Brentano argued indeed. *Secondarily*, we are rather *content*-oriented, and the intensity of a sensation relates to the codification of relevant information in the content of a psychic phenomenon.[25]

The problem of *measuring the act* should therefore be broken down into its constitutive elements, which varyingly regard the *Spannung* of the attention (the limits set by the beginning and end of the act), and therefore the structure of the time of presentness, the emotional aspects that may or may not be present, the presence or otherwise of the content of the act (there may be also acts without content), and so on.

From an *experimental* point of view, only Külpe and the Würzburg school, on the one hand, and some studies by Benussi on the other, undertook an analysis of *acts without objects*. And indeed, the affinity between Brentano's theory of intentional reference and the Würzburgians is well underlined by the criticism raised by Wundt against Bühler's resurrection of the theory of the medieval *actus purus* of Thomistic origin.[26]

From a *theoretical* point of view, by contrast, it was Meinong and Husserl (and Daubert) who most directly addressed the problem of the act.[27]

Generally speaking, by 'acts' or 'functions' are meant some sort of original component of representation which is always *concrete, temporally existent* and *actual.* Acts are dynamic phenomena, spatio-temporally extended, of a continuous nature, and they are of two kinds depending on their more 'perceptual' or 'mental' content.[28]

A different level of analysis concerns the representations of *objects*, variously intended as *configurations* (*Gebilde*) of *activities* (*Handlungen, Tätigkeiten*) or *processes* (*Vorgänge*).[29] This kind of analysis is to be found in Twardowski, in a paper written in 1912, where he develops a naturalist theory of *concrete linguistic signs*, (verbs, substantives, etc.) as a rejoinder to the accusations of psychologism brought by Husserl against his previous theory.[30] In this article Twardowski observes that:

> Psychologism originates from a confusion between actions and psychic products and from the fact that psychic products may have properties whose relations (i.e. relations concerning truth and falsity) can be defined a priori, without any care of psychology.[31]

Moreover,

> Psychophysical products which express certain mental products are also termed 'signs' of those mental products, and the mental products themselves are termed their respective 'meanings.' Thus any mental product which bears to a psychophysical product the relation of being expressed by the latter is a meaning.[32]

However, Twardowski's theory is rather complex, since in it there can be envisaged:

(i) a theory of expression (*Ausdruck*);
(ii) a theory of meaning (*Bedeutung*);
(iii) a theory of *linguistic* expression and of *linguistic* meaning;
(iv) a theory of non-independent configurations (*Gebilde*).

The different aspects should not be confused, since historically they sign a very important step towards the development of the original theory of Brentano's intentional reference and, for example, one of the main point of departure of the analytic interpretation of Brentanism itself.

From a systematic point of view, instead, the question concerns the intrinsic temporality (*Vergänglichkeit*) of the psychic functions. In fact, these seem to have a sort of non instantaneous 'duration' in time if they are *expressed* in *psychophysic* configurations.[33]

An expressed psychic configuration, according to Twardowski, becomes *meaningful* when it is shared by other people, as happens in a piece of musical art, which prompts the same psychic configuration in the listener. The *linguistic* expressions of psychic configurations and meanings, consequently, are the best *tools for giving stability* to the products of psychic functions.

According to the later Twardowski, then, *meaning* is given by the *content* of the *linguistic sign*, but only as *a set of well defined properties*, which is a further example of the shift from aspects relative to the *notification* of an actual psychic content to aspects relative to its linguistic *expression* through the *translating* and *fixing* of the content in *symbolic* or *propositional aspects*.

A similar idea of 'objects' as 'products of action' is proffered by Benussi when he speaks of psychic functions as *objectual* functions and reality *transformers*. Thus the function of *Stellvertretung* of representations, already mentioned, acts by *objectifying* or reifying a form of *psychic energy* or function.[34]

The continuity and the passage between the level of *pre*sentations to the level of *re*presentations, the role played in this process by numerical and/or visual procedures is still an area of enquiry for the cognitive sciences.[35]

f. Considering the complex structure of the representation, and its dual outer and inner boundary, onto what type of *spatio-temporal* continuum can the structure of the representation be mapped? This is a very important question for cognitive science and scientists working in AI. In fact, although Brentano's space of sensations, as far as the intensity of light is concerned, can be approximated by Marr's primal sketch, one must still decide how to reconcile the various kinds of superimposed continua which constitute the multiple and multiform qualitative fillings of our ordinary experience. For example, how one can map sound, volume, pressure, temperature, etc., which require different formal tools and different depictions according to the nature of their primitives.[36]

g. Finally, what is the role of the linguistic module, and what different aspects of the representation are manifested in the functions of *notification* (*Kundgabe*) and of *expression* (*Ausdruck*)? The two terms, in fact, are often used synonymously by authors, but sometimes one observes the passage from a prevalent use of one term to the other, as in the case of Bühler (1918: where the accent is on *Kundgabe*, *Auslösung* (triggering), *Darstellung*; 1934: where it is on *Ausdruck, Appell* (appeal), *Darstellung*). In this change is stressed the passage from information (*Kundgeben*) to meaning (*Bedeuten*), from presentation (*Vorstellen*) to representation (*Darstellen*) and in this passage an essential role is played by *language* as *re-presentative function*.

One notes here the various terminological shifts (the pair *Kundgabe-Ausdruck* and the pair *Auslösung-Appell*), due at least in part to Meinong's influence after publication of Bühler's *Die Krise der Psychologie* in 1927.[37]

As regards the pair *Kundgabe-Ausdruck* in particular, corresponding to the semantic shift is a different emphasis on stages of information processing (that is, notification of a *mental content* and its *linguistic* expression, and moreover the passage from a psychological set-up to a linguistic and logical one).[38]

Nowadays the *theoretical* complexity of these questions is sometimes simplistically summed up as the bi-polar opposition that divides the theory of representation in cognitive studies, namely the 'most subtle question': *Is perception direct or hypothetical?*

As I have observed elsewhere,[39] two main positions are taken in the debate on the theory of representation, the one led by Gregory which argues a constructivist point of view, and the one led by Gibson which sustains a theory of direct (or ecological) reference of perception.[40] In this case too, the contraposition leaves many problems unresolved relative to the 'format' of the representation, to the prevalence of what are dichotomously referred to as top-down or bottom-up processes in the construction of experience, and to the existence of phases in presentation, mainly restricted to the role played by the mnestic traces, which Gregory sustains but Gibson denies. A feature worth noting is that both approaches to the theory of representation embrace the idea that perception proceeds in a single direction from stimuli to meaning, *from the exterior to the interior*.

Among the key issues in the debate are (i) the 'data driven' or 'concept driven' nature of information processing by the cognitive processes, (ii) the role and nature of 'phases' of representation and, even more crucially, the question whether (iii) construction of the world goes 'from the outside to the inside' or vice versa. Classical psychophysics in its entirety, in fact, assumed the former hypothesis.

And yet Fechner himself distinguished between an 'external psychophysics' and an 'internal psychophysics,' which manifested panpsychistic traits. Brentano's criticism of Fechner focused in particular on certain aspects of inner psychophysics, like the concept of order and the qualitative character of phenomenal experience, which do not coincide with the 'physics' of the external world and therefore cannot be reduced to it.[41] This is one of the most important questions, not only for the internal dynamics of Fechner's theory, but mainly because it singles out and guarantees the *ontological non reducibility* of the *phenomenal* or *perceptive level* in the broad sense on the physical one. This was also the legacy left by Brentano to Husserlian phenomenology and Gestalt, which respectively analysed the question from the theoretical and experimental points of view.

Finally, the question also closely concerns a fundamental philosophical stance, that of immanentist realism. In fact, provided that the reality of the world is not questioned (realist assumption), but at the same time is assumed to be mediated by inner reference, the problem is wheter the priority is given to the construction of a psychophysics from outside, or vice versa. If one opts for its construction from inside, metaphysically what I usually call a sort of 'open monadology' follows, the *windows* of which are the acts of presentation, and in particular of perceptual presentation in the 'moment-now'.

It is worth noting that a reply to some of the issues currently debated were provided in its time by the school of Graz.[42]

Höfler, Witasek and Benussi, for example, assumed that perceptual aspects and cognitive integrations are connected '*in the various phases* of the field *of actual presentation*,' thereby rejecting in principle the current dichotomy in the cognitive sciences, but did it assuming 'a phased structure of the actual perception,' in which *hypothetical* unconscious mnestic contents do *not* operate. Moreover, the temporal structure of actual perception does not correspond to the linear and a-symmetrical structure of a *before* and *after*.[43]

Unlike the Berliners, who in general analysed the object of the concrete presentation on the basis of its already-formed and relatively stable (or multi-stable) structures, the Graz researchers also examined the *categorial dynamic formation* of these structures, and therefore also the structure of the *Vorstellung* understood as an *act of completion* of the perceived forms.

On this basis the Brentanians, and the Meinongians as well, distinguished among different types of presentations. Generally speaking, Brentano distinguished *two* different types of presentation:

1. *unitary presentations*, which comprise both intuitions and generalisations of intuitions;
2. *attributive* unitary presentations, which derive from direct and indrect identification of the intuitive presentations.

Meinong in turn distinguished between:

1. *intuitive* perceptions, like the presentations of actually given sounds;
2. *relative* presentations, like the presentations of temporal similarities and differences between sounds, whose objectual correlates are 'states of affairs' or 'objectives'.

To those distinctions corresponds that between *simple objects* of presentations, like musical notes or trees, and *complex* or *higher-order objects*, like melodies or forests.[44]

According to the theory of *intentional reference* of Brentanian origin, this process, which is akin to the intellective processes of the apprehension of form, concerns the various 'directions' of the mind addressed to the apprehension of various types of objectuality, be these colours, musical notes, a melody, a group of dots, a phrasing, amodal presences, or more generally, the apprehension of a difference in distance or position among phenomena. All these cases involve *different types* of *objects* or objectualities and they have diverse *psychological foundation*.[45]

In order to explain the conception, we may very simply refer, for example, to the difference between the presentation of 'several dots' and the presentation of '9 dots'; or the difference between 'a mere succession of black and white' and the presentation of 'the difference between black and white.'[46] For the Meinongians, the presentations of form, like number, melody, difference, and so on, are new and real psychic configurations, so-called *produced* configurations founded on intuitive and primary presentations. Corresponding to presentations produced *a parte objecti* are the founded or higher-order objects, which do not necessarily 'exist' in a strict sense: I can always *think of* an angel or of a golden mountain.[47]

Many of these *inner* aspects of the dynamics of presenting have been entirely ignored, or at least not clearly distinguished, by current debate on the nature of the representation. For example, there are some aspects of Meinongian theory which have sunk into oblivion and which concern a particular feature of the representation: assumptions (*Annahmen*).

Assumptions are neither judgements nor hypotheses, in the sense of a comparison among several and successive presentations in terms of stored mental representations. Assumptions derive from the mere *apprehension of the mode of being* (*So-sein*) of an object, and in Meinong's sense of the term they lack any kind of 'truth assertion' concerning the object itself.[48]

Although the theme of assumptions has been given only semantic development, within Brentanism, this aspect of the theory may be of relevance to cognitive inquiry. For example, the cases of amodal surfaces can be recast on the basis of Meinongian assumptions. In fact, they involve the *apprehension of the perceptually given mode of being* of that particular 'object'. In the Berlin School and in Koffka in particular, the *frame of reference* (*Bezugsystem*) plays a similar role.[49]

Endly, the introduction of particular psychic acts like assumptions, lying midway between presentations and hypotheses, may avert the risk of excessive reductionism, but without excluding the presence of *constructive* aspects of cognitive activity.[50]

That the perception of form is *a-sensory* (*außersinnliche*), in fact, does not automatically entail that at the bottom this is a theory of *mosaic psychophysical depiction* or that it is based on the hypothesis of the constancy of stimuli.

In a Gestaltist configuration, granted the specificity of the Berliners' and the Grazers' analyses, in fact, 'top down' and 'bottom up' processes act *simultaneously*, because perception is viewed as a *whole* of numerous processes with differing degrees of complexity and salience. These processes, in fact, are not hierarchically ordered in a 'before' and 'after' temporal relation, but interact as *forces of a field* in the actual duration, which has traits of different phenomenic salience.[51]

3. BASE FORMAT ONE: THE BOUNDARY OF REPRESENTATION

The subtlety of the distinctions drawn by the Graz school within the general framework of the theory of production highlights the relationship between some sort of *base format* of the representation (acts of concrete presentation of primitive factors relative to visual, auditory or tactile aspects of the perceptual continuum) and *founded formats* dependent on the former and founded on lateral presentations (so called 'objectives') and mnestic representations in the strict sense. A number of examples were given by the analysis of the perception of different temporal intervals, whose final outcome depends on the different relative presentations involved in the process (position of intervals, salience of pauses, different filling of intervals, intensity of sounds, etc.).

In other words, the Graz school stressed the presence of an *inner temporal stratification of the morphogenesis of the representation*. Current research on the auditory stream could indeed benefit from reactualising the theory and inserting all these apparently homeless phenomena in a theoretical framework.[52]

It is now only necessary to take one further step back to the foundational level, or to the *spatio-temporal perceptual* structure of the base format. And this is in fact what the Grazers did in their experiments.

Generally speaking, researchers currently distinguish at least *two* levels in the perceptual process: one which can be defined as *pre-attentive*, and one which is more *cognitive* and involves the identification and categorisation of the *objects* in the field.[53] Nevertheless, these definitions are still too loose for analysis of the base format. In the base format there are numerous phenomenal differences among the particulars, even at the minimal level of cues: for example, *form* completion (which involves phenomena of segmentation, grouping of fragments, orientation, type of occluder, edge assignment, etc.)

apparently does not simultaneously imply forms of *content* filling (like brightness, colour, texture, depth, etc.); or else evaluation of an empty space between two points seems to involve greater activation of the forces of the perceptual field than does evaluation of a full line, etc.

The Grazers, for their part, stressed that *from a dynamic point of view* the theory of representation does not begin with *objects*, but rather with *events* or processes as the primitive phenomena of *change* internal to an actual duration.[54] But, in this sense, qualitative change is explained as a relationship among the *various phases* of *what is* to *other phases* of *what is*, which are *in different stages of enfoldment*. The qualitative change given in duration has an *implicit order* in terms of a series of *interprenetrating elements* in different grades of enfoldment, characterised by the fact that they are *all co-present*, like the notes in a melody. The same holds for every actual perception of events or of a series of events which tend towards a certain pattern, which amounts to claiming the dynamic co-presence of 'lower-order' and 'higher-order' objects in the duration.

According to the analysis conducted by the Graz school there are, in the base format of representation, moments *internal to the presenting* constituted by forms of cognitive completion, whether these be phenomena of assimilation, collections of parts (similar to those labelled as *Bindung* in Matthei or *Verbindung* in Husserl),[55] or those of amodal completion first identified by Benussi and Musatti and subsequently analysed by Kanizsa.[56]

These *internal determinants*, as Benussi called them, which also comprise analytical and synthetic personal attitudes (*Einstellungen*), are conditions which influence perceptual performance. In fact, with an analytic attitude the elements tend to be perceived as separate and independent, with a synthetic attitude the figure is mostly perceived as unitary.[57]

Internal conditions *work together* with other types of conditions or *external determinants* like the difference in brightness between figure and ground, the particular hue of a colour, the influence of the secondary lines in the Müller – Lyer illusion, and so on.[58] In other words, the format of the representation is a *dynamic whole* with an *external* and *internal* border.

From this point of view, and considering the fact that in order for the cognitive modules to be activated a certain temporal extension is required, the field of *perceptual duration* provides a testing ground for the analysis of the concept of 'representation.' The principal area of the temporal analysis of duration is constituted by sounds; but also lights, colours and even their particular states (like transparency or the perception of whiteness) offer countless examples of the fact that the object of presentation is a *structure dependent* on the entity, dimensions and order of the '*perceived*' *temporal intervals*.[59] It is this intuition that underlies stroboscopic movement, for example.[60]

As von Ehrenfels, Witasek, Ameseder and Benussi showed, the time of the duration is largely reducible to the time of a *qualitative change*. This is perception of a change in the *position*, *direction velocity*, *pureness* and *degree* of the forms perceived, which varyingly depend on the phenomenal salience, texture, clarity, and so on, of the events of a duration, and which affect the perception of a sequence of stimuli. One cannot even say that the time of the duration is composed of a series of *quanta* (Stroud's perceptual moments)[61] because, for example, according to the pitch, timbre and qualitative similarity of the sounds in a triplet which fall within one of these moments, these sounds may be perceived – and are thereby *recombined* by the perceptual processes – in *a manner different* from their sensory *order*.[62]

The phenomenon is apparent not only in sounds but in other perceptible materials like colours. In fact, corresponding to the phenomenon of temporal inversion in the auditory field are the phenomena of inversion in the visual field,[63] and likewise the phenomena of transparency, when the colour dissolves and the colour of the surface beneath is seen *through* that superimposed on it.[64] Involved in this case, too, is the *double-face* structure of perception, because on the one hand there is *a single* linear chain of stimuli, while on the other there are two perceptual 'events,' in this case separated in space but coincident in time.

In this sense, the colour and sound perception analysed by the Brentanians showed both:

(i) that the perception of sounds and of coloured shapes is a product of *multiple dependence* even at the phenomenic level;[65]

(ii) that in sounds and colours as *elementary observables*, there operate *functionally* different 'bottom up' and 'top down' cognitive processes, a feature underlined particularly by Husserl.[66]

The same holds for phenomena of amodal completion, analysis of which seems to allow different interpretations *according to the duration of exposure*. For stimulus onset asychronic duration (SOA) of times of more than 200msec, in fact, occluded surfaces show priming effects identical to those of fully visible, *complete surfaces*. At SOAs around 100msec, on the contrary, occluded surfaces tend to show priming effects more similar to those of *mosaic surfaces*. We may therefore conclude that there are *two parallel processes* which *compete* to gain access to consciousness. These two processes reflect *two different perceptual modes* always active and operating in parallel.[67]

The analysis of the genesis of perceptions shows that the form of percepts is stabilised in the duration. It is topologically extended and flattened in a field of forces created by the perceptual contours. The visual or auditory percept thus produced is therefore the outcome of an equilibrium of primitives or of *formative factors* (i.e., not of *direct* factors as in Fechner' sense) in the complexity. The form of a visual object, therefore, seems to be the *resultant* of forces in which there operate *spatial primitives* consisting of the points, surface lines, directions, angles and parallelograms of a space which still does not possess the features of a Euclidean space, but comprises tactile and kinaesthetic qualities, movement, velocity, and tension towards a form. It is therefore an essentially *dynamic structure* of the phenomena of vision. In fact, to use an expression of phenophysics, there arise *temporal structures mapped on to a space of qualities.*[68]

This outcome should raise at least some questions for researchers in AI concerning the formalisation of perceptual primitives for coding in algorithms for further implementation.

4. Base Format Two: The Pieces and Moments of Representation

One of the general assumptions of the classical theory of Gestalt is that a whole is greater than its parts. There exist, moreover, different types of wholes: aggregate (a heap of sand), integral (Socrates) and essential (the whole by definition).[69] A second distinction concerns the fact that the parts of an aggregate can also exist independently (the individual grain of sand), while in the case of integral and essential wholes, the existence of the whole precedes the existence of the parts, which are such only in relation to the whole to which they pertain. There are therefore at least *two laws of the composition* of wholes which we may label, descriptively, as:

(i) bottom up;
(ii) top down.

In both cases there arises a *relation of ontological dependence*: in the first case, i.e. a sum, the relation is *from the parts* to the whole; vice versa, in the other case, i.e. Gestalt, the relation is *from the whole* to its parts.[70]

This is an essentially phenomenological theory: in fact, despite the shortcomings of his *Third Logical Investigation*, Husserl's achievement was to elaborate a theory of wholes and parts.[71] Husserl distinguished in the phenomenological object:

(1) *separable parts* or *fractions (Stücke)*;
(2) *non separable moments (Momente).*

Parts in their term are distinguished into:

1.1. *substantial* (parts of the object reciprocally external to each other, for example an arm);
1.2. *non substantial* (parts of the object which co-penetrate each other), for example the blood circulation system;

2.1. *immediate moments* (properties of the non-substantial parts and forms of connection among the substantial parts), for example 'red';
2.2. *mediate moments* (properties of the moments of objects), for example the 'intensity' of red.

Subsequent mereological analyses, however, have underestimated or indeed neglected certain fundamental features of the original theory, from the point of view of both wholes and parts.[72]

It also seems that the primitives with which these attempts at formalisation begin are still too poorly defined conceptually, on the one hand, or on the other excessively formalised without preliminary reflection on the *type* of formalisation to adopt as the one that sheds clearest light on the particular phenomenon under observation.

The predominance of the parts in the *genesis* of the form, an assumption also shared by Benussi and by Musatti, is nowadays also apparent in Marr's research on the *phase structure* of vision which, from a phenomenological point of view, seems to obey the fundamental form of 'sense-data' processing known in Husserl's theory as *segmentation* (*Sonderung*), and which constitutes the counterpart to the fusion (*Verschmelzung*) of 'phenomenal objects' in perception.[73]

As it is well known, the first phase, in Marr's analysis, involves an initial two-dimensional ordering of the visual data which passes through two stages:

1. elaboration of punctiform *elements* (2 dimensional sketch or grey level description) dependent on variations in light intensity;

2. elaboration of *surfaces* (2.5 dimensional sketch) due to the fact that computation of local variations in intensity consists in measuring the *perceptual gradients* of vision or in plotting *angles* which originate on the *edges* of objects, *indices of depth*, *segments*, *patches*, characteristic notes of the *texture* and *shading* of objects.

The corners and edges of objects are perceptually continuous. Hence even primary-level computation involves the classic principles of Gestalt, namely the laws of proximity, similarity, continuity, closure, etc. In fact, the 'thing' that is computed is in the final analysis but *the figure on a ground*. This level of description is *primary* and depends on the subject and his/her *point of view*: in fact, for example, it does not represent the *hidden surfaces* of objects. The perceptual base of this stage of representation nowadays is mantained by Jackendoff's theory of representation, but it is worth noticing that these aspects, like recognition of the forms of objects on the basis of boundaries, chromatic contrast, filling intensity of the sensory qualities, etc., are all already present, however from a *descriptive* point of view, in Brentano's dictations on space, time and perceptual continua.[74]

3. The second phase involves the elaboration of *objects* or *parts* of objects at a three-dimensional level (3 dimensional sketch).

In this phase, from the plotting of 'things' as 'objectual figures against a ground' one moves to the *spatial and volumetric* three-dimensional level of '*objects*.' The process is actualised in canonical representations which follow the *volumetric primitives* (generalised cylinders) which move along axes of direction (the system of co-ordinates centred on the object). Then the *parts* of an object may be located in this space.

Of the three sketches, the first two are mainly bottom-up processes, while the third, involving the extraction of canonical volumetric forms, suggests the intervention of specific top-down processes.

One of the still-unresolved problems concerning this theory of vision is the *potential point of view* of the object. This is dependent on the observer but cannot always be assumed to be so. Consider the *canonical aspects* (Husserl's *Abschattungen*) of the frontal or lateral view of the facade of a house analysed by Koenderink and van Doorn.[75] Marr and Nishihara, for example, tackle the problem by adopting the system of the object based on cylinders and axes of direction, but the *subjective* problem of the representation of the point of view is still unsolved.[76]

Since the very structure of the *potential point of view* corresponds to the structure of Husserl's *noema*, a return to phenomenological analysis also in this case might help to clarify the perceptual layers from a descriptive point of view.[77]

According to Husserl, in fact, the perceptual field is spatially constituted by *limiting forms* which we identify as dividing the perceptual whole into parts. In order to define the nature of the perceptual field, organised as a system of places (limiting forms) and of markers (oppositions like light-dark, centre-periphery, figure background), however, one must be able to formulate formal concepts and axioms, using concepts like 'point', 'line', 'surface', 'distance', 'direction', etc. And this aspect immediately rises the question of how to *represent* and *modelize* the perceptual continuum which is not immediately a mathematical one.[78]

From a phenomenological point of view, a spatial multiplicity, as an aggregate of *points of lights*, becomes a spatial surface when a multiplicity relative to *distance* (which is unidimensional and linear) is connected with a multiplicity relative to a *change of perspective* (which is two-dimensional and cyclical).[79]

5. A 'VON UNTEN OBEN & VON OBEN UNTEN' SCHEMA FOR REPRESENTATION

In order to depict this type of complexity of the *format of representation*, in 1925 Benussi formulated a scheme which summarised its features of:

(i) *temporal structure;*
(ii) *dynamic whole.*

Benussi's scheme depicts the *genesis* and the *passage* in actual perception from a *pre*sentative format to a *re*presentative one; that is, from a phase of *perceptual presence* to a phase of *mental presence*, to use terminology first coined by Benussi and then adopted by Musatti and finally disseminated by Kanizsa.

By means of experiments Benussi showed that the *dynamics* of the representation moves structurally through three stages:

1. initially, a content of an *act* of presentation is shaped in a duration, according to a succession of *content-features* which follow a *consecutive* order as if they were parts of a temporal process. These content-features are *not* seen as 'well defined elements,' a conception which Langer and Shanon afterwards would share too;[80]

2. once the *act* is completed, the final holistic content is *non-distributed*, in the sense that the parts are no longer arranged in consecutive order. Moreover, in this phase, the now 'sensorily absent' components of the whole are presented as if they were still present, but only 'mentally.' Consider again the notes of a melody, in which all the notes are 'mentally' present but 'sensorily' absent. This is a situation of mental *simultaneous contemporaneity* which preserves traits of temporal distributive signs, like 'precedent' or 'successive.'[81]

It is at this level that the transition takes place from a sensory modality with, so to say, *high 'perceptual gradients'* to a cognitive modality with *high 'cognitive gradients'*: in short, the initial steps of transition from *pre*sentation to *re*presentation – for example from 'seeing' to 'thinking' in Kanizsa's sense.

In other words, the presentational and the representational define the *two poles* of cognitive activity.[82] A clearcut case of *re*presentational cognitive expression is a set of well structured expressions subjected to computational operations as happens in formal logics. But, more generally, all *pre*sentational cognitive expressions can be built into *re*presentational terms, i.e. *language oriented.*[83]

3. the third stage of representation concerns *memories* – that is when the *presentation content* is re-activated by memory. In this case the whole is reactualised by the *composition* and *recomposition of its parts* in an *order* which is *not* that in which it was originally constituted. Obviously *cognitive* integrations are very powerful here. For example, a sequence *a, b, c,* can be re-presented as *b, c, a.*[84]

This last aspect sheds light on a conception of memory seen no more as a storehouse of representations, but as a new-presentation able to replace the old state.

In the second stage, which marks the passage from presentation to representation, there also occurs the origin of *meaning* and of *substantivity* in the strong sense. On the basis of the formation of non-distributed content, in fact, we have a noun that denominates (not the parts but) the final holistic 'object' of representation (*noema*). This aspect corresponds to the *Benennung* in the Brentanists' theory of language, that is, to the nominal semantic category.[85] Linguistic description, indeed, confirms Benussi's findings: experiments show that when a series of figures are described, the emphasis is placed first on the formal characteristics of objects which respond to a criterion of the *segregation of the form*.

Benussi's scheme, which seems to point out very important steps towards a *theory of imagery*, suggests further conclusions, including:

(i) the existence of *phases* in the act of actual presentation, which are *not* in a asymmetric succession;[86]

(ii) it specifies that this is a series of *interprenetrating elements* in different grades of enfoldment, *all co-present* in a duration. Thus it is not a mere top-driven hypothesis; probably it is a situation of data-driven *plus* cognitive completions due to *internal* determinants; in other words, the differences between bottom-up and top-down processes are more of *degree* than of kind;

(iii) it highlights the distinction and the passage from *pre*sentative aspects (phase 1) to *re*presentative ones (phase 2, 3), i.e. the passage from a certain type of *objectual quality* (perceptual presence) to another type of *objectual* *quality* (mental presence), to which corresponds different types of functions and of schemes;[87]

(iv) finally, it shows the *modification* of the whole that comes about when one operates on the basis of mnestic materials of even a few milliseconds ago.

On this point a major step was taken by the *early* Stout, who argued that the *contents* of presentations are:

> *sense impressions* and *images by which* we perceive or imagine material things and their qualities, states and processes. A special term is required to designate contents of immediate experience which thus fulfil, or are capable of fulfilling, the function of presenting or *introducing* objects which are not themselves contents of immediate experience. The term selected for this purpose by the group of writers I am dealing with (Meinong, Husserl, &c.), is '*Inhalt*.'[88]

However, strictly speaking Stout's contents correspond neither to those of Meinong nor to those of Husserl, which in the first instance are phenomena 'encountered' or 'lived' immediately in the actual presentation. These in fact are 'things' which have form, extension and properties like something being blue, hot, heavy, etc., which no '*mental state*' can have, as Findlay points out in his commentary on Stout.[89]

The things of the actual presentation, in fact, *do not* have any *indicative* or *symbolic* significance. They are *directly perceived forms* (*gegenständliche Bilder*) which 'immediately *present*' their meaning on the basis of their *So-sein*; they are *not signs* that *represent* (*vertreten*) a meaning.

The problem is that when we are 'living through' them we do not pay attention to them. It is only when we have to *express* them or *communicate* them that in some manner we 'objectify' the qualities of phenomenally experienced things and speak of them in terms of the 'content of a presentation of redness' or the 'content of a presentation of heat,' and so on. In this way the indubitable data phenomenally 'encountered' in actual perception assume the appearance of *hypothetical* entities through their transformation into *mental contents* and *linguistic expressions*.

For decades, the same assumption has been used to ignore a series of theoretical and experimental findings, concerning the *original structural aspects* of perception, and hastily put apart as 'unconscious' mnestic traces.[90]

Possibly, the *later* Stout's conception of sense-data approaches this problem, since both sense and physical objects are 'of a piece,' as he puts it, i.e. both are *material*: but assuredly they are not *qualia* of a *definitional* type.[91]

As regards contemporary philosophy, the assumption that it is not possible to achieve the *conscious constructions of percepts* starting from original primitive elements like *acts of presentation* – and in general that analysis of both *Gestalten* and *meanings* must begin with *mental contents* rather than *direct perceptual information* – is present in Føllesdal's interpretation of the Husserlian concept of *noema*, a highly defective 'mentalistic' interpretation, which however became predominant in those realms of AI interested in phenomenology.[92] We could also argue that Føllesdal's interpretation of Husserl's noema is strictly reminiscent of Helmholtz's position, according to which sense data (*hylè*) are transformed into perceptions via unconscious inferences (*noesis*).[93] On the contrary, Husserl very soon abandoned this first initial conception and indeed, to cite only the crux of the problem, according to him every *perception* is primary:

> a web of partial intentions, fused in the unity of a completive intuition. Correlate of this last one is the *thing*, while correlates of those partial intentions are *parts and moments of the thing*.[94]

which concerns the already mentioned problem of the 'object' revealing itself from different sides in the process of perception. Moreover, one should not forget that the concept of noema in Husserl is a complex issue which concerns, among the various aspects:

(i) *sense*, or the noematic object considered in the *how* of its determinations;
(ii) *noematic nucleus*, or 'qualitative positional character' (*matter*) of the act, as opposed to the *quality* of the act;
(iii) *pure x*, or bearer of predicates, as the noematic aspect of what is the moment now in consciousness.

Generally speaking, there has been a main misunderstanding of Husserl's conception by analytic philosophers, because the noema is not a list of a descriptive type, as they maintain, but a *hierarchy of rules*, in a cognitive sense.

In Husserl's conception of noema, in fact, *the mental activity plays a major role in referentiality* and also in determining the *type* of objects to which we are directed. From that point of view, a 'transcendental' analysis is but a regressive analysis, concerning the ultimate sources of all psychic configurations (*Gebilde*). For example, the noema of a facade implies ordered appearances of the other perspectives of the house, which means that the representational content is factually a rule to synthesise the 'multiplicity of experience' in a very Kantian sense.

At various levels, Searle's, Dennett's and Davidson's accounts show the same reductionist interpretation of the theory of intentionality, based on a series of shifts, for example in the interpretation of the original conception. Among these shifts, there are the interpretation of 'act' as 'attitude,' 'act matter' as 'propositional attitude' and, in the specific case of Searle, of 'act-quality' as 'illocutionary force.'[95]

The difficulty encountered adopting an analytic approach to the problem of intentionality is admitted by Dennett himself in the *Oxford Companion to Mind.*[96]

6. A METAPHYSICALLY CORRECT CONSTRUCTIVISM

From what has been said, the crucial theoretical feature of the choice of a particular 'format' of representation – for example, the choice of 'externally' directed aspects rather than 'internally' directed ones – consists firstly in *which primitives* of the description are selected, and secondly which *direction* is considered metaphysically primary in psychophysics.[97]

From the foregoing observations concerning the theories in question, it seems that the arguments put forward by the authors considered lead to a moderate constructivism. This does not seem to be, or not entirely, a radical constructivist conception, as in Varela and Maturana,[98] because it involves not *solely* a 'top-down' process but one which is both 'bottom-up' and 'top-down,' and in which *self-reference* does not create a world separate from the reality of transphenomenal things, a world beyond the *mind*'s capacities of apprehension. The *phases* of the actual perception of something red, something quadrangular, something loud, in the multiform continuum of experience of Brentano's metaphysics, in fact, are partial presentations of dynamic wholes, of which the *perceptor* is a *non-independent part* as a sort of 'active agent tuned to objects.'[99]

Self-reference, in fact, understood in the Brentanian sense of proprioception, involves not only the mind but also the *psyche* in the Aristotelian sense of a *complex and internally stratified form* able to apprehend the entities of various *irreducible* but *ontologically connected layers* of reality.

All the authors to whom I referred to in my contribute worked on the *boundary* between internal and external psychophysics, from Fechner's direct factors to phenomenological forms to mental contents to ideal objects, without any hypothesis of reductionism.

Assume for a while that metaphysics is no longer 'off limits.' Is it therefore so inadvisable to consider a '*third* manifesto for naïve physics' on a Brentanian basis, i.e., a theory of scientifically-founded, even neurophysiological, intentionality?[100] This is the hidden question that prompted me to consider a book like the present one to elicit discussion with researchers in the cognitive sciences.

Dipartimento di Sociologia e Ricerca Sociale
Università di Trento
Via Verdi, 26
I-38100 Trento
e-mail: liliana.albertazzi@soc.unitn.it

[1] For the first approach see the classic Newell, Shaw and Simon 1958; Fodor 1968; Johnson-Laird 1983. For the second see for example Fodor and Pylyshyn 1988; Anderson 1995. For a recent survey of the dynamic components of connectionism see Port and van Gelder 1996.
[2] Cf. for example Marcel 1983a, and b.
[3] Cf. Lakoff 1987; Langacker 1983; Talmy 1985. On common-sense see Davis 1990, on naïve physics Hayes 1985a, and 1985b. On this different approach to semantics see Albertazzi, forthcoming e.
[4] Cf. Langacker 1991, 535-6.
[5] On this see Cantwell Smith 1996.
[6] On the classical approach to representation in psychophysics cf. Luccio, this volume.
[7] Cf. Köhler 1929.
[8] Cf. Metzger 1986, esp. iv and v.
[9] Cf. Kant 1790. On the concept of *Vorstellung* from Wolff to Kant cf. Knüfer 1911; Hafner 1974. As to the linguistic expressions used by criticism cf. Cassirer 1906, vol. 3, ch. 5.
[10] Cf. Bernhardi 1801.
[11] Cf. Knüfer 1911, 57 ff.
[12] Cf. Brentano 1874, B.2, § 5.
[13] Cf. Helmholtz 1896, 10.
[14] Cf. Palagy 1902, esp. chap. 4; Morscher 1973; Sebestik 1992.
[15] Cf. Nehrlich 1996, and forthcoming.
[16] Cf. Marty 1908.
[17] Cf. Fodor 1968; Pylyshyn 1981a, and b; Kosslyn 1983. For the debate on imagery cf. Shepard and Cooper 1982; Finke 1989; Tye 1991. Cf. also Introduction, this volume. An antirealistic position similar to those of Fodor and Pylyshyn is shared by van Fraassen and Putnam.
[18] Brentano 1982.
[19] On the problem of the 'I' cf. Koffka's and Dennett's idea of self, in Koffka 1935 (set of memory traces) and Dennett 1991 (an abstraction related to a sort of 'center of narrative gravity').
[20] On the parts of the psychological whole cf. Albertazzi 1994.
[21] Höfler 1890, § 6, emphasis mine.
[22] Cf. Tetens 1777. There are various similarities in Tetens' and Brentano's vocabulary (for example the concepts of *Empfindungsvorstellungen*, *Phantasmata*, *Apperzeption*, etc.), but they obviously do not concern the problem of a 'classification' of psychic activities, which was truely Brentanian. A subsequent approach to the Brentanian idea of 'energy' may be found in Wertheimer 1924. Nowadays it shows similarity with the position expressed in Palmer 1995.

[23] Cf. Brentano 1905. On this cf. Albertazzi 1995/1996.
[24] Cf. Calabresi 1930.
[25] Cf. Poli 1998.
[26] Cf. Wundt 1907. On this cf. Kusch, this volume.
[27] Cf. Külpe 1920; Meinong 1917; Husserl 1900–1, Fifth Investigation.
[28] On the extensity (*Extensität*) of psychic events cf. Brentano 1976. On the topic cf. Albertazzi forthcoming c.
[29] The text of this article, originally written in German, has been translated by Twardowski himself into Polish with the title 'O czynnosciach wytworach.' A shortened version of the Polish text has been translated into English with the title 'Actions and products,' in Pelc 1979, 13–27. The integral translation of the Polish text is in Twardowski 1999, 103-132. As to the terminology, the German text has 'functions' (*Funktionen*) and 'configurations' (*Gebilde*), the Polish text has 'activities' and 'products.' The German text (*Funktionen und Gebilde*) has been recently edited by J. Brandl in *Conceptus* 36, 1996, n. 75, 157–189, with an explanatory Introduction by the same author. Cf. Brandl 1996. On Twardowski cf. Cavallin, this volume.
[30] Twardowski 1912.
[31] Twardowski 1912, 271.
[32] Twardowski 1912, 19–20.
[33] Idem, §§ 30, 33.
[34] Cf. Benussi 1925, Part I, 4 ff. On this cf. Albertazzi forthcoming a.
[35] Cf. for example Marr 1982; Jackendoff 1987; Kosslyn 1994.
[36] On this point cf. Albertazzi forthcoming c.
[37] Cf. Meinong's letter of 26.8.1919.
[38] Brentano's 1956 terminology is *Kundgeben, Aufdecken* (to reveal) and *Ausdruck des Begriffsapparates*. Twardowski's 1894's is *Kundgeben, Ausdecken* (to disclose) and *Nennen*. Marty's 1908 is *Bedeuten* (i.e. *Darstellen*), *Aufdecken* and *Nennen*. Meinong's 1894 is *Ausdruck, Nennen, Bedeuten*. Bühler borrowed the concept of the 'representative function of language' from Marty. Cf. Kiesow 1988, 54, n. 11. Bühler also drew on Külpe 1920 for the concept of the coordination (*Zuordnung*) at the basis of signifying and on Hilbert's projective geometry. Cf. Bühler 1909. Note finally that *Benennung* or nomination consists in *singling out a unitary percept* and corresponds to an Aristotelian *tode ti* endowed with a certain *typicity*.
[39] Albertazzi 1996, and 1999b.
[40] Cf. Gregory 1981. The perception as hypothesis thesis maintains that sensory information translates in a language of the mind. From this point of view it has a lot of point in common with Fodor 1968 and Pylyshyn 1981a, and b. For a keen analysis of Gibson's point of view cf. Shanon 1993, 315 ff., distinguishing between an (i) external week computational type and an (ii) external intensional one.
[41] Cf. Brentano 1874. Panpsychism, as a theory of consciousness, is nowadays present in some trends of quantum mechanics. Cf. Penrose 1989; Lockwood 1989.
[42] Cf. Albertazzi forthcoming b.
[43] The symmetry thesis is argued by Rock 1983, and Sekuler and Palmer 1992. From a phenomenological point of view, however, 'before' and 'after' are terms related to a phantasy representation or recollection.
[44] Cf. Meinong 1899.
[45] Cf. Albertazzi 1995.
[46] The examples are taken from Witasek 1907, § 26, but a number of them is to be found in Benussi's work on the perception of temporal intervals. Cf. Benussi 1913.

[47] Cf. Meinong 1899. Witasek distinguishes three types of *produced* presentations: (i) *Gestaltvorstellungen* (presentations of form) (*Raum*), (ii) *Vergleichsvorstellungen* (presentations of comparison) and *Verbindungsvorstellungen* (presentations of connection). Cf. Witasek 1907, 233 ff.
[48] Cf. Meinong 1910. On the 'modes of being' intended as tertiary qualities see Metzger 1941, ch. 2.
[49] Cf. Koffka 1935. I owe this observation to Alf Zimmer.
[50] For a reductionist constructive hypothesis see Roth 1995, esp. 75.
[51] The two different positions are maintained respectively by Rock 1983, and Gibson 1950.
[52] Cf. Michon and Jackson 1985; Sekuler and Palmer 1992.
[53] Cf. Neisser 1967.
[54] Cf. Eilan, McCarthy and Brewer 1993.
[55] Cf. Husserl 1891; Matthei 1929. The 'collection' is a *constructive piece* of the psychic wholes, like 'connection,' 'cognition,' etc.
[56] Cf. Kanizsa 1991.
[57] Benussi's point of view seems to be very similar to Witkin's cognitive styles, e.g. field dependent vs field independent, or levelling vs sharpening. Cf. Witkin 1983. I owe this remark to Alf Zimmer.
[58] On internal and external determinants cf. also Benussi 1922–23; Matthei 1929. Among the 'internal' determinants there are also Sander's and Krueger's structures of 'totality' and of 'parts.' Cf. Sander 1930. Among 'external' determinants are also Ehrenfels' *Überschaubarkeit* and Köhler's *physical Gestalten*, and the *grouping* of forms analysed by Wertheimer. Cf. Ehrenfels 1890; Köhler 1929; Wertheimer 1922-23. On the 'rediscovery' of *Zeitverschiebung* (temporal displacement) phenomena by the cognitivists cf. Michon and Jackson 1985.
[59] Cf. Pierantoni 1996, 440. Also chromatic sensations and perceptions are characterised by the temporality of the computation originated by the neuro-biochemical algorithm.
[60] Cf. Wertheimer 1922-23.
[61] Cf. Stroud 1949.
[62] Cf. Vicario 1973, and 1993.
[63] Cf. Legrenzi 1971.
[64] Cf. Gerbino 1988.
[65] Cf. Albertazzi 1994, and 1996.
[66] Cf. Husserl 1966.
[67] Cf. Bruno 1997.
[68] Cf. Thom 1980.
[69] Cf. Aristotle, *Metaphysics* B.10, 1023 b26–28.
[70] Cf. Albertazzi 1994, and 1995.
[71] Cf. Husserl 1900–1, and 1939.
[72] Cf. Goodman 1951.
[73] Cf. Husserl 1900-1. On this cf. Petitot 1994. From a psychological point of view cf. Kanizsa 1956, and 1991.
[74] Cf. Jackendoff 1987; Brentano 1928, and 1976.
[75] Cf. Koenderink and van Doorn 1977.
[76] Cf. Marr and Nishihara 1978.
[77] Cf. Husserl 1966.
[78] Cf. Albertazzi forthcoming c.
[79] Cf. Husserl 1966, ch. 3, § 31; Husserl 1939 (1973), 255.
[80] Cf. Langer 1988; Shanon 1993.

[81] This is the so called phase of immediate or consecutive memory.
[82] As Shanon puts it, 'cognitive expressions are characterised by different profiles along the representational-presentational continuum'. Cf. Shanon 1993, 286. Also Husserl's theory of intentionality, consequently, in principle would be readable according to Shanon's point of view.
[83] On this point cf. Simon's distinction between personal and scientific representation of reality in Simon 1981. Cf. also Murray 1995, 173 ff.
[84] The same holds in states of suggestion, where the subject of experiment must evaluate, for example, a series of points. Suggestion, according to Benussi, allows the exclusion of some conditions of consciousness which might influence the state under observation because they are *part* of it. Cf. Benussi 1925. On this topic cf. Albertazzi forthcoming a. As to Husserl, instead, also primary memory is part of the perception itself, differently from phantasy and recollection. Cf. supra, note n. 43.
[85] Cf. supra n. 38. Cf. also for a comparison Shanon 1993, 285.
[86] Similar analyses were conducted by Calabresi 1930, and Bonaventura 1929.
[87] Cf. also Musatti 1926, ch. 10, 80.
[88] Cf. Stout 1896, 355.
[89] Cf. Findlay 1933, 28.
[90] Cf. Husserl 1966.
[91] Cf. Stout 1931, and 1952; Findlay 1933, 28.
[92] Cf. Husserl 1900–1, Sixth Logical Investigation, § 10; Føllesdal 1969. For a contrary view see Gurwitch 1964, and Dreyfus 1984b. On the whole topic cf. Schuhmann 1991; Albertazzi 1989.
[93] On this problem, which involves also Kraus' defective interpretation of Brentano, cf. Zimmer 1998.
[94] Husserl 1900–1, 2nd ed., Sixth Logical Investigation, § 13.
[95] For a similar opinion cf. Mohanty 1997.
[96] Cf. Dennett in Gregory 1987, 383–386.
[97] See Nadia example, in Pierantoni's paper, this volume.
[98] Cf. Maturana and Varela 1982.
[99] Cf. Swain 1991. I take this observation from Zimmer 1998.
[100] For a similar approach cf. Cantwell Smith 1996.

REFERENCES

Albertazzi, L.: 1989, *Strati*. Trento, Reverdito.

Albertazzi, L.: 1994, 'The Psychological Whole. I: The Temporal Parts of Presentation,' in Albertazzi and Libardi, eds., 1994, 145–175.

Albertazzi, L.: 1995, 'Forms of Completion,' *Grazer Philosophische Studien*, **50**, 321–340.

Albertazzi, L.: 1995/1996, 'Die Theorie der indirekten Modifikation', *Brentano Studien*, **6**, 263-282.

Albertazzi, L. ed.: 1996, *The Philosophy of Alexius Meinong*, special issue of *Axiomathes*, **7**.

Albertazzi, L. ed.: 1999a, *Shapes of Form. From Gestalt Psychology and Phenomenology to Ontology and Mathematics*. Dordrecht, Kluwer.

Albertazzi, L.: 1999b, 'Form Metaphysics', in Albertazzi, ed., 1999a, 257-305.

Albertazzi, L.: Forthcoming a. 'Transformers and Conductors of Reality. The Case of Real Psychic Analysis,' *Proceedings of the Meeting 'Sign Processes in Complex Systems,'* Dresden, October 6–11, 1999.

Albertazzi, L.: Forthcoming b. 'Presentation and Production', in Albertazzi, Jacquette and Poli, eds.

Albertazzi, L. ed.: Forthcoming c, *Unfolding Perceptual Continua*. Amsterdam, Benjamins Publishing Company.

Albertazzi, L. ed.: Forthcoming d, *Meaning and Cognition: A Multidisciplinary Approach*. Amsterdam, Benjamins Publishing Company.

Albertazzi, L.: Forthcoming e, 'Which Semantics?', in Albertazzi, ed., forthcoming d.

Albertazzi, L. and Libardi, M. eds.: 1994, *Mereologies*, special issue of *Axiomathes*, **5**.

Albertazzi, L., Cimino, G. and Gori-Savellini, S. eds.: 1999, *Francesco De Sarlo e il laboratorio fiorentino di psicologia*. Bari, Edizioni Giuseppe Laterza.

Albertazzi, L., Jacquette, D. and Poli, R. eds.: Forthcoming. *The School of Alexius Meinong*. Aldershot, Ashgate.

Anderson, J.A.: 1995, *An Introduction to Neural Networks*. Cambridge, Mass., MIT Press.

Baumgartner, P. and Payr, S. eds.: 1995, *Speaking Minds: Conversations with Twenty Eminent Cognitive Scientists*. Princeton, NJ, Princeton University Press.

Benussi, V.: 1913, *Psychologie der Zeitauffassung*. Heidelberg, Winter.

Benussi, V.: 1922–23, *Lezioni tenute dal Prof. Vittorio Benussi nell'anno accademico 1922–23, redatte dal Dott. Cesare Musatti*, typescript.

Benussi, V.: 1925, *La suggestione e l'ipnosi come mezzi di analisi psichica reale*. Cappelli, Bologna.

Bernhardi, A.F.: 1801, *Sprachlehre*. Berlin, Frölich. Repr. 1973, Hildesheim, Olms.

Biedermann, I.: 1985, 'Human Image and Understanding: Recent Research and a Theory,' *Computer Vision, Graphics, and Image Processing*, **32**, 29–73.

Block, N. ed.: 1981, *Readings in the Philosophy of Psychology*, vol. 2, Cambridge, Mass., MIT Press.

Bonaventura, E.: 1929, *Il problema psicologico del tempo*. Milan, Soc. An. Istituto Editoriale Scientifico.

Brentano, F.: 1874, *Psychologie vom empirischen Standpunkte*. Leipzig, Duncker & Humblot. En. tr. 1973, *Psychology from an Empirical Standpoint*, by A. C. Rancurello, D. B. Terrell and L. McAlister. London, Routledge & Kegan Paul. Repr. 1995.

Brentano, F.: 1905, *Untersuchungen zur Sinnespsychologie*. Leipzig, Duncker & Humblot.

Brentano, F.: 1928, *Vom sinnlichen und noetischen Bewusstseins*, ed. by O. Kraus, Leipzig, Meiner. En. tr. 1981, *Psychology from an Empirical Standpoint, vol. III: Sensory and Noetic Consciousness*, by L. McAlister and M. Schättle. London, Routledge & Kegan Paul.

Brentano, F.: 1956, *Die Lehre vom richtigen Urteil*, ed. by F. Mayer-Hillebrand, Bern, Francke.

Brentano, F.: 1976, *Untersuchungen zu Raum, Zeit und Kontinuum*, ed. by S. Körner and R.M. Chisholm. Hamburg, Meiner. Eng. tr. 1988, *Philosophical Investigations on Space, Time and Continuum*, by B. Smith, ed. by S. Körner and R. M. Chisholm, London, Croom Helm.

Brandl, J.: 1996, 'Kazimierz Twardowski über Funktionen und Gebilde. Einleitung zu einem Text aus dem Nachlaß', *Conceptus*, **36**, 145-156.

Bruno, N.: 1997, 'Amodal Completion of Partly Occluded Surfaces: Is There a *Mosaic* Stage?,' *Journal of experimental psychology: Human perception and performance*, **5**, 1–15.

Bühler, K.: 1909, 'Rezension von A. Marty 1908, *Untersuchungen zur Grundlegung der allgemeinen Grammatik und Sprachphilosophie*,' vol. I (*Göttingische gelehrte Anzeigen*, **171**), Niemeyer, Halle, 947–79.

Bühler, K.: 1918, 'Kritische Musterung der neuen Theorie des Sätzes', *Indogermanisches Jahrbüch*, **6**, 1–20.

Bühler, K.: 1934, *Sprachtheorie. Die Darstellungsfunktion der Sprache*. Jena, Fischer. Repr. 1965, Stuttgart, Fischer.

Calabresi, R.: 1930, *La determinazione del presente psichico*. Florence, Bemporad.

Cantwell Smith, B.: 1996, *On the Origin of Objects*. Cambridge, Mass., MIT Press.

Cassirer, E.: 1906, *Das Erkenntnisproblem in der Philosophie und der Wissenschaft der neueren Zeit*. Berlin, Dümmler.

Davis, E.: 1990, *Representations of Commonsense Knowledge*. San Francisco, Ca., Morgan Kaufmann.

Dennett, D.: 1991, *Consciousness Explained*. Boston, Little, Brown.

Dennet, D.: 1987, 'Intentionality', in Gregory, ed., 1987, 383-386.

Dreyfus, H.L. ed.: 1984a, *Husserl, Intentionality and Cognitive Sciences*. Cambridge, Mass., MIT Press, 2nd ed.

Dreyfus, H.L.: 1984b, 'Husserl's Perceptual Noema,' in Dreyfus, ed., 1984a, 97–124.

Ehrenfels, Ch. von : 1890, 'Über Gestaltqualitäten,' *Vjerteljahrschrift für wissenschaftliche Philosophie*, **14**, 242–292. En. tr. in Smith, ed., 1988, 82-117.

Elliston, F. and McCormick, P. eds.: 1997, *Husserl: Expositions and Appraisal*. Notre Dame, Ind., Notre Dame University Press.

Eilan, N., McCarthy, R. and Brewer, B. eds.: 1993, *Spatial Representation*. Oxford, Blackwell.

Eschbach, A.: 1988, *Karl Bühler's Theory of Language*. Amsterdam, Benjamins Publishing Company.

Fechner, Th.: 1860, *Elemente der Psychophysik*. Leipzig, Breitkopf & Härtel.

Findlay, J.N.: 1933, *Meinong's Theory of Objects*. London, Humphrey Milford, Oxford University Press.

Finke, R.A.: 1989, *Principles of Mental Imagery*. Cambridge, Mass., MIT Press.

Fodor, J.A.: 1968, *The Modularity of Mind: An Essay on Faculty Psychology*. Cambridge, Mass., Harvard University Press.

Fodor, J.A. and Pylyshyn, Z.W.: 1988, 'Connectionism and Cognitive Architecture: A Critical Analysis, *Cognition*, **28**, 3-71.

Føllesdal, D.: 1969, 'Husserl's Notion of Noema,' *Journal of Philosophy*, **66**, 680–7.

Gerbino, W.: 1988, 'Models of Schematic Transparency: A Theoretical Analysis,' *Gestalt Theory*, **10**, 5–20.

Gibson, J.J.: 1950, *The Perception of the Visual World*. Boston, Houghton-Mifflin.

Gibson, J.J.: 1979, *The Ecological Approach to Visual Perception*. Boston, Houghton–Mifflin.

Goodman, N.: 1951, *The Structure of Appearance*. Dordrecht, Reidel.

Gregory, R.: 1981, *Mind in Science. A History of Explanations in Psychology and Physics*. Cambridge, Cambridge University Press.

Gregory, R.L. ed.: 1987, *The Oxford Companion to the Mind*. Oxford, Oxford University Press.

Gurwitch, A.: 1964, *The Field of Consciousness*. Pittsburg, Duquesne.

Hafner, B.J.: 1974, *Darstellung. Die Entwicklung des Darstellungsbegriff von Leibniz bis Kant und sein Anfang in der antiken Mimesis und der mittelalterlichen Repraesentatio*. Dissertation, University of Düsseldorf.

Hayes, P.J.: 1985a, 'The Second Naive Physics Manifesto', in Hobbs and Moore, eds., 1985, 1-36.

Hayes, P.J.: 1985b, 'Naive Physics I: Ontology for Liquids', in Hobbs and Moore, eds., 1985, 71-107.

Hauske, G. and Butenandt, E. eds.: 1978, *Kybernetik*. München, Oldenbourg.

Helmholtz, H.: 1896, *Handbuch der physiologischen Optik*. Hamburg, Voss, 2nd ed.

Hobbs, J.R. and Moore, R.C. eds.: 1985, *Formal Theories of the Commonsense World*. Noirwood, Ablex.

Husserl, E.: 1891, *Philosophie der Arithmetik: Psychologische und logische Untersuchungen*. Halle, Niemeyer. Repr. 1970, Husserliana XII, The Hague, Nijhoff.

Husserl, E.: 1900-1, *Logische Untersuchungen*. Halle: Niemeyer. Repr. 1975. Vol. I ed. by E. Holenstein, Husserliana XVII, Den Haag, Nijhoff 1975. Repr. 1984. Vols. I and II ed. by U. Panzer, Husserliana XIX , *ivi*, 1, 2. En. tr. 1970, *Logical Investigations*, ed. by J.N. Findlay, London, Routledge & Kegan Paul.

Husserl, E.: 1939, *Erfahrung und Urteil. Untersuchungen zur Genealogie der Logik*, ed. by L. Landgrebe, Prag, Academia Verlag. En. tr. 1973, *Experience and Judgement*, by J. Churchill and K. Ameriks, Evanston, Northwestern University Press.

Husserl, E.: 1966, *Analysen zur passiven Synthesis*, ed. by M. Fleischer, Husserliana XI, Den Haag, Nijhoff.

Jackendoff, R.: 1987, *Consciousness and the Computational Mind*. Cambridge, Mass., MIT Press.

Johnson-Laird, P.N.: 1983, *Mental Models. Towards a Cognitive Science of Language, Inference, and Consciousness*. Cambridge, Cambridge University Press.

Kanizsa, G.: 1956, 'Il fattore empirico nella percezione,' *Rassegna di Psicologia Generale e Clinica*, **1**, 9–15.

Kanizsa, G.: 1991, *Vedere e pensare*. Bologna, Il Mulino.

Kant, I.: 1781, *Kritik der reinen Vernunft*. Riga, Johann Friederich Hartknock.

Kant, I.: 1790, *Kritik der Urteilskraft*. Berlin und Libau, Lagarde und Friederich.

Kiesow, K.F.: 1988, 'Aussageninhalt bei Gomperz, Bühler und Popper,' in Eschbach, ed., 1988, 349–367.

Koffka, K.: 1935, *Principles of Gestalt Psychology*. New York, Harcourt, Brace & Co. Repr. 1963, London, Kegan Paul.

Koenderink, J.J. and van Doorn, A.J.: 1977, 'How an Ambulant Observer Can Construct a Model of the Enviroment from the Geometrical Structure of Visual Inflow,' in Hauske and Butenandt, eds., 1977, 244-47.

Köhler, W.: 1929, 'Ein altes Scheinproblem,' *Die Naturwisseschaften*, **17**, 395-401. En. tr. in Köhler 1971, 125-141.

Köhler W.: 1971, *The Selected Papers of Wolfang Köhler*, ed. by M. Henle, New York, Liveright.

Kosslyn, S.: 1893, *Ghosts in the Mind's Machine*. New York, Norton.

Kosslyn, S.: 1994, *Image and Brain*. Cambridge, Mass., MIT Press.

Knüfer, C.: 1911, *Grundzüge der Geschichte der Begriffs Vorstellung von Wolff bis Kant*. Berlin, Dissertation.

Külpe, O.: 1920, *Vorlesungen über Psychologie*. Leipzig, Hirzel.

Lakoff, G.: 1987, *Women, Fire and Dangerous Things*. Chicago, University of Chicago Press.

Langacker, R.: 1987, *Foundations of Cognitive Grammar. Theoretical Prerequisites, I*. Stanford, Stanford University Press.

Langacker, R.: 1991, *Foundations of Cognitive Grammar. II: Descriptive Application*. Stanford, Stanford University Press.

Langer, S.: 1988, *Mind: An Essay on Human Feeling*. Baltimore, John Hopkins University Press.

Legrenzi, P.: 1971, 'Un fenomeno di dislocazione temporale in campo visivo,' *Atti dell'Accademia Patavina SS LL AA*, Classe di scienze matematiche e naturali, **83**, 135–182.

Lockwood, M.: 1989, *Mind, Brain and the Quantum: The Compound 'I'*. Oxford, Basil Blackwell.

Marcel, A.J.: 1983a, 'Conscious and Unconscious Perceptions: An Approach to the Relations between Phenomenal Experience and Perceptual Processes,' *Cognitive Psychology*, **15**, 238-300.

Marcel, A.J.: 1983b, 'Conscious and Unconscious Perceptions: Experiments on Visual Masking and Word Recognition,' *Cognitive Psychology*, **15**, 197-237.

Marr, D.: 1982, *Vision*. San Francisco, Freeman & Co.

Marr, D. and Nishihara, H.K.: 1978, 'Representation and Recognition of the Spatial Organization of Three-dimensional Shapes,' *Proceedings of the Royal Society of London*, **200**, 269-294.

Marty, A.: 1908, *Untersuchungen zur Grundlegung der allgemeinen Grammatik und Sprachphilosophie*, vol. I. Niemeyer, Halle.

Matthei, R.: 1929, *Das Gestaltproblem* (*Ergebnisse der Psychologie*, **29**). Münich, Bergmann.

Maturana H.R. and Varela F.J.: 1987, *The Tree of Knowledge: The Biological Roots of Human Understanding*. Boston, New Science Library.

Masin, S.C. ed.: 1993, *Foundations of Perceptual Theory*. Amsterdam, North Holland.

Meinong, A.: 1894, *Beiträge zur Theorie des psychischen Analyse. Psychologisch-etische Untersuchungen zur Wert-Theorie*. Festschrift der K.K-Karl-Franzens-Universität zur Jahresfeier am 15. November 1894. Repr. in *Gesamtausgabe*, ed. by R.M. Chisholm, R. Kindinger and R. Haller, 7 vols. Graz, Akademische Druck- und Verlaganstalt 1968-1978, vol. 1, 307-388.

Meinong, A.: 1896, 'Über die Bedeutung des Weber'schen Gesetzes. Beiträge zur Psychologie des Messens,' *Zeitschrift für Psychologie und Physiologie der Sinnesorgane*, **11**, 81–133, 230–285, 353–404. Repr. in Meinong 1971, 215-480.

Meinong, A.: 1899, 'Über Gegenstände höherer Ordnung und ihre Verhältniss zu inneren Wahrnehmungen,' *Zeitschrift für Psychologie und Physiologie der Sinnesorgane*, **21**, 181–271.

Meinong, A.: 1910, *Über Annahmen*. Barth, Leipzig. Repr. in *Gesamtausgabe*, cit., vol. 4. En. tr. 1976 by J. Heanue, Berkeley, University of California Press.

Meinong, A.: 1917, *Über emotionale Präsentation*. Barth, Leipzig. Repr. in *Gesamtausgabe*, cit., vol. 3. En. tr. 1972, *On Emotional Presentation*, Evanston, Northwestern University Press.

Meinong, A.: 1971, *Abhandlungen zur Erkenntnistheorie und Gegenstandstheorie*, ed. by R. Haller, Graz, Akademische Druck- u. Verlagsanstalt.

Metzger, W.: 1941, *Psychologie: die Entwicklung ihrer Grundannahmen seit der Einführung des Experiments*. Dresden, Steinkopff.

Metzger, W.: 1986, *Gestaltpsychologie*, ed. by M. Stadler and H. Crabus, Frankfurt a.Main, Verlag Waldemar Kramer.

Michon, J.A. and Jackson, J. eds.: 1985, *Time, Mind and Behaviour*. Berlin, Springer.

Mohanty, J.N.: 1997, 'Husserl's Theory of Meaning,' in Elliston and McCormick, eds., 1997, 18-37.

Morscher, E.: 1973, *Das logische An-Sich bei Bernard Bolzano*. Salzburg, Anton Pustet.

Murchison, C.: 1930, *Psychologies of 1930*. Worchester, Mass., Clark University.

Murray, D.J.: 1995, *Gestalt Psychology and the Cognitive Revolution*. New York, Harvester/Wheatsheaf.

Musatti, C.: 1926, *Analisi del concetto di realtà empirica*. Città di Castello, Il Solco. Repr. 1964, as *Condizioni dell'esperienza e fondazione della psicologia*. Florence, Editrice Universitaria.

Nehrlich, B.: 1996, 'Sprachliche Darstellung als Prozess: Die Pragmatisierung eines Begriffs von Kant bis Bühler,' *Zeitschrift für Semiotik*, **18**, **(4)**, 423–440.

Nehrlich, B.: Forthcoming, 'Linguistic 'Representations' as *Darstellung*: From Bernhardi to Bühler'.

Neisser, U.: 1967, *Cognitive Psychology*. New York, Apple-Century-Croft.

Newell, A., Shaw, J.C and Simon, H.A.: 1958, 'Elements of a Theory of Human Problem Solving,' *Psychological review*, **65**, 151–66.

Norman, D.A. and Rumelhart, D.E. eds.: 1975, *Explorations in Cognition*. San Francisco, Ca., Freeman.

Palagy, M.: 1902, *Kant und Bolzano. Eine kritische Parallele*. Halle, Niemeyer.

Palmer, R.: 1975, 'Visual Perception and World Knowledge: Notes on a Model of Sensory-Cognitive Interaction,' in Norman and Rumelhart, eds., 1975, 279–307.

Palmer, S.E.: 1995, 'Gestalt Psychology Redux,' in Baumgartner and Payr, eds., 1995, 156-176.

Pelc, J. ed.: 1979, *Semiotics in Poland 1894–1969*. Warsaw, Polish Scientific Publishers, Reidel.

Penrose, S.E.: 1989, *The Emperor's New Mind*. Oxford, Oxford University Press.

Petitot, J.: 1985, *Morphogénèse du sens*. Paris, Presses Universitaires de France.

Petitot, J.: 1994, 'Passion de formes. Dynamique qualitative sémiophysique et intellegibilité,' Fontenay-St. Cloud, Ens Editions.

Pierantoni, R.: 1996, *La trottola di Prometeo. Introduzione alla percezione acustica e visiva*. Bari, Laterza.

Poli, R. ed.: 1998, *The Brentano Puzzle*. Aldershot, Ashgate.

Poli, R.: 1998, 'Antonio Aliotta e la misura dei fenomeni psichici,' in Albertazzi, Cimino and Gori-Savellini, eds., 1999, 369–392.

Port, R.F. and van Gelder, T. eds.: 1995, *Mind as Motion*. Cambridge, Mass., MIT Press.

Pylyshyn, Z.W.: 1981a, 'Psychological Explanations and Knowledge-dependent Processes,' *Cognition*, **10**, 267–274.

Pylyshyn, Z.: 1981b, 'Imagery and Artificial Intelligence,' in Block, ed., 1981, 170–194.

Rock, I.: 1983, *The Logic of Perception*. Cambridge, Mass., MIT Press.

Roth, G.: 1995, *Das Gehirn und seine Wirklichkeit*. Frankfurt, Suhrkamp, 2nd ed.

Sander, F.: 1930, 'Structure, Totality of Experience and Gestalt,' in Murchison, ed., 1930, 188–204.

Schuhmann, K.: 1991, 'Intentionalität und intentionale Gegenstände beim frühen Husserl,' *Phänomenologische Forschungen*, **24/25**, 46–75.

Sebestik, J.: 1992, *Logique et mathématique chez Bernard Bolzano*. Paris, Vrin.

Sekuler, A.B. and Palmer, S.E.: 1992, 'Perception of Partly Occluded Objects: A Microgenetic Analysis,' *Journal of Experimental Psychology: General*, **121**, 95–11.

Shanon, B.: 1993, *The Representational and the Presentational. An Essay on Cognition and the Study of Mind*. New York, Harvester/Wheatsheaf.

Shepard, R.N. and Cooper, L.: 1982, *Mental Images and their Tranformations*. Cambridge Mass., MIT Press.

Simon, H.A.: 1981, *The Sciences of the Artificial*. Cambridge, Mass., MIT Press, 2nd ed.

Smith, B. ed.: 1988, *Foundations of Gestalt Theory*. Munich, Philosophia.

Stout, G.F.: 1896, *Analytic Psychology*. London, Swan, Sonnenschein.

Stout, G.F.: 1931, *Matter and Mind*. Cambridge, Cambridge University Press.

Stout, G.F.: 1952, *God and Nature*, ed. by A. K. Stout, Cambridge, Cambridge University Press.

Stroud, J.M.: 1949, 'The Psychological Moment in Perception,' in van Foerster, ed., 1949, 27–63.

Swain, M.J.: 1991, 'Promising Directions in Active Vision,' *Active vision workshop*, University of Chicago, cit. in Zimmer 1998, 113.

Talmy, L.: 1985, 'Force Dynamics in Language and Cognition,' *Cognitive Science*, **12**, 49–100.

Tetens, J.N.: 1777, *Philosophische Versuche über die menschlische Natur und ihre Entwickelung*. Leipzig.

Thom, R.: 1980, *Modèles matématique de la morphogénèse*. Paris, Borurgois.

Twardowski, K.: 1894, *Zur Lehre vom Inhalt and Gegenstand der Vorstellungen: Eine psychologische Untersuchung*. Vienna, Hölder. Repr. 1982, Munich, Philosophia. En. tr. 1977 by R. Grossmann. The Hague, Nijhoff.

Twardowski, K.: 1912, 'Actions and Products,' part. En. tr. in Pelc 1979, 13–27. Integral tr. in Twardowski 1999, 103-132.

Twardowski, K.: 1999: *On Actions, Products and other Topics in Philosophy*, edited by J. Brandl and J. Wolenski, Rodopi, Amsterdam.

Tye, M.: 1991, *The Imagery Debate*. Cambridge, Mass., MIT Press.

van Foerster, H. ed.: 1949, *Conference on Cybernetics. Transactions of the Sixth Conference*. New York, Macy Foundations.

Varela, F., Thompson, E. and Rosch, E.: 1992, *Embodied Mind*. Cambridge, Mass., Harvard University Press.

Vicario, G.B.: 1973, *Tempo psicologico ed eventi*. Florence, Giunti.

Vicario, G.B.: 1993, 'On Experimental Phenomenology,' in Masin, ed., 1993, 197–219.

Wertheimer, M.: 1922–23, 'Untersuchungen zur Lehre vom der Gestalt,' *Psychologische Forschung*, **1**, 47–58; **4**, 301–350.

Witasek, S.: 1907, *Grundlinien der Psychologie*. Leipzig, Meiner.

Witkin, A.P.: 1983, 'Scale Space Filtering', *Proceedings of the Eighth International Joint Conference on Artificial Intelligence*, **2**, 1019-1022.

Wundt, W.: 1907, 'Über Ausfragexperiment und über die Methoden zur Psychologie des Denkens,' *Psychologische Studien*, **3**, 301–60.

Zimmer, A.: 1998, 'On Agents and Objects: Some Remarks on Brentanian Perception,' in Poli, ed., 1998, 93–113.

MARTIN KUSCH

THE POLITICS OF THOUGHT: A SOCIAL HISTORY OF THE DEBATE BETWEEN WUNDT AND THE WÜRZBURG SCHOOL

1. INTRODUCTION

The early twentieth-century German debate over the thought psychology of the Würzburg School is mentioned in most standard textbooks on the history of psychology and in several studies on the historical roots of cognitive science. Not only is there general agreement amongst historians that this controversy was a key event in the history of experimental psychology; at least one prominent historian of 'the cognitive revolution,' Howard Gardner, has even wondered whether the Würzburg School's work has ever been surpassed.[1]

Despite the long-standing, general agreement as to the importance of the *Streit um die Denkpsychologie* (fight over thought psychology), it has not, until recently, attracted much detailed historical scholarship. In my book *Psychological Knowledge: A Social History and Philosophy*, I have sought to fill this lacuna.[2] In this paper, I want to summarise the main results of my book.

As the title of my paper already indicates, my approach to the history of psychology and philosophy is informed by recent work in the sociology of scientific knowledge.[3] I wish to show that more was at stake in the debate then whether or not humans ever have so-called 'imageless thoughts,' whether laws of association suffice to explain the train of thoughts, or whether or not retrospection under experimental conditions is reliable. At issue were also the following questions: Under what social arrangements is the human mind best able to think? How should cognitive and social authority be divided between experimenter and experimental subject? How should psychological research schools be organised and run? Where should the discipline of psychology be located institutionally? What is the correct way to think about the relationship between the individual and the collective in general, or the citizen and the state in particular? And, which confession – Catholic or Protestant – should define the role of the Christian in early-twentieth century Germany? In what follows,

L. Albertazzi (ed.), The Dawn of Cognitive Science. Early European Contributors, 61–88.

I shall go through these issues one by one. I begin with a brief summary of the Würzburgers' main contentions.

2. THE WÜRZBURGERS' INNOVATIONS

The controversy was first provoked by a number of methodological and theoretical writings coming out of the Psychological Institute of the University of Würzburg between, roughly, 1900 and 1907. As the members of that institute moved on to positions elsewhere, thought psychology spread, too. Soon thought-psychological studies as well as their theoretical interpretations and philosophical defences were also produced at the universities of Bonn, Munich, and Königsberg. To avoid clumsy circumscription, I shall speak of the thought psychologists as 'the Würzburgers' even when their institutional location was Bonn, Munich, or Königsberg. Key Würzburgers in this broader sense were Oswald Külpe, the director of the Würzburg institute from 1896 until 1909, Narziss Ach, Karl Bühler, Karl Marbe, August Messer, Otto Selz, and Henry Jackson Watt. Each one of these men developed ideas that were controversial even within the school itself, but here I do not have time to explain the school-internal debate. For our purposes, we can reduce the Würzburgers' innovations to three ideas.

First, the Würzburgers claimed to have discovered new kinds of mental contents, in addition to the traditional trio of sensations, feelings, and presentations (*Vorstellungen*). The Würzburgers called these new, irreducible mental contents 'situations of consciousness' (*Bewusstseinslagen*), 'awarenesses' (*Bewusstheiten*), and 'thoughts' (*Gedanken*). These new contents were experiences of an imageless, non-depictive (*anschauungslos*) knowing.[4]

Second, in their attack on the sufficiency and exhaustiveness of the traditional trio of sensations, feelings, and presentations, the Würzburgers also rejected associationism. Over and above associations, they argued, one had to assume 'determining tendencies' (*determinierende Tendenzen*) that linked presentations and thoughts together in a goal-oriented fashion, the goal being the solving of a problem or 'task' (*Aufgabe*).[5]

Third, and finally, the Würzburgers believed that a retrospective self-observation (*Selbstbeobachtung*) of one's own thought processes was possible and reliable provided it occurred under controlled experimental conditions. The most important of these conditions was the presence of a second person. Successful retrospection demanded an experimental setting with two persons such that the 'retrospectionist' as an experimental subject was assisted in his or her retrospection by an experimenter. The experimenter would first set a task for the subjects, a task that would demand thinking to solve it. The subjects would try to solve the task and give their answer as soon as they felt satisfied with their solution. Immediately afterwards, the subjects would furnish a complete account of all that took place in their mind between hearing the problem and providing the solution.[6]

3. RECLUSE VERSUS INTERLOCUTOR

The Würzburg psychological institute was one of the four main institutes of its kind in Germany during the first decade of the twentieth century. The others were Wilhelm Wundt's laboratory in Leipzig, Georg Elias Müller's institute in Göttingen, and Carl Stumpf's institute in Berlin. Several of the Würzburg psychologists had trained with Wundt or Müller; Külpe had worked under both. The competition and rivalry between these 'schools' was often intense, and it reached its climax during the Thought Psychology Controversy. My main focus in this paper will be on the dispute between the Würzburgers and Wundt. But it must be mentioned, at least in passing, that the Würzburgers' work was discussed very widely indeed. Highly critical writers included Anschütz, von Aster, Deuchler, Henning, G.F. Lipps, Meumann, Moskiewicz, G.-E. Müller and Reichwein; sympathetic commentators included the Phenomenologists Husserl, Scheler and Stumpf; the Neokantians Cassirer, Hönigswald, and Rickert; and the Neothomists Baeumker, Geyser, Grabmann and Gutberlet.[7]

Wundt sympathised with the Würzburgers' attack on associationism, but he opposed their new mental entities and their new method. According to Wundt, the traditional trio of presentations, sensations, and feelings was sufficient. What the Würzburgers took to be *Bewußtseinslagen* or *Gedanken* were, according to Wundt, simply conceptual feelings. Talk of pure thoughts, Wundt maintained in his 1907 – criticism of Bühler, was an unfortunate resurrection of the Scholastics' *actus purus*. As concerned the Würzburgers' methods, Wundt claimed that thought and other higher mental processes could not be investigated experimentally. Experimental psychology could only reach the 'lower' psychological processes of sensation and perception, but to investigate the 'higher' processes of thinking, one needed to resort to non-experimental collective psychology, that is, *Völkerpsychologie*. Moreover, perfect psychological experiments were such that they could be carried out on one's own. After all, Wundt took it to be well established that the presence of another person was almost always a disturbance. This was so in particular when the experimental subject – or anyone else for that matter – was engaged in thinking. Therefore, the ideal psychological experimental subject as well as the thinker in general, had to be something of a social isolate, a 'recluse.'[8]

Wundt's and the Würzburgers' different views on the mind and the psychological experiment must be seen in the context of the very different social orders in the two respective institutes. Let's begin with Wundt. In his own institute Wundt played the role of a remote 'sage': He controlled his research school from a carefully orchestrated distance; he was inaccessible to his students and subordinates; he reacted to their attempts at independent work 'somewhat arrogantly and without any attempt to understand them,'[9] and he tried to pre-empt rather than stimulate their thinking.[10] Moreover, Wundt did not participate in the experimental work of his famous laboratory.[11]

Wundt's preferred model of social order – a hierarchical structure with a recluse at the top – was reflected in his psychological theory of the human mind, and his conception of the internal structure of psychology as a field of knowledge. In his theory of the human mind, Wundt made a sharp distinction between the 'lower' processes of, for example, sensation and perception and the 'higher' processes of, first and foremost, thinking. Although the higher processes presupposed the lower, the two realms were governed by different laws. The attribute 'higher' in 'higher processes' was not just a convenient spatial metaphor, it was also meant to signal a difference in value. Wundt made this point explicitly. The term also expressed something 'above experimentation.' Because of their complexity, higher-thinking-processes could not be controlled and manipulated under experimental conditions. It was for this reason that their study had to be left to *Völkerpsychologie*.[12]

Moreover, since *Völkerpsychologie* studied the 'higher' processes, it was itself 'higher' with respect to experimental psychology. At one point Wundt went even further, equating *Völkerpsychologie* with psychology proper: He wrote of experimental psychology as 'the *Vorschule* [nursery school] for the [!] psychologist.'[13] Given that Wundt saw experimental psychology as the nursery school for the psychologist, it is not surprising that his students were expected to learn the skills of experimental work. What *is* surprising is the fact that Wundt made little effort to train any of them in the 'higher' area of psychology. Indeed, he reserved the 'higher' and more important area of psychology, and thus the study of thought for the exclusive attention of a single individual: himself.

This then was Wundt's hierarchical recluse model for the social-cum-cognitive order of psychology. The internal structure of a psychological institute consisted of a lower stratum of experimental psychologists and a higher stratum consisting of the solitary master psychologist. The latter owed his dominance and solitude to his subject matter: the thinking mind. Since the study of the 'higher' mental processes demanded a 'higher' form of psychology, the master psychologist rightly controlled the 'lower,' experimental study, directed at the 'lower' mental processes, that was carried out by the 'lower' ranks of students and assistants. And since thinking itself demanded solitude – this proposition being supported by experimental work and by common sense – the master psychologist had better safeguard such solitude for himself as well.

The ways in which – in Wundt's case – social order and psychological order were inextricably intertwined makes it difficult or even impossible to distinguish between the Würzburgers' rejection of Wundt's psychological theories and their repudiation of Wundt's preferred social order in the psychological institute. This is all the more obvious as the Würzburgers worked in a very different kind of social setting from their Leipzig colleagues. Külpe would always have time, endlessly listen to students' and colleagues' ideas, and assist in developing them. Moreover, the major fruits of Külpe's psychological research were not Külpe's own publications; these fruits were generously left to his students in discussions, supervisions, and introspective reports. Furthermore, Külpe never imposed his ideas upon his students. The style of leadership in the Würzburg research school was thus very egalitarian. Groping for a label for Külpe's style of leadership, one obvious term suggesting itself is 'interlocutor.'[14]

The term 'interlocutor' also serves well to set apart the Würzburgers' preferred social model of psychological experimentation. Time and time again, they pointed to the equality, reciprocity, indeed even the 'love' between experimental subject and experimenter. Wundt's recluse model of experimentation – that is, the idea that the experimental subject must be isolated from the experimenter – was close to incomprehensible to the Würzburgers.[15]

At least for one key member of the Würzburg school, Bühler, the egalitarian spirit of the Würzburg laboratory and its emphasis on the equality of experimenter and experimental subject went hand in hand with an egalitarian model of the human mind. Bühler denied Wundt's claim that thought processes were higher than the processes of sensation, feeling, and presentation. As Bühler saw it, all of these processes could be more or less complex, but in principle a thought could be just as simple as a sensation. In this way, Bühler rejected Wundt's main argument for a sharp division between higher and lower processes in the mind. By rejecting the Wundtian assumption of thought processes as complicated and inaccessible to experimental psychology, Bühler weakened – at least to the satisfaction of himself and other Würzburgers – Wundt's argument that the study of higher processes called for a special kind of psychology, that is, collective psychology. At least implicitly he also denied that one had to leave the study of thinking to those who had first undergone training in the experimental psychology of sensation, feeling, and presentations 1908.[16] Later the Würzburg school also attacked Wundt's collective psychology more directly.[17]

It fits nicely with my argument that Bühler flatly rejected Wundt's claim according to which the presence of another person impeded thinking: "How does Wundt know that such interference *had to* occur in our case [...] it is no more than an unproved assumption on Wundt's part."[18]

4. PSYCHOLOGISM VERSUS LOGICISM

Wundt's and the Würzburgers' opposite viewpoints on the question of non-depictive thought was also connected to their respective views concerning the relationship between psychology and philosophy. Their disagreement in this area was of course part of a wider debate in German philosophy at the time, that is, the debate over psychologism, *der Psychologismusstreit.*[19]

In this many-sided controversy, Husserl and the Neokantians advocated what we might call a 'logical' or 'philosophical purism:' They insisted that experimental psychology could not contribute anything of value to logic and epistemology. Experimental psychology studied mental and psychophysical phenomena, logic investigated the realm of meanings, or pure, transcendental consciousness.[20]

Most philosophical purists were purists only in one direction, however. That is to say, while they denied that experimental psychology had anything to give to philosophy, they in turn tried to contribute to psychology. This was particularly true of Husserl. Husserl envisaged a new philosophical discipline, which would, amongst other things, 'prepare the ground for *psychology as an empirical science.*'[21] Husserl believed that experimental psychology needed his help to be able to rise from its current '*de facto* unscientific' level. Experimental psychology found itself in this sad state because it had neglected the tasks of a 'direct and pure' descriptive analysis of consciousness.

Wundt's view on the relation between logic and psychology was something of a mirror image of Husserl's logical purism. Wundt never tired of deploring the intrusion of logic into psychology; and he believed that logic could do with a little help from psychology.[22]

It was of course the last-mentioned idea that led many 'pure' philosophers to brand Wundt a proponent of psychologism.[23] And indeed, Wundt was committed to the idea that the psychological study of logical thought was the necessary first step in the development of any scientific logic. He emphasised that only a psychological investigation into thought could identify the special and unique features of logical thinking. These features were spontaneity, self-evidence, and universality. Moreover, Wundt believed that because logical laws of thought were identified within human thinking, and because logical thought was usually intertwined with other kinds of mental activities, the formulations and explanations of logical laws would always and inevitably contain psychological concepts.[24]

Despite his advocacy of these ideas, Wundt did not accept that his position amounted to some sort of 'psychologism.' Instead he accused Brentano and Husserl of being amongst the worst '*Psychologisten*,' Husserl because of his reliance on the concept of self-evidence in his criticism of psychologism, and Brentano because of his theory of judgements.[25]

Finally, just as Husserl and other 'pure' philosophers tried to cleanse logic of psychological ideas, so also Wundt aimed to exorcise logical concepts and conceptions from the realm of psychology. So while Husserl was fighting 'psychologism,' Wundt saw himself as combating 'intellectualism' and 'logicism.'[26] Wundt was most hostile towards the alleged intellectualism and logicism of Brentano, Stumpf, and Husserl. And of this trio, Husserl was the most dangerous. Brentano and Stumpf's intellectualism was based on oversight and carelessness; but to Husserl Wundt attributed, as it were, a logical imperialism of sorts: "His programme is not just to ban psychologism from logic; his programme is to ban psychologism from psychology." Wundt suspected that Husserl wanted "to turn psychology into logic," or "into an applied logic." Wundt also spoke of "Husserl's [...] recklessly developed logicism that seeks to turn psychology into a reflective dissection of words and concepts."[27]

Wundt showed no inclination to accept Husserl's claim that phenomenological analysis clarified important concepts for both psychology and logic. The concepts that Husserl selected for his study were:

> the scientifically unchecked concepts of vulgar psychology, and these concepts do not become in the least bit more psychologically scientific by this detour through purely logical conceptual analysis. Husserl's phenomenology was a psychology without psychology.[28]

The Würzburgers' thought-psychological studies started to appear just one year after Husserl's famous onslaught on psychologism in the *Prolegomena* to the *Logische Untersuchungen*. Faced with the choice between Wundt's psychological purism concerning logic, on the one hand, and Husserl's logical purism *vis-à-vis* psychology on the other hand, a number of key Würzburgers tended to side with the latter. That is to say, they allowed – *pace* Wundt – for the possibility that logic and 'pure' philosophy might influence experimental psychology, while insisting that logic owed nothing to psychology. No wonder then that, as mentioned above, phenomenologists, Neokantians, and Neothomists were so positive in their reaction to the Würzburgers.

The relationship between logic and psychology was explicitly the topic of the first famous Würzburg study on thought psychology, Karl Marbe's *Experimentell-psychologische Untersuchungen über das Urteil: Eine Einleitung in die Logik* (Experimental-psychological Investigations into the Judgment: An Introduction to Logic), published in 1901. Marbe set himself the task of testing the psychological claim that there was a specific quale that characterised all judgements. Wundt was amongst those who had made proposals on what this quale might be like. Marbe's subjects were unable to find these qualia in cases where, by the standards of logic, they had formed judgements, that is, in cases where they had said or done something to which the concepts of true and false could be applied. Marbe therefore concluded by calling for a greater independence of logic from psychology:

> Currently logic is to a considerable extent nothing but a non-methodological psychology of judgement. In the future, logic will have to construe itself as non-psychologically as possible.[29]

There can be no doubt that Wundt regarded Marbe's book as something of a betrayal. Here was, after all, a former Leipzig student and experimental psychologist, who sided with *the* enemy of experimental psychology, that is, Husserl, and accused Wundt of psychologism. The extent of Wundt's anger can be measured by his reactions. In the fifth edition of his *Grundzüge* he called Marbe's book a 'maltreatment of psychology.' The judgements formed by Marbe's subjects were not 'original judgements' but mere 'artefacts of the experiment.'[30] And Marbe was of course one of the two main targets of Wundt's famous 1907 criticism, together with Bühler. Wundt was active also behind the scenes. For instance, he forbade Meumann to put Marbe on the editorial board of the *Archiv für Psychologie*.[31] And Wundt repeatedly let Marbe's superior, Külpe, know how little he thought of the author of the *Experimentell-psychologische Untersuchungen über das Urteil*.[32]

It is significant that Wundt wrote his extensive criticism of the Würzburg experiments only after Bühler had published the first of his three articles on thoughts in 1907.[33] Indeed, Wundt concentrated most of his attack on Bühler. This should not come as a surprise: For with Bühler's papers, the very 'scholastic reflection psychology of Brentano' and Husserl, a type of psychology that Wundt had wanted to keep out of the psychological literature, entered the *Archiv für Psychologie*. And it did so with a vengeance.

Bühler began his first paper with praise for Husserl's methodology in the *Logical Investigations* and went on to claim that his study of 'the *hic et nunc* of what is experienced while thinking' would prove Husserl's assumptions true. In the main body of his text, Bühler made repeated use of Husserl, Brentano, and Stumpf. He used Husserl's terminology for analysing parts and moments of thoughts, adopted Husserl's classification of acts, and used Husserl's distinction between meaning-intention and meaning-fulfilment.[34] Indeed, his reliance upon Husserl was so substantial that one contemporary critic commented that "Bühler's [...] experiments are [...] an attempt to check and confirm Husserl's phenomenology in an experimental way."[35]

Even if Bühler was crucial in introducing phenomenology into Würzburg, he was no doubt helped by Messer. There were only a few references to Husserl's work in Messer's famous 1906 study on thought, but Messer did adopt, for instance, Husserl's categories of meaning-intention and meaning-fulfilment.[36] Within two years of completing his thought-psychological study, Messer had converted to phenomenology almost completely: Messer's book *Sensation and Thought*[37] was as much an introduction to Husserl's *Logical Investigations* as it was an introduction to modern psychology. Not surprisingly, Messer also warned of the dangers of psychologism and adopted Husserl's brand of antipsychologism.[38]

As I have shown in my earlier book *Psychologism: A Case Study in the Sociology of Philosophical Knowledge*,[39] the debate over the systematic ties between logic and psychology was at the same time a debate over the question of whether the new, experimental psychology still belonged within philosophy departments. Husserl and other pure philosophers wanted the experimental psychologists out, and they publicly complained whenever 'yet another' philosophical chair had 'fallen' to an experimentalist. The ultimate provocation was the appointment, in 1912, of Müller's student E. Jaensch, to H. Cohen's chair in Marburg. Natorp, Husserl, Rickert, Windelband, the Riehl and Eucken drew up a petition, collected signatures for it, and submitted it to all German-language universities and ministries of education. The petition was signed by 107 philosophers in Germany, Austria and Switzerland, demanding that no more philosophical chairs go to experimental psychologists.[40]

Interestingly enough, the leader of the Würzburg School, Külpe, in fact sympathised with the pure philosophers' view on professorial chairs. Külpe held the view that 'the combination of psychology (as a special science) with philosophy is beyond the capacity for work, the talent and the inclination of a *single* human being,' and he went so far as to express understanding for the pure philosophers' opposition to the 'invasion of specialists.' Külpe suggested that experimental psychology should lobby for chairs in the medical faculty.[41]

Read in this light, it was actually surprising that Külpe did not sign the philosophers' petition – as Messer did. Indeed, the pure philosophers were themselves startled. Rickert sent Külpe the text of the petition with the following note attached to it: "Dear colleague! It would be great if you too could sign. I hope it is clear that we are not trying to say anything *against* psychology."[42] Rickert also referred to the proximity between Külpe's and the pure philosophers' positions in the daily *Frankfurter Zeitung*, and expressed his astonishment about the fact that Külpe had not signed.[43] Perhaps the explanation lay with the fact that Külpe and some other experimentalists were working on their own petition. In Külpe's *Nachlaß* of around the same time I have found a draft petition – to be signed by members of the *Gesellschaft für experimentelle Psychologie* (Society for Experimental Psychology) – demanding separate chairs for experimental psychologists.[44]

Although Wundt had often argued for an institutional independence for psychology during earlier decade, he was quite unwilling to give in to the demands of Husserl and the Neokantians. His pamphlet *Die Psychology im Kampf ums Dasein*[45] (Psychology in the Struggle for Survival) attacked both the pure philosophers and Külpe. Wundt now insisted that philosophy needed psychology and that fundamental psychological issues were close to problems in metaphysics and the philosophy of mind.

5. COLLECTIVISM VERSUS INDIVIDUALISM

Different stances regarding the new thought psychology were also informed by different social, ethical, and political philosophies. Broadly speaking, Wundt was a 'collectivist,' whereas Bühler, Külpe, Marbe, and Messer were 'individualists' of sorts. Wundt's opposition to introspection was part and parcel of his general hostility towards individualism; the Würzburgers' advocacy of self-observation was part of a conscious rejection of collectivism.

To begin with, Wundt's *Völkerpsychologie* was of course a collective form of psychology, it studied collective phenomena like language, myth and custom. *Völkerpsychologie* dealt both with small collectivities such as the family, the group, the tribe, and the local community, and with large communities such as nations, peoples, states, or humankind as a whole. Amongst collectivities of various size and character, however, the *Volk* had a special position: "In the overall development of mental life [...] *Volk* is the main unifying concept. All other [collectivities] are related to it. Families, classes, clans, and groups exist only within the *Volk*."[46]

No doubt, the stress upon the Volk in Wundt's psychological thinking was an expression of the political aspirations that were common in the third quarter of the nineteenth century. This was the period when many German intellectuals supported German national unification. Wundt himself was actively involved in the politics of his home state of Baden, and from 1866 until 1868 he even served as representative for a Heidelberg district in the Baden diet.[47]

In Wundt's system, *Völkerpsychologie* was closely connected to ethics. *Völkerpsychologie* was the 'portico of ethics.'[48] Wundt's *Ethik* (consisting of three volumes that went through several editions), advocated a political collectivism according to which individual happiness was irrelevant and the collective production of *Bildungsgüter* (educational achievements), organised by the nation-state, was all important. The state served 'ideal purposes of absolute value, purposes with respect to which the life of the individual has no value at all.'[49] One of the key means for this production was nationalism. The creation of ever more, and ever better, cultural works presupposed the competition between peoples and their states. To strive for a single 'world-state' was thus ultimately immoral.[50]

For Wundt, individualism was intertwined with intellectualism. In line with its exaggerated emphasis on reasoning, intellectualism treated every social phenomenon as something that had to be planned, debated, and agreed upon. It therefore was prone to construe the state or other collectivities as emerging from contracts between pre-social individuals. In Wundt's view, "no conception of the state misunderstands its essence more radically than the individualistic theory according to which the state emerged from a real or fictive social contract." If the state was nothing but an agreement amongst self-interested parties then there was no reason to deny the possibility or desirability of a world-state. But such a world-state would be so thin, so restricted in its rights vis-à-vis its citizens, that it was tantamount to having no state at all. And that was anarchism.[51]

In Wundt's view, individualism was *the* central premise of a host of politically dangerous positions: Marxism, social democracy, utilitarianism, and capitalism. In the case of Marx and Lassalle, Wundt deplored their psychology as well as their ethical stance. Predictably, their materialistic psychology led them to embrace determinism and individualism. Their determinism made them construe the human being as a 'machine that reacts [...] to external influences.' Their psychological individualism was most perspicuous in their internationalism; after all, internationalism denied the importance of collectives such as the family, the tribe, or the nation. And psychological individualism led to ethical and political individualism. The 'key motive' of Marxism and social democracy was 'egoism;' both aimed exclusively for the 'satisfying of the natural needs of life' of the individual.[52]

Finally, different forms of individualism – socio – ontological, metaphysical, psychological, ethical, political – were all interconnected. Wundt made this point most forcefully in comments on Descartes and Herbart. Descartes' metaphysical ideas 'reflected' the ethical views of his time: 'Descartes' atomistic concept of the soul is the true reflection of the psychological and ethical individualism of the Enlightenment.'[53] And Herbart's assumption of a conflict-free, collective soul was influenced by 'the higher and educated German middle classes of his time, the conservatively thinking civil servants and the academics.'[54]

Given the ways he viewed individualism, Wundt had no choice but both to conceive of the Würzburgers as advocates of individualism, and regard their *psychological* theories as *politically* dangerous.

As far as Wundt's likely perception of the Würzburgers is concerned, their denial of the need for a collective psychology must have appeared as a very clear symptom of individualism. The Würzburgers tried to study thinking with the help of the very method – introspection – that Wundt had marked as the primary tool of *experimental* psychology. And experimental psychology was of course the *heartland* of *individual* psychology. In advocating introspection as *the* decisive method for thought psychology, the Würzburgers in fact insisted that all of psychology could throughout be based upon individualistic premises. There can be no doubt that Wundt himself made the link between this thought-psychological individualism and political individualism in its full sense. For instance, in his autobiography he reported that trying to understand thinking on the basis of individual consciousness had always seemed to him as hopeless as the attempt to understand the state 'as a purely individual discovery.'[55]

Was Wundt right about the Würzburgers' commitment to individualism? And what did the Würzburgers make of Wundt's collectivism in psychology, ethics, and political philosophy? It is difficult to provide a straightforward answer to these questions. Very few of the Würzburgers wrote about social and political issues, and even those few who did usually expressed their views only long after the main thought-psychological studies had been published. Nevertheless, and with these provisos in mind, it can be shown that at last some key Würzburgers rejected both Wundt's social ontology and aspects of his ethical and political thinking. One might therefore suspect that their individualistic way of approaching the study of thinking was linked to a more principled individualistic philosophical commitment. And this suspicion is not new. At least one observer of the German psychological scene, the Swiss C. Sganzini, reported in 1913 that the '*Funktionspsychologie*' of Külpe and the Würzburg school now represented the main alternative to Wundt's interpretation of 'collective-mental processes.' For the 'younger schools of psychologists and sociologists,' there existed only 'individual-psychological processes.' They accepted that some of these individual-psychological processes were 'social.' But to call an individual-psychological process 'social' did not mean for them that it was part of some larger collective process or soul; a psychological process was 'social' only if it was 'caused by stimuli coming from other humans.'[56]

Turning to the texts of the Würzburgers themselves, Bühler rejected Wundt's concept of *Volksseele* in his *Habilitationsthesen*[57] and Marbe clearly formulated an individualistic social ontology in his book *Die Gleichförmigkeit in der Welt* (The Uniformity of the World).[58] There Marbe argued that the Wundtian *Volksseele* was 'actually a farce,' and that collective phenomena needed to be explained starting from individuals and their interaction.[59] Marbe also challenged Wundt's nationalism. One central line of Marbe's criticism concerned the adequacy of the emphasis on *peoples* in Wundt's *Völkerpsychologie*. The force with which Marbe belaboured this point deserves attention, especially if we remember that Marbe's book was published halfway through World War I, that is, at a time when Wundt and others were busy arguing for the superiority of the German *Volksseele* over all others. Indeed, the relevant passage is worth quoting here:

> Do we not also find within political parties a peculiar connection of presentations and feelings, just like we find within the whole *Volk*? [...] Is it not true that great numbers of persons have placed in the past, and are placing now, the international consciousness of social democracy above the national consciousness? And is it not true that in their case the international consciousness is more developed than the national one?[60]

Külpe and Messer commented directly on Wundt's ethical and political writings. Messer shared with Wundt the concern that social democracy might increasingly endanger the state, and he also feared that 'ethical individualists' might undermine belief in 'the moral significance of the state.' Unlike Wundt, however, Messer sided 'with Kant and our classic literature' in maintaining that in cases of conflict between the demands of the state and the morality of humankind the latter should always be placed above the *raison d'etat*.[61]

Külpe wrote that:

> a revolt against the state can have moral significance in circumstances where the state follows tendencies that harm culture, that are egoistic, and that run counter to the comprehensive interests of humankind.[62]

And, directly contradicting Wundt, Külpe insisted that: "cultural values are not *per se* moral values. And it would be wrong to say that it is primarily, or exclusively, the morality of a people that enables it to create cultural values."[63]

6. PROTESTANTISM VERSUS CATHOLICISM

It remains for me to argue for the importance of a fourth social variable in the debate over the Würzburgers' thought psychology: religion. To begin with, we need to remember that *Kulturkampf* (cultural war) and *Konfessionenstreit* (confessional struggle) had repercussions within academia in general, and within philosophy in particular. For instance, Catholic philosophers had better chances of being hired in Catholic Bavaria than in Protestant Prussia; and Protestant philosophers were more likely to find employment in Protestant rather than in Catholic lands. Moreover, the very existence of half a dozen chairs for Catholic philosophy caused a lot of anger amongst Protestant philosophers.[64] Such chairs were usually the outcome of compromise formulas at the end of the *Kulturkämpfe*. Such 'affirmative action' seemed uncalled for in the eyes of Protestants. As many of them saw it, the future of German philosophy lay in developing further German Idealism rather than Neothomist realism.

These two equations 'Protestant philosophy equals German Idealism' and 'Catholic philosophy equals Neothomist realism' were widely accepted by philosophers and theologians of both confessions.[65] Perhaps more important for my concerns however, was a further dichotomy that was linked to the confessional philosophical divide. This was the opposition between voluntarism and intellectualism. To mention just one influential contemporary source, for the militant Protestant Friedrich Paulsen, it was voluntarism – an emphasis on the will as the most important element of the human soul – that linked Luther to Kant. Luther's great achievement had been the overcoming of Scholastic intellectualism, that is, a 'false unity of belief and knowledge' and Kant's emphasis on 'the primacy of practical reason' had renewed the voluntarist programme. Interestingly enough, Paulsen added a third name to the list of important voluntarists: Wilhelm Wundt.[66]

Like Paulsen, Wundt, throughout his career, attacked Catholicism as outdated and backward, and praised Protestantism as the religion of progress and science. The very existence of Catholicism in his own time and age proved to Wundt that too many people were still steeped in '*superstition.*'[67] As far as scholasticism was concerned, Wundt singled out Thomas Aquinas for special criticism. Thomas's work was without originality, and its essence was compilation and eclecticism. Moreover, Wundt chastised Thomas also for having been one of the earliest advocates of the contract theory of the state.[68]

With respect to contemporary Neoscholasticism, Wundt distinguished between 'official' and 'free' versions. The official version was the Neothomism advocated by the Pope and faithfully repeated by the so-called 'Catholic philosophers' of the universities. By 'free Neoscholasticism' Wundt meant the philosophy of Brentano and his followers. I have already mentioned that Wundt disliked their 'logicism' and 'intellectualism;' now we can add to this that Wundt also wished to emphasise a *religious* divide between himself and the Brentano school. Brentano 'was quite unable to develop an independent philosophy,' and his resurrection of mediaeval Catholic thinking was clever but not deep. As Wundt saw it, the ideological function of Brentano's philosophy was clear enough: It tamed critical Catholics into making their peace with the Pope.[69]

Like Paulsen, Wundt too saw deep parallels between Luther and Kant. Indeed, 'Luther [had] created less a new *religion than a new ethics.*' According to this proto-Kantian ethics, activity was more important than contemplation, and the concept of duty was primary with respect to the concept of virtue.[70] Luther had been the first to 'raise the moral conscience of the individual to the position of a judge over his thinking and doing.' And the final philosophical codification of this 'viewpoint of moral freedom' had come with 'Kantian Idealism.' Moreover Luther deserved credit for having emphasised the importance of the religious community.[71]

The clearest link between Wundt's Protestantism and his psychology was undoubtedly his pronounced 'voluntarism.' Voluntarism for Wundt was the correct alternative to psychological logicism and intellectualism; and intellectualism was most characteristic of the Brentano school, that is, of 'free Neoscholasticism.' As Wundt saw it, the Würzburgers followed Brentano, Stumpf, and Husserl; thus they 'had to' be committed to intellectualism, be opposed to voluntarism, and thus be direct or indirect supporters of religious sentiments that Wundt disliked. No wonder, therefore, that the link between Bühler's non-depictive thought and the Scholastics' *actus purus* figured so centrally in Wundt's criticism.

Wundt himself called his psychology 'voluntaristic' because it was based on the idea that the will was the key to understanding higher processes of thinking, the essence of the self, and the endpoint of the development of emotional life. The will made systematic thinking possible. According to Wundt's account of thinking, associative processes always produced a number of presentations, but it was the human will, that is, apperceptive processes, that chose amongst those presentations.[72] Moreover, Wundt held that the only experienced continuity in inner experience was the 'continuum [...] of acts of willing.' From this Wundt concluded that 'from the viewpoint of psychology, *the will is the I.*'[73]

The will also figured centrally in Wundt's 'inductive metaphysics,' that is a metaphysics that was to speculatively build upon the best-supported scientific theories of the day. Not surprisingly, in Wundt's metaphysical treatises, his own psychology was treated as one such best theory. And since his psychology was voluntaristic, so was his metaphysics. Thus the metaphysical essence of the individual was the 'pure will,' and the history of the world was the history of the all-encompassing world-will. Finally, as Wundt saw it, treating the 'pure will' as the essence of soul precluded talk of a permanent individual soul-substance.[74]

It strengthens my argument for the 'Protestant qualities' of Wundt's philosophy and psychology that he was repeatedly the target of attacks coming from Neothomist philosophers. To begin with, as they saw it, Wundt's voluntarism constituted something of a category mistake. Rather than adopting a neutral soul-substance as the bearer of all mental activities and states, Wundt identified the bearer with one kind of mental content, that is, with volition. This came out clearly when Wundt equated the will with the ego-subject.[75]

In arguing for an individual soul-substance, Neothomists often relied on arguments about the unconscious; an entity or realm the existence of which Wundt denied but the Würzburgers affirmed. The Neothomist Josef Geyser reasoned that conscious events had to have causes, and that such causes were not introspectively accessible. These causes therefore had to be unconscious. But they could not be material: Mental and material events were *toto genere* different, and this precluded the possibility that mental events were determined by material events. Thus the causes of mental life had to be unconscious mental causes.[76] Other Neothomists agreed.[77] Grünholz submitted that Wundt had no good arguments against the unconscious. 'The serious and honest thinker Oswald Külpe' had shown this to Grünholz' satisfaction.[78]

Wundt's adherence to psychophysical parallelism – a topic not yet mentioned and one that I cannot go into here in more detail – was also unacceptable to his Neothomist critics. To Gutberlet it was so obvious that presentations influenced bodily processes and *vice versa* that Wundt's advocacy of parallelism could be explained only as being due to 'self-deception' and 'persistent blindness.'[79]

And finally, the Neothomists deeply disliked Wundt's voluntaristic metaphysics. Gutberlet found Wundt's theory of God as the will of the world 'nebulous and incomprehensible,' and he admitted that he had not been able to make much sense of it all, despite 'multiple rounds of reading and thinking and comparing different passages.'[80]

Turning from Wundt's Protestantism to the Würzburgers' Catholicism, the figure that most deserves our attention is Külpe. This is because Külpe was in fact a Protestant! Does this destroy my argument according to which different religious commitments were influential in the debate over thought psychology? Fortunately the answer is 'no.' As it turns out, and as both friends and foes agreed, the contents of Külpe's philosophy and psychology were more befitting of a Catholic than a Protestant. On the one hand we might note that Külpe was never attacked by, and never attacked, any of the Neothomists of the time. On the other hand, it deserves mentioning that of the calls to professorial chairs that Külpe received between 1904 and 1914, most came from universities in predominantly Catholic areas.[81]

What then were Külpe's views that brought him the sympathy of the Neothomists? And what was his criticism of Wundt's Protestant traits?

To begin with Külpe defended the assumption of an individual soul-substance against Wundt. To the argument that the soul-substance could not be observed, Külpe replied that atoms were not observable either. To the objection that the link between the soul-substance and particular mental processes was dubious, Külpe retorted that the link between mental and physical phenomena in Wundt's psychophysical monism was no less unclear. And finally, Wundt was wrong to claim that the assumption of a soul-substance was incompatible with mental change and development. Wundt's argument was plausible only if one assumed an unchanging soul-substance.[82]

A second area of contention in which Külpe sided with the Neothomists against Wundt was the mind-body problem. Like the Neothomists, and unlike Wundt, Külpe defended interactionist dualism. He wrote that dualism fitted 'much better with natural science and psychology than materialism and spiritualism.'[83]

A third issue on which Külpe and the Neothomists saw eye to eye was their assessment of Paulsen's and Wundt's voluntarism. As Külpe judged the case, there were no good reasons to assume that 'the will is the primary function of the soul.' Neither did such good reasons exist now, nor was it likely that such reasons would ever be found.[84]

Although Külpe refused to replace voluntarism by intellectualism, his warm advocacy of the intellect in some of his writings on the one hand, and his linking of the intellect to Scholasticism on the other hand, were telling of his sympathies. In one place he wrote that 'acceptance of the intellect, and emphasis on the spiritual independence of our soul has always given wings to thinking;' and he went on to mention, as one of his prime examples, 'the intellectualistic perspective of the culmination period of mediaeval Scholasticism.'[85]

Külpe's emphasis upon the intellect brings us to the fourth, and final issue on which Külpe and the Neothomists could agree wholeheartedly. I mean the issue of epistemological realism. To many philosophers in turn-of-the-century Germany, to endorse epistemological realism was to side with Aristotle and St. Thomas against Kant, German Idealism, and the Neokantians. Despite Wundt's anti-Kantian stance in epistemology, Külpe added him to the list of anti-realists.[86]

According to Külpe, his own 'critical realism' could be distinguished from various other epistemological stances by means of the several questions. For our purposes, the most important question was the first. It asked whether science was ever justified in positing anything as a real entity, that is, as an entity that was not given in original, immediate experience. Realist epistemologies answered 'yes,' 'conscientialism' (*Konszientialismus*) answered 'no.' 'Conscientialism,' a term invented by Külpe, thus covered various forms of 'immanent' philosophy, positivism, idealism, and – at least within the realm of the psychological – Wundt.[87] Indeed, Külpe regarded it as a major inconsistency in Wundt's thinking that Wundt wished to be a realist with respect to natural science but a conscientialist with respect to psychology. Here it was of course Wundt's actuality theory of the soul, that is, his rejection of the unconscious mental and any underlying soul substance and that committed him to 'psychological conscientialism.'[88]

For Külpe a sound epistemological realism was inseparable from taking on board the results of Würzburg thought psychology. Realism presupposed a human faculty by means of which knowers could free themselves from being confined to their intra-mental world of sensations, presentations, and feelings. As long as knowers remained on the level of sensations, presentations, and feelings, they never had reason to posit anything 'beyond,' anything 'real.' Obviously, the needed faculty was thinking. But not just any account of thinking would do. If thinking itself was nothing but a sequence of sensations, feelings, and presentations, then nothing was gained. Only a thinking with its own independent laws, only a thinking for which sensations, feelings, or presentations were accidental vehicles at best, only such thinking could justify the assumption of a mind-independent, real world.[89]

Külpe also located his own realistic epistemology historically. Basically, there were three ways of thinking about categories: the Aristotelian, the Lockean, and the Kantian. For Aristotle, categories were objective: 'predicates that apply to beings.' For Locke, categories were subjective: 'mere givens of consciousness, formed according to psychological laws.' Finally, for Kant, categories were both objective and subjective. They were objective insofar as they were not formed by the individual, and subjective insofar as they were not true of things-in-themselves. Külpe sided with Aristotle.[90]

Neothomist philosophers were quick to capitalise on Külpe's realist leanings. The leading German Neoscholastic journal, *Philosophisches Jahrbuch der Görres-Gesellschaft*, published a number of highly positive papers on Külpe's work, and Martin Grabmann published two booklets on parallels between Külpe's philosophy and psychology on the one hand, and Thomas Aquinas's thought on the other.[91] To make his case, Grabmann picked out various passages from Külpe's oeuvre and showed that either St. Thomas or else his modern-day followers had said pretty similar things. He also used Messer as an additional witness. In his obituary for Külpe, Messer had written that his former colleague had wanted 'to prove that the Kantian epistemological revolution was without justification;' Külpe therefore had attempted to return to 'the *pre*-Kantian conception of knowledge, a conception central to Aristotle and the Scholastic.'[92]

All in all, Grabmann found seven 'points of contact' between Külpe and the medieval schoolmen. First, Külpe's use of 'immanent criticism' reminded Grabmann of Aristotle and Thomas Aquinas. Second, both Külpe and the Scholastics took their starting point from the empirical sciences. Third, Külpe's advocacy of the method of self-observation would have delighted the likes of Aristotle, Augustine, and St. Thomas. The fourth parallel between Külpe and the scholastics was that both took a keen interest in the problem of realism. Fifth, Thomas agreed with Külpe that the object of knowledge was extramental. Sixth, both men had emphasised the importance of the concept of 'intention,' and seventh, and finally, both had spoken of the 'dignity and task of thinking;' this task was 'a deep and complete knowledge of reality.'[93]

Up to this point, I have concentrated on the opposition between Wundt's Protestantism and Külpe's Catholicism. To conclude let me at least comment briefly on three other key Würzburg psychologists. Bühler, Marbe, and Messer were all reared in the Catholic religion. Interestingly enough, as young students they all were in contact with E. Krebs, a priest in Freiburg who nowadays is remembered primarily because of his friendship with the young Heidegger almost two decades later. After Heidegger had informed Krebs of his intention to leave the Catholic church in 1919, Krebs entered into his diary the following comment:

> Having qualified to lecture on Catholic philosophy, Heidegger will get himself into a lot of trouble for now changing sides. He is growing away from Catholic thinking, going the same way I saw Bühler going.

To the apostate Bühler, Krebs added a list of further names of one-time Catholic philosophers, amongst them Marbe and Messer.[94]

Little can be said of Marbe in this respect. He himself reported having received 'a strict religious (Catholic) education,' but added that already from his gymnasium days he 'had not used it any more.'[95] Bühler did his M.D. at the University of Freiburg in 1903; this presumably was the time of his close contact with Krebs. The following year Bühler went to Strassburg to do a Ph.D. with the Neothomist philosopher C. Baeumker. His Ph.D. was a somewhat unambitious summary of 'the system' of the eighteenth-century Scottish philosopher Henry Home (better known as Lord Kames).[96] From

Strasburg Bühler moved on to Berlin, were he acted as Stumpf's assistant in the academic year 1905-6. And from working with this 'free Neothomist,' Bühler then went on to Würzburg to begin his work on his *Habilitationsschrift*. Some sources report that evidence that Bühler was a 'fallen Catholic' did not emerge until 1916, when Karl Bühler married Charlotte, who was from an assimilated Jewish family and who had been reared a Protestant. The Bühlers married in a Protestant ceremony.[97]

Finally, Messer's relationship with the Catholic church was complicated. On the one hand, he always insisted on his 'high regard and affection for this religion.'[98] On the other hand, he found unacceptable the church's position that religious doubt was a sin, and that independent thinking in religious matters was acceptable only if it ended up confirming official church doctrines.[99]

Messer had strong sympathies for the leading German modernist, the Würzburg theology professor H. Schell. Messer's sympathies for Schell were obvious from the fact that he wrote about Schell in his *Die Philosophie der Gegenwart in Deutschland,*[100] that he sought a dialogue with Schell when he worked in Würzburg during the academic year 1904-5, and that he left the Catholic church over Pius X's condemnation of modernism.[101]

As far as Messer's epistemological and metaphysical views were concerned, these largely coincided with Külpe's. Messer was more outspoken, however, on the relationship between his (and Külpe's) epistemological views and Neothomism:

> Neothomist philosophy does not deserve the contemptuous treatment to which it is usually subjected [...] Neothomism is right to base knowledge of reality upon the combined activity of (outer and inner) experience on the one hand, and the interpretation and explanation of the material of experience by rational thinking, on the other hand.[102]

7. CONCLUSION

In this paper, I have tried to show that the debate over thought psychology, especially as far as the controversy between Wundt and the Würzburgers was concerned, was intertwined with a number of social, political, cultural, and religious issues. Stances for and against imageless thought were influenced by differing views on how a research school should be organised, on how psychology related to philosophy both intellectually and institutionally, on what was the appropriate way to think about the relationship between the individual and the state, and on which religion best befitted the modern world.

In conclusion I would like to emphasise that I do not wish to claim that the technical content of the controversy over *Denkpsychologie* can be *reduced* to the social institutions, social visions, or social interests that I brought into the story above. Nor do I wish to deny that we can still ask 'who was right, Wundt or the Würzburgers.' But I do believe that situating the debate into its social and cultural context enriches our understanding of what all the debate was about, and why it was fought over with so much passion. Equally crucially, coming to realise and appreciate the interplay of various social orders on the one hand, and the 'psychological order' of theories about the mind, on the other hand, teaches us something important about the mind itself. Perhaps it indicates that the complete defeat of Wundtian psychological collectivism involved a loss – a loss from which cognitive science has not yet recovered.

Martin Kusch
Department of History and Philosophy of Science
University of Cambridge
Free School Lane
Cambridge CB2 3RH
Email: mphk2@cam.ac.uk

NOTES

[1] Gardner 1987.
[2] Kusch 1999.
[3] E.g. Barnes, Bloor, Henry 1996.
[4] Ach 1905, 1910, 1964a; Bühler 1907a; Marbe 1901, 1964.
[5] Ach 1905, 1910, 1964b; Koffka 1912; Watt 1905, 1964.
[6] Ach 1905, 1910; Bühler 1907a; Marbe 1901.
[7] Anschütz 1911, 1912a, 1912b, 1913a, 1913b; von Aster 1908; Deuchler 1909; Henning 1919; G.F. Lipps 1906; Meumann 1907a, 1907b; Moskiewicz 1910a, 1910b; G.-E. Müller 1911, 1913, 1917, 1919; Reichwein 1910; Husserl 1913 (1950); Scheler 1922; Stumpf 1906; Cassirer 1910; Hönigswald 1913; Rickert 1904; Baeumker (1913) 1921; Geyser 1908a, 1908b, 1914; Grabmann 1913, 1916; Gutberlet 1908.
[8] Wundt 1907.
[9] Hellpach 1948, 175.
[10] Tawney 1921, 181.
[11] Judd 1921, 174.
[12] Wundt 1897, 41; 1903, 341–45; 1908, 202, 209; Blumenthal 1970, 14–5; Weimer 1974, 245.
[13] Wundt 1908, VIII.
[14] Bühler 1922, 249; Hellpach 1948, 175; Kiesow 1930, 167; Marbe 1961, 198; Messer 1922, 11–13.
[15] Ach 1905, 9, 26–27; Bühler 1907a, 299–300, 309, 313; Külpe 1914, 1229.
[16] Bühler 1908a.
[17] E.g. Marbe 1916.
[18] Bühler 1908a, 90.
[19] Kusch 1995.
[20] Husserl (1900) 1975; Cohen 1902; Natorp 1887; Rickert 1904, 1913; Windelband 1884.

[21] Husserl (1911) 1987, 18.
[22] Wundt, e.g. 1906, 1, 11.
[23] Dubs 1911, 119; Geyser 1909, 259; Heidegger (1913) 1978, 79–90; Husserl (1900) 1975, 69, 80, 184; Moog 1918, 360, 1919, 101, 1922, 110–129; Natorp 1901, 277; Spranger 1905, 14; Ssalagoff 1911, 161; Windelband 1884, 248.
[24] Wundt 1906, 75–6, 84–5, 88–9.
[25] Wundt 1910, 526–36, 612.
[26] Wundt 1910.
[27] Wundt 1910, 516–9, 569.
[28] Wundt 1910, 579.
[29] Marbe 1901, 98.
[30] Wundt 1903, 580.
[31] Bringmann and Ungerer 1980, 54.
[32] Hammer 1994, 244.
[33] See Bühler 1907a, 1908b, 1908c.
[34] Bühler 1907a, 299, 329, 347.
[35] Aster 1908, 62.
[36] Messer 1906, 85.
[37] Messer 1908.
[38] Messer 1906, 179.
[39] Kusch 1995.
[40] Kusch 1995, 190–93.
[41] Külpe 1912b.
[42] Külpeana V, 13.
[43] Külpe 1912b, 2.
[44] Külpeana V, 13.
[45] Wundt 1913a.
[46] Wundt 1913b, 5.
[47] Diamond 1980, 42; Oelze 1991, 80–1.
[48] Wundt 1912 I, iii.
[49] Wundt 1912 III, 102.
[50] Wundt 1897, 634, 657–8, 661.
[51] Wundt 1917, 336.
[52] Wundt 1912 II, 246, 1912 III, 81, 251, 257.
[53] Wundt 1912 III, 33.
[54] Wundt 1912 II, 215.
[55] Wundt 1920, 218.
[56] Sganzini 1913, 235.
[57] Bühler1907b.
[58] Marbe 1916.
[59] Marbe 1916, 115.
[60] Marbe 1916, 117.
[61] Messer 1912, 146, 156, 158.
[62] Külpe 1910a, 321.
[63] Külpe 1915a, 31.
[64] E.g. Hensel 1909.
[65] E.g. Erhard 1902; Eucken 1901; Gutberlet 1911; Hensel 1909; Paulsen 1901a, 1901b.
[66] Paulsen 1901a, vi–vii, 13, 97, 391.

[67] Wundt 1912 III, 219–20.
[68] Wundt 1897, 612.
[69] Wundt 1910, 599–600.
[70] Wundt 1912 II, 90–91.
[71] Wundt 1920, 352.
[72] E.g. Wundt 1897, 41.
[73] Wundt 1914, 121.
[74] Wundt 1897, 379.
[75] Grünholz 1913, 313.
[76] Geyser 1908, 83.
[77] Gutberlet 1896, 17.
[78] Grünholz 1913, 317.
[79] Gutberlet 1913, 147, 155.
[80] Gutberlet 1891, 357.
[81] Külpeana V, 18.
[82] Külpe 1910a, 277–83.
[83] Külpe 1910a: 203–4.
[84] Külpe 1910a: 287.
[85] Külpe 1912b, 1107; cf. also 1912a, 33–34.
[86] Külpe 1910a, 191.
[87] Külpe 1910b, 1027.
[88] Külpe 1910a, 191.
[89] Külpe 1910b, 1029.
[90] Külpe 1915a, 4, 52–58.
[91] Grabmann 1913, 1916.
[92] Quoted from Grabmann 1916, 356.
[93] Grabmann 1916, 357–67.
[94] Krebs in Ott 1993, 109.
[95] Marbe 1945, 49.
[96] Bühler 1905.
[97] Gardner and Stevens 1992, 141.
[98] Messer 1924, 121.
[99] Friedwalt (alias Messer) 1905; 1921, 172; 1922, 15; 1924, 128; 1927, 3.
[100] Messer 1927.
[101] Messer 1922, 15.
[102] Messer 1927, 7.

UNPUBLISHED SOURCES: CONSULTED ARCHIVES

Bayrische Staatsbibliothek München, Handschriftenabteilung, Külpeana.

REFERENCES

Ach, N.: 1905, *Über die Willenstätigkeit und das Denken*. Göttingen, Vandenhoeck & Ruprecht.

Ach, N.: 1910, *Über den Willensakt und das Temperament: Eine experimentelle Untersuchung*. Leipzig, Engelmann.

Ach, N.: 1964a, 'Awareness,' in Mandler and Mandler, eds., 1964, 152–67.

Ach, N.: 1964b, 'Determining Tendencies,' in Mandler and Mandler, eds., 1964, 201–8.

Anschütz, G.: 1911, 'Über die Methoden der Psychologie,' *Archiv für die Gesamte Psychologie*, **20**, 414–98.

Anschütz, G.: 1912a, 'Spekulative, exakte und angewandte Psychologie. Eine Untersuchung über die Prinzipien der psychologischen Erkenntnis II,' *Archiv für die Gesamte Psychologie*, **24**, 1–30.

Anschütz, G.: 1912b, 'Tendenzen im psychologischen Empirismus der Gegenwart: Eine Erwiderung auf O. Külpes Ausführungen "Psychologie und Medizin" and "Über die Bedeutung der modernen Denkpsychologie," *Archiv für die Gesamte Psychologie*, **25**, 189–207.

Anschütz, G.: 1913a, *Die Intelligenz: Eine Einführung in die Haupttatsachen, die Probleme und die Methoden zu einer Analyse der Denktätigkeit*. Osterwieck, Zickfeldt.

Anschütz, G.: 1913b, 'Einige Bemerkungen zu meiner Kritik von O. Külpes Ausführungen "Psychologie und Medizin" and "Über die Bedeutung der modernen Denkpsychologie," *Zeitschrift für Psychologie*, **66**, 155–60.

Aster, E. v.: 1908, 'Die psychologische Beobachtung und experimentelle Untersuchung von Denkvorgängen,' *Zeitschrift für Psychologie*, **49**, 56–107.

Baeumker, C.: 1921 (1913), *Anschauung und Denken*. Paderborn, Schöningh.

Barnes, B., Bloor, D. and Henry, J.: 1996, *Scientific Knowledge: A Sociological Analysis*. London, Athlone Press.

Blumenthal, A.: 1970, *Language and Psychology*. New York, Wiley.

Bringmann, W. G. and Ungerer, G.: 1980, 'Experimental vs. Educational Psychology: Wilhelm Wundt's Letters to Ernst Meumann,' *Psychological Research*, **42**, 57–73.

Bühler, K.: 1905, *Studien über Henry Home*. Dissertation. Strassburg, Bach.

Bühler, K.: 1907a, 'Remarques sur les problèmes de la psychologie de la pensèe,' *Archives de Psychologie*, **6**, 376–86.

Bühler, K.: 1907b, 'Tatsachen und Probleme zu einer Psychologie der Denkvorgänge, I: Über Gedanken,' *Archiv für die Gesamte Psychologie*, **9**, 297–365.

Bühler, K.: 1907c, *Theses quas ad veniam legendi impetrandam defendet Carolus Bühler*, VII Junii MCMVII. Wirceburgi, Stürtz.

Bühler, K.: 1908a, 'Nachtrag: Antwort auf die von W. Wundt erhobenen Einwände gegen die Methode der Selbstbeobachtung an experimentell erzeugten Erlebnissen,' *Archiv für die Gesamte Psychologie*, **12**, 93–123.

Bühler, K.: 1908b, Review of Stumpf (1906), *Archiv für die Gesamte Psychologie*, **11**, 1–5.

Bühler, K.: 1908c, 'Tatsachen und Probleme zu einer Psychologie der Denkvorgänge. II. Über Gedankenzusammenhänge,' *Archiv für die Gesamte Psychologie*, **12**, 1–23.

Bühler, K.: 1908d, 'Tatsachen und Probleme zu einer Psychologie der Denkvorgänge. III. Über Gedankenerinnerungen,' *Archiv für die Gesamte Psychologie*, **12**, 24–92.

Bühler, K.: 1922, 'Oswald Külpe,' *Lebensläufe aus Franken*, **2**, 244–55.

Cassirer, E.: 1910, *Substanzbegriff und Funktionsbegriff*. Berlin, Cassirer.

Cohen, H.: 1902, *Logik der reinen Erkenntnis*. Berlin, Cassirer.

Deuchler, G.: 1909, 'Beiträge zur Erforschung der Reaktionsformen,' *Psychologische Studien*, **4**, 353–430.

Diamond, S.: 1980, 'Wundt before Leipzig,' in Rieber, ed., 1980, 3–70.

Dubs, A.: 1911, *Das Wesen des Begriffs und des Begreifens: Ein Beitrag zur Orientierung in der wissenschaftlichen Weltanschauung*. Halle, Niemeyer.

Ehrhard, A.: 1902, *Der Katholizismus und das zwanzigste Jahrhundert im Lichte der kirchlichen Entwicklung der Neuzeit*. Stuttgart, Roth, 3rd ed.

Eucken, R.: 1901, 'Thomas von Aquino und Kant: Ein Kampf zweier Welten,' *Kantstudien*, **6**, 1–18.
Friedwalt, A.: (=A. Messer) 1905, *Katholische Studenten*. Stuttgart, Greiner & Pfeiffer.
Gardner, H.: 1987, *The Mind's New Science: A History of the Cognitive Revolution*. New York, Basic Books.
Gardner, S. and Stevens, G.: 1992, *Red Vienna and the Golden Age of Psychology, 1918–1938*. New York, Praeger.
Geyser, J.: 1908a, 'Die Vorzüge und Schwächen der neueren Untersuchung der Denkvorgänge durch das Ausfrageexperiment,' *Philosophisches Jahrbuch der Görres-Gesellschaft*, **21**, 90–102.
Geyser, J.: 1908b, *Lehrbuch der allgemeinen Psychologie*. Münster, Schöningh.
Geyser, J.: 1909, *Einführung in die Psychologie der Denkvorgänge: Fünf Vorträge, gehalten im April 1909 auf dem pädagogischen Kursus in Cöln*. Münster, Schöningh.
Geyser, J.: 1914a, *Die Seele: Ihr Verhältnis zum Bewusstsein und zum Leibe*. Münster, Schöningh.
Grabmann, M.: 1913, *Der Gegenwartswert der geschichtlichen Erforschung der mittelalterlichen Philosophie*. Wien, Herder.
Grabmann, M.: 1916, 'Der kritische Realismus Oswald Külpes und der Standpunkt der aristotelisch-scholastischen Philosophie,' *Philosophisches Jahrbuch der Görres-Gesellschaft*, **29**, 333–69.
Grünholz, E.: 1913, 'Eine kritische Untersuchung über das Denken im Anschluß an die Philosophie Wilhelm Wundts,' *Philosophisches Jahrbuch der Görres-Gesellschaft*, **26**, 305-27.
Gutberlet, C.: 1891, 'W. Wundt's System der Philosophie,' *Philosophisches Jahrbuch der Görres-Gesellschaft*, **4**, 281-96, 341–59.
Gutberlet, C.: 1908, 'Der gegenwärtige Stand der psychologischen Forschung,' *Philosophisches Jahrbuch der Görresgesellschaft*, **21**, 1–32.
Gutberlet, C.: 1911, 'Religionspsychologie,' *Philosophisches Jahrbuch der Görresgesellschaft*, **24**, 147–76.
Gutberlet, C.: 1916, 'Rudolph Eucken,' *Philosophisches Jahrbuch der Görres-Gesellschaft*, **29**, 113–37.
Hall, S.T. et al.: 1921, 'In Memory of Wilhelm Wundt,' *The Psychological Review*, **28**, 153–88.
Hammer, S.: 1994, *Denkpsychologie–Kritischer Realismus: Eine wissenschaftshistorische Studie zum Werk Oswald Külpes*. Frankfurt a. Main, Lang.
Heidegger, M.: 1978 (1913), 'Die Lehre vom Urteil im Psychologismus: Ein kritisch-positiver Beitrag zur Logik,' in Heidegger, *Frühe Schriften*, ed. by F.-W. von Herrmann, *Gesamtausgabe*, vol. 1. Frankfurt a. Main, Klostermann, 59–188.
Hellpach, W.: 1948, *Wirken in Wirren*, 2 vols., vol. 2. Hamburg, Wegner.
Henning, H.: 1919, 'Assoziationslehre und neuere Denkpsychologie,' *Zeitschrift für Psychologie*, **82**, 219–26.
Hensel, P.: 1909, 'Die Aussichten der Privatdozenten für Philosophie,' *Frankfurter Zeitung*, 29. 7. 1909, 1–2.
Hönigswald, R.: 1913, 'Prinzipien der Denkpsychologie,' *Kantstudien*, **18**, 205–45.
Husserl, E.: 1975 (1900), *Logische Untersuchungen. Erster Band: Prolegomena zur reinen Logik*. Husserliana XVIII, ed. by E. Holenstein. The Hague, Nijhoff.

Husserl, E.: 1984 (1901), *Logische Untersuchungen. Zweiter Band: Untersuchungen zur Phänomenologie und Theorie der Erkenntnis*, Husserliana XIX, ed. by U. Panzer. The Hague, Nijhoff.

Husserl, E.: 1987 (1911), 'Philosophie als strenge Wissenschaft,' in Husserl, *Aufsätze und Vorträge (1911–1921)*, Husserliana XXV, ed. by T. Nenon and H. R. Sepp. Dordrecht, Nijhoff, 3–62.

Husserl, E.: 1950 (1913), *Ideen zu einer reinen Phänomenologie und phänomenologischen Philosophie Erstes Buch: Allgemeine Einführung in die reine Phänomenologie*. Husserliana III, ed. by W. Biemel. The Hague, Nijhoff.

Judd, C. H.: 1921, 'In Memory of Wilhelm Wundt,' in Hall *et al.*, 1921, 'In Memory of Wilhelm Wundt,' *Psychological Review*, **3**, 153–88, 173–8.

Kiesow, F.: 1961 (1930), 'F. Kiesow,' in Murchison, ed., 1961, 163–90.

Koffka, K.: 1912, *Zur Analyse der Vorstellungen und ihrer Gesetze: Eine experimentelle Untersuchung*. Leipzig, Quelle & Meyer.

Külpe, O.: 1910a, *Einleitung in die Philosophie*. Leipzig, Hirzel, 5th ed.

Külpe, O.: 1910b, 'Erkenntnistheorie und Naturwissenschaft,' *Physikalische Zeitschrift*, **23**, 1025–35.

Külpe, O.: 1912a, *Die Realisierung: Ein Beitrag zur Grundlegung der Realwissenschaften*. Leipzig, Hirzel.

Külpe, O.: 1912b, 'Psychologie und Medizin,' *Zeitschrift für Psychopathologie*, **1**, 187–267.

Külpe, O.: 1914, 'Über die Methoden der psychologischen Forschung,' *Internationale Monatsschrift für Wissenschaft, Kultur und Technik*, **8**, 1053–70, 1219–32.

Külpe, O.: 1915a, *Die Ethik und der Krieg*. Leipzig, Hirzel.

Külpe, O.: 1915b, *Zur Kategorienlehre* (Sitzungsberichte der Koeniglich Bayerischen Akademie der Wissenschaften, Philosophisch-philologische Klasse, 5. Abhandlung, Vorgetragen am 6. Februar 1915). München, Verlag der Koeniglich Bayerischen Akademie der Wissenschaften.

Kusch, M.: 1995, *Psychologism: A Case Study in the Sociology of Philosophical Knowledge*. London, Routledge.

Kusch, M.: 1999, *Psychological Knowledge: A Social History and Philosophy*. London, Routledge.

Lipps, G.F.: 1906, 'Review of Ach, 1905,' *Literarisches Zentralblatt*, **20** (May 12th), 677–8.

Mandler, J. M. and Mandler, G. eds.: 1964, *Thinking: From Association to Gestalt*. New York, Wiley & Sons.

Marbe, K.: 1901, *Experimentell-psychologische Untersuchungen über das Urteil: Eine Einleitung in die Logik*. Leipzig, Engelmann.

Marbe, K.: 1916, *Die Gleichförmigkeit in der Welt: Untersuchungen zur Philosophie und positiven Wissenschaft*, vol. 1. Munich, Beck.

Marbe, K.: 1945, *Selbstbiographie des Psychologen Geheimrat Prof. Dr. Karl Marbe in Würzburg*. Deutsche Akademie der Naturforscher, Halle a. d. Saale, Leopoldina.

Marbe, K.: 1961, 'Autobiography,' in Murchison, ed.,1961, 181–213.

Marbe, K.: 1964, 'The Psychology of Judgements,' in Mandler and Mandler, eds., 1964, 143–8.

Messer, A.: 1906, 'Experimentell-psychologische Untersuchungen über das Denken,' *Archiv für die Gesamte Psychologie*, **8**, 1–224.

Messer, A.: 1908, *Empfindung und Denken*. Leipzig, Quelle & Meyer.

Messer, A.: 1912, *Das Problem der staatsbürgerlichen Erziehung historisch und systematisch behandelt*. Leipzig, Nemnich.

Messer, A.: 1921, *Einführung in die Erkenntnistheorie*. Leipzig, Meiner, 2nd ed.

Messer, A.: 1922, 'August Messer,' in Schmidt, ed., 1922, vol. 3. (pages are not numbered consecutively).

Messer, A.: 1924, *Glauben und Wissen: Geschichte einer inneren Entwicklung*. München, Reinhardt, 3rd. ed.

Messer, A.: 1927, *Die Philosophie der Gegenwart in Deutschland*. Leipzig, Quelle & Meyer, 6th ed.

Meumann, E.: 1907a, Review of Wundt, 1907b, *Archiv für die Gesamte Psychologie*, **10**, 117–34.

Meumann, E.: 1907b, 'Über Assoziationsexperimente mit Beeinflussung der Reproduktionszeit,' *Archiv für die Gesamte Psychologie*, **9**, 117–50.

Moog, W.: 1918, 'Die Kritik des Psychologismus durch die moderne Logik und Erkenntnistheorie,' *Archiv für die Gesamte Psychologie*, **37**, 301–62.

Moog, W.: 1919, *Logik, Psychologie und Psychologismus*. Halle, Niemeyer.

Moog, W.: 1922, *Die deutsche Philosophie des 20. Jahrhunderts in ihren Hauptrichtungen und Grundproblemen*. Stuttgart, Enke.

Moskiewicz, G.: 1910a, Review of Messer, 1908, *Zeitschrift für Psychologie*, **54**, 545–50.

Moskiewicz, G.: 1910b, 'Zur Psychologie des Denkens (Erste Abhandlung),' *Archiv für die Gesamte Psychologie*, **18**, 305–99.

Müller, G. E.: 1911, *Zur Analyse der Gedächtnistätigkeit und des Vorstellungsverlaufes, I. Teil* (*Zeitschrift für Psychologie*, suppl. vol. 5). Leipzig, Barth.

Müller, G. E.: 1913, *Zur Analyse der Gedächtnistätigkeit und des Vorstellungsverlaufes, III. Teil* (*Zeitschrift für Psychologie*, suppl. vol. 8). Leipzig, Barth.

Müller, G. E.: 1917, *Zur Analyse der Gedächtnistätigkeit und des Vorstellungsverlaufes, II. Teil* (*Zeitschrift für Psychologie*, suppl. vol. 9). Leipzig, Barth.

Müller, G. E.: 1919, Review of Selz, 1913, *Zeitschrift für Psychologie*, **82**, 102–20.

Murchison, C. A. ed.: *A History of Psychology in Autobiography*, 7 vols., vol. III. New York, Russell & Russell.

Natorp, P.: 1887, 'Über objective und subjective Begründung der Erkenntnis I,' *Philosophische Monatshefte*, **23**, 257–86.

Natorp, P.: 1901, 'Zur Frage der logischen Methode. Mit Beziehung auf Edm. Husserls "Prolegomena zur reinen Logik," *Kantstudien*, **6**, 270–83.

Oelze, B.: 1991, *Wilhelm Wundt: Die Konzeption der Völkerpsychologie*. Münster, Waxmann.

Ott, H.: 1993, *Martin Heidegger: A Political Life*. En. tr. by A. Blunden. New York, HarperCollins.

Paulsen, F.: 1901a, *Einleitung in die Philosophie*. Berlin, Hertz, 7th ed.

Paulsen, F.: 1901b, *Philosophia militans: Gegen Klerikalismus und Naturalismus*. Berlin, Reuther & Reichard.

Reichwein, G.: 1910, *Die neueren Untersuchungen über Psychologie des Denkens nach Aufgabestellung, Methode und Resultaten übersichtlich dargestellt und krititisch beurteilt*. Dissertation. Halle a. S., Kaemmerer.

Rickert, H.: 1904, *Der Gegenstand der Erkenntnis: Ein Beitrag zum Problem der philosophischen Transcendenz*. Tübingen, Mohr, 2nd, rev. ed.

Rickert, H.: 1913a, *Die Grenzen der naturwissenschaftlichen Begriffsbildung*. Tübingen, Mohr, 2nd, rev. ed.
Rickert, H.: 1913b, 'Zur Besetzung der philosophischen Professuren mit Vertretern der experimentellen Psychologie, *Frankfurter Zeitung*, 4 March 1913, 1–2.
Rieber, R. W. ed.: 1980, *Wilhelm Wundt and the Making of a Scientific Psychology*. London, Plenum Press.
Scheler, M.: 1922, 'Die deutsche Philosophie der Gegenwart,' in Witkop, ed., 1922, 127–224.
Schmidt, R. ed.: 1922, *Philosophie der Gegenwart in Selbstdarstellungen*, 7 vols., vol. 3. Leipzig, Barth.
Spranger, E.: 1905, *Die Grundlagen der Geschichtswissenschaft*. Berlin, Reuther & Reichard.
Ssalagoff, L.: 1911, 'Vom Begriff des Geltens in der modernen Logik,' *Zeitschrift für Philosophie und philosophische Kritik*, **143**, 145–90.
Stumpf, C.: 1906, 'Erscheinungen und psychische Funktionen,' *Aus den Abhandlungen der königlichen Preussischen Akademie der Wissenschaften vom Jahre 1906*. Berlin, Verlag der königlichen Akademie der Wissenschaften.
Tawney, G. A.: 1921, 'In Memory of Wilhelm Wundt,' in Hall *et al.*, 1921, 'In Memory of Wilhelm Wundt,' *Psychological Review*, **3**, 153–88, 1178–81.
Watt, H. J.: 1905, 'Experimentelle Beiträge zu einer Theorie des Denkens,' *Archiv für die Gesamte Psychologie*, **4**, 289–436.
Watt, H. J.: 1964, 'Experimental Contribution to a Theory of Thinking,' in Mandler and Mandler, eds., 1964, 189–201.
Weimer, W. B.: 1974, 'The History of Psychology and Its Retrieval from Historiography, I: The Problematic Nature of History,' *Science Studies*, **4**, 235–58.
Windelband, W.: 1884, 'Kritische oder genetische Methode?,' in Windelband 1884, 247–79.
Windelband, W.:1884, *Präludien*. Freiburg i. Br., Mohr.
Witkop, P. ed.: 1922, *Deutsches Leben der Gegenwart*. Berlin, Wegweiser-Verlag.
Wundt, W.: 1897, *System der Philosophie*. Leipzig, Engelmann, 2nd rev. ed.
Wundt, W.: 1903, *Grundzüge der physiologischen Psychologie*, 3 vols., vol. 3. Leipzig, Engelmann, 5th ed.
Wundt, W.: 1906, *Logik: Eine Untersuchung der Prinzipien der Erkenntnis und der Methoden wissenschaftlicher Forschung*, 1st vol. Stuttgart, Enke, 3rd ed.
Wundt, W.: 1907, 'Über Ausfrageexperimente und über die Methoden zur Psychologie des Denkens,' *Psychologische Studien*, **3**, 301–60.
Wundt, W.: 1908, *Logik: Eine Untersuchung der Prinzipien der Erkenntnis und der Methoden wissenschaftlicher Forschung*, 3 vols., vol. 3. Stuttgart, Enke.
Wundt, W.: 1910, 'Psychologismus und Logizismus,' in Wundt 1910, 1st vol., 511–634.
Wundt, W.: 1910, *Kleine Schriften*. Leipzig, Engelmann.
Wundt, W.: 1912, *Ethik: Eine Untersuchung der Tatsachen und Gesetze des sittlichen Lebens*, 3 vols. Stuttgart, Enke, 4th ed.
Wundt, W.: 1913a, *Die Psychologie im Kampf ums Dasein*. Leipzig, Engelmann.
Wundt, W.: 1913b, *Elemente der Völkerpsychologie*. Leipzig, Kröner, 2nd. ed.
Wundt, W.: 1914, *Sinnliche und übersinnliche Welt*. Leipzig, Kröner.
Wundt, W.: 1917, *Völkerpsychologie: Eine Untersuchung der Entwicklungsgesetze von Sprache, Mythus und Sitte, vol. 8: Die Gesellschaft*, Part 2. Leipzig, Kröner.
Wundt, W.: 1918, *Völkerpsychologie: Eine Untersuchung der Entwicklungsgesetze von Sprache, Mythus und Sitte, vol. 9: Das Recht*. Leipzig, Kröner.
Wundt, W.: 1920, *Erlebtes und Erkanntes*. Stuttgart, Kröner.

RICCARDO LUCCIO

REPRESENTATION IN PSYCHOPHYSICS

1. INTRODUCTION

To speak about representation in psychology with reference to psychophysics might appear somewhat paradoxical. If by 'representation' is meant (in a broad sense) a mental construct in the absence of (or at least independent from) actual sensory stimulation, then psychophysics is paradigmatically the exact contrary of the psychology of representation. In a certain sense, psychophysics is the branch of psychology which studies the relation between stimulation and sensation, and does so relatively regardless of mental contents. I believe that this image is only partly true, or least it is only adequate. In order to gain thorough understanding of the historical evolution of the concept of representation in psychology, careful examination of some aspects of psychophysics is recommended.

One of the key problems of scientific psychology when it was endeavouring to establish itself as a natural science was the problem of measuring mental facts, and also of identifying a physical parameter, variations in which could be related to variations in mental facts. This was the customary procedure of all the natural sciences, from physics to chemistry, from astronomy to geology, and it was argued that only in this way could psychology, too, achieve the status of a natural science. Of decisive importance in this regard was research on reaction times, in particular the research conducted using the so-called 'subtractive method' by the Dutch physiologist and ophthalmologist Frans Cornelis Donders.[1] In the same period, however, mental events were also being measured from an entirely different direction, with the beginning of psychophysics. In this case too, the pioneer in this area of research was a scientist working in a field apparently very distant from psychology: the physicist Gustav Th. Fechner (1860).

L. Albertazzi (ed.), The Dawn of Cognitive Science. Early European Contributors, 89–101.

2. THE WEBER-FECHNER LAW

A professor at Leipzig from 1834 to 1839, Fechner was a leading physicist in the first half of the century until a serious illness constrained him to long years of inactivity. There is something mysterious about Fechner's illness: according to one version credited to Fechner himself,[2] the illness affected his eyes, the cause possibly being over-long fixation on the sun to study retinal after-images.[3] However, in the same period he also suffered from mental disturbances. Fechner was also influenced by Indian philosophy and developed a personal philosophical theory with mystical aspects that today appears irremediably outdated but at that time undoubtedly had a certain appeal – one remembers that Schopenauer, for instance, belongs to the same period. Roughly speaking, Fechner (1851) thought that the 'soul' was a characteristics of the organisation of matter, and that all matter, however simple, was endowed with it. Hence not only men and animals, but also plants and the inorganic world, and the systems into which they were organised, from planets to stones, had a soul. Note that Fechner's doctrine was an utterly monistic and materialistic doctrine: it was an original solution to the 'problem of materialism' then preoccupying the scientific world, particularly in Germany.[4]

The specification of a relation, expressed in a clear mathematical form with universal value, which could define a stable relationship between the world of the soul and the world of matter was of enormous importance. According to Fechner, it was just as important as Newton's law on universal gravitation or Coulomb's on electrical charges. And Fechner believed that he had found such a law, deriving it from a relation discovered some years earlier by a friend a little younger than him (whose brother succeeded Fechner in the chair of physics at Leipzig), the physiologist and anatomist Ernst Heinrich Weber.

In 1834, while studying touch and hearing, Weber made a discovery of great relevance to all future developments of psychology (although he was unaware of doing so). He noticed that if in a certain sensory modality (touch, hearing, sight, etc.), a subject is presented with a stimulus of a given intensity R, and if he tried to evaluate the degree of variation in this stimulus needed to perceive the variation, this just noticeable difference (ΔR) was not constant, but depended linearly on the initial value of R. What was constant was the ratio k (called Weber's constant) between (ΔR and R):

$$k = \frac{\Delta R}{R}$$

In other words, if we begin with a stimulus of intensity, say, 10, and if in order to appreciate its variation, we must pass from 10 to 11 ($\Delta R = 1$, $k = 0.1$), with the same kind of stimulus and beginning with 20 we should increase its intensity by 2, with 30 by 3, and so on.

Fechner decided not to consider a discrete scale of noticeable intervals among intensities of stimuli, but to examine the possible variations in sensation and the continuous variation in the intensity of the stimulation; in other words, the continuous increase for an infinitesimal value dR of the stimulus of intensity R. One could do this by substituting ΔR with the infinitesimal, increasing and integrating. The result is Fechner's famous law:

$$E = \frac{1}{k} \log R + c,$$

where c is the constant of integration. In other words, according to this law the sensation increases with the logarithm of the intensity of the stimulus. Research carried out on all the sensory modalities has shown the validity of Fechner's law to a highly satisfactory level of approximation, except for the higher and lower values of the scales of intensity.

Note that a logarithmic law of the relation between psychological values and objective values was not entirely a novelty. Over a century before, in 1713, the great physicist and mathematician Nicole Bernoulli had presented a series of problems to another mathematician, his cousin P. R. de Montmort. Among them was the famous fifth problem on relations between probability and value. Another great Bernoulli, Daniel (1733), answered this problem with his famous 'paradox of St. Petersburg,' in which he demonstrated that, in gambling, moral hope (that is, the utility of the gain), unlike mathematical hope, is proportional to the logarithm of the ratio between the capital that one obtains with the gain and the initial capital.[5] Fechner was well aware of Bernoulli's work, and he cites it explicitly on two occasions in the *Elemente*.[6]

Apart from Fechner's philosophical ideas, one can safely say that by formulating this law he opened a major chapter in modern science, that of psychophysics, which has seen a huge number of applications, from industry to medicine. By way of example, the most widely used scale of the intensity of sound, the scale in decibels (dB), is a logarithmic scale based on Fechner's law.

Psychophysics immediately set about determining the value of Weber's constant for the various sensory modalities, and the minimum and maximum values of intensity of the stimuli that could be perceived. These measurements are called 'values of threshold,' which may be differential and absolute.

The differential threshold is what was denoted above with the symbol ΔR;

we may say that it is the difference of value that enables subjects to perceive 50% of times the change of intensity of a stimulus. By 'absolute threshold' is meant the value of a stimulus that allows in absolute to perceive it 50% of times. A lower threshold (the minimum value) is distinguished from a higher one (the maximum value): note, however, that an excessively high stimulus can no longer be perceived, or sometimes perceived as belonging to another sensory modality; for instance, when a sound is too high, its perception is no longer acoustic, but painful.

3. ASTRONOMERS AND PSYCHOPHYSICS

One notes in parenthesis that psychophysics was born on terrain already largely prepared for it. Weber himself had at least two forerunners: two physicists, the French Pierre Bouguer (1760) and the Belgian Charles Delezenne (1827). The former, known as the founder of photometry, had estimated the differential threshold for the intensity of illumination. Of course, Bouguer did not use the term 'threshold,' but referred to the "strength that a light must have in order that a fainter one disappear."[7] The latter measured the 'discernibility' (again, the differential threshold) between two pitches.

More interesting is the contribution by astronomers, with a formulation practically alike to Fechner's one which appeared a decade before his publication, and about two years after his 'sudden enlightenment' of 22nd October 1850 recounted by Fechner in the *Elemente*.[8] It is worth recalling the decisive role played by astronomy in the birth of scientific psychology. The study of the reaction times had in fact originated in research conducted by Bessel[9] on so-called 'personal equation.' The reason for the role played by astronomers is simple: lacking recording and photographical apparatus, for a reliable estimate of their observations and to avoid subjective biases, they were forced to construct their own psychophysics (which psychologists had not yet created).

Given that the important problem was the magnitude and luminosity of astronomical bodies, what was needed was a scale of subjective magnitude and luminosity. Yet the ancient Greeks had already proposed scales of this kind, beginning with Hipparchus, and until the end of the eighteenth century astronomers had most frequently used Ptolomy's. But the problem now arose of the stars whose luminosity varied, a problem that could not be solved with such rough methods.

An interesting solution was offered by Dawes, in 1851. Notice that to solve the problem earlier J. Herschel had used a property of the astronomical mirrors: in conditions of equality of illumination the relative luminosity of two mirrors was given by the ratio between their surfaces. He was consequently able to determine, by observations carried out between 1834 and 1838 in South Africa,[10] that a 1 to 100 ratio between the areas of two mirrors corresponded to a subjective equalisation of luminosity between a star of first magnitude (maximum luminosity) and one of sixth magnitude.

According to Dawes, following Herschel, the subjective luminosity of the stars increased by an arithmetical factor, while that of objective luminosity increased by a geometrical factor. Dawes thus proposed a 'light ratio' of 4 for each class of magnitude. However, as pointed out by Pogson (1856), although it is true that the difference in objective luminosity, as noted by Herschel, is 100 between the first and the sixth class, the 'light ratio' must be the fifth root of 100, that is, 2.512. This is called 'Pogson's ratio,' which is still used in astronomy today.

Note that if M_1 and M_2 are the subjective luminosities of two stars (their classes of magnitude), and I_1 and I_2 are their respective luminous intensities, we have:

$$M_1 - M_2 = 2.512 \log \frac{I_1}{I_2}$$

Which is exactly Fechner's law on the difference between sensations.

4. THE DEVELOPMENT OF PSYCHOPHYSICS IN THE NINETEENTH CENTURY

Fechner was a far more complex psychologist than the textbooks would have us believe. With a rich sense of humour (under the pseudonym 'Dr Mises' he published satirical papers, as well as writing on more serious philosophical and sociological arguments),[11] he also conducted a number of experiments in the psychology of perception and the psychophysiology of seeing. Of particular importance was his research on the perception of light and colour, and on the phenomena of constancy and contrast.[12]

Numerous criticisms were brought against Fechner's ideas. Some of these criticisms, like Bergson's,[13] denied that the scientific study of sensations was possible. According to Bergson, in fact, the error of psychophysicists, that were misdirected by an illusion of the 'common sense,' consisted in confusing sensation with 'excitation:'

> It was therefore inevitable that the moment would arrive when science, accustomed to this confusion between sensation and excitation, tried to measure the former in the same way as it measured the latter: this was the object of psychophysics [...] If we acknowledge that one sensation may be stronger than another, and that this inequality resides in the sensations themselves [...] it is natural that one should search for a quantitative ratio between their intensities.[14]

For Bergson, however, a sensation was not an extensive magnitude, but a 'pure quality,' and any attempt to measure it was nonsensical.

There were also some scholars who sought to improve Fechner's law, this being the case of the Belgian Delboeuf (1872) or Plateau (1872), or the great Austrian philosopher Alexius Meinong (1896).

I shall confine my discussion to Meinong (1896). His criticism was closely linked with his theory of objects (*Gegenstandstheorie*), of which his work on psychophysics is perhaps the first complete formulation.[15] Broadly speaking, Meinong distinguished between two psychological levels, one lower (*founding*) and one higher (*founded*). To these, on the psychological side corresponded presentations at the lower level, and thoughts, desires and volitions at the higher level; and on the objectual side, lower-order objects and higher-order objects (objective, dignitative and desiderative). For Meinong, Ehrenfels' Gestalt qualities (*Gestaltqualitäten*) were higher-order objects. Note that lower-order objects possess all the properties that define them: a specific chair has all the properties relative to its spatial dimensions, to its colour, etc.; while a higher-order object is constituted only by an incomplete set of properties.

As to the problem of measurement, and thus of psychophysics, of central importance for Meinong was the concept of 'difference' (*Verschiedenheit*), which is itself an object of higher order, being an incomplete object. When measuring an object psychologically, we measure its 'difference' respect to other objects belonging to the same category relatively to one or more properties, but not to all the properties of such objects. Indeed, we simultaneously establish a relation of difference between stimuli and a relation of difference between sensations, which concerns, however, only the selected properties of both. These relations are precisely relations of difference.

Stated thus, Meinong's criticism is not incompatible with Fechner's theory. But it is incompatible with it when he states that the relations of difference can be only relations of ratio, and not of subtraction. Fechner, in contrast, is concerned with subtraction, and his dR, which allows integration of Weber's law, is a subtraction. If the problem is framed in terms of ratio, matters change, and Meinong shows that the fundamental psychophysical law, instead of being a logarithmical function, becomes a power function (for the sake of brevity, the relative mathematical steps are omitted):

$$E = cR^k.$$

As we shall see, about half a century later S. S. Stevens, with his 'new' psychophysics, proposed a function of the same kind but starting from radically different premises.

5. THURSTONE AND THE LAW OF COMPARATIVE JUDGEMENTS

A major advance in psychophysics was achieved by Thurstone in the 1920s. We have seen that according to Fechner there is a continuous correspondence between each minimum increase in the intensity of stimulation and the sensation produced. In other words, Fechner's law states a point-to-point correspondence between the continuum of the stimuli and the continuum of the sensations. This correspondence, however, is by no means certain, and it appears plausible that the judgement of sensations behaves as a continuous random variable that distributes itself according to a certain probability function around the points identified as the absolute threshold and the differential threshold. In effect, the same value of physical intensity in the stimulus does not always correspond to the same sensation: a fact which emerges with particular clarity if we conduct comparison between pairs of stimuli.

Let us suppose that we have only three stimuli, A, B, and C, in order of intensity of stimulation, and that we present them in pairs to the subject, asking him to judge which is the more (or less) intense in each pair. If the stimuli are not distinctly spaced along the physical continuum, it will be very unlikely that in 100 % of cases the first is judged as more intense than the second, the second more than the third and the first more than the third, while in a small number of cases the first is judged more intense than the second, the second more than the third, and the third more than the first, with the paradoxical result of the non-transitivity of the asymmetric more/less relation. We infer that the judgmental process does not always have the same 'strength' for the same stimulus.

The problem was addressed in 1927 by L. L. Thurstone, who propounded the law of the comparative judgement and refined the relative psychophysical methods. According to Thurstone, the 'discriminal process' which corresponds to each stimulus distributes itself normally. Let us assume two stimuli, a and b, to which correspond two discriminal processes with means of respectively Ma and Mb corresponding to scaling values. According to Thurstone, the sensation due to the difference between these means is given by the product of the z value, which corresponds to the distance between these two scaling values (recall that the z value is the ratio of the difference between the means and the standard error) times the standard error of the difference between means:

$$M_a - M_b = z_{a-b}\sqrt{\sigma^2_{Ma} + \sigma^2_{Mb} - r_{ab}\sigma_{Ma}\sigma_{Mb}},$$

where r_{ab} is the correlation coefficient between the two distributions. This is Thurstone's famous law of comparative judgements.

The unknowns of this law, however, are too numerous for it to be used practically . But there is a specific case in which one can determine the distances between all the discriminal processes, and therefore the scaling values for each stimulus intensity. This case (Thurstone's case V), requires two assumptions: the first is that the two discriminal processes are not correlated, with rab = 0; the second, that the two distributions have equal variance. If these two conditions are satisfied, the law reduces to:

$$M_a - M_b = z_{a-b}\sigma\sqrt{2}.$$

Both assumptions are plausible in most cases. In fact, there is no reason to assume either that the two discriminal processes must be correlated or that discriminal processes on the same continuum lead to different distributions. However, a counter-proof is impossible, although one can gain a good approximation by using the estimated variances, given that the correlation is zero (Thurstone's cases III and IV).

Thurstone's law of comparative judgements was extended in the 1950's by Torgerson to categorical judgements. In this case, one assumes that, in parallel to the continuum of the judgements in which the discriminal processes proceed there is a continuum of categories to which the stimuli of different intensity can be assigned. One assumes, however, that the boundaries between categories are not fixed but instead behave as continuous and normally distributed random variables. The problem, then, is analogous to the previous one, except that what is evaluated is not the distance between judgements relative to pairs of stimuli, but between a stimulus and the boundary of a category.

6. MORE RECENT DEVELOPMENTS IN PSYCHOPHYSICS

Since Thurstone, psychophysics has seen a number of attempts at further development. After the Second World War, three main approaches were followed: S. S. Stevens' 'new psychophysics' (1951), Helson's adaptation level theory (ALT) (1964), and signal detection theory (SDT).[16] Today, Stevens' psychophysics clearly predominates, having almost completely ousted Fechner's theory. SDT has been given significant development in various areas of research, like memory and social psychology, but ALT is now only of historical importance.[17]

In the 1930's, S. S. Stevens (perhaps the greatest theoretician of measurement this century – also beyond the strictly psychological domain) profoundly criticised Fechner's psychophysics. According to Stevens the methods used in classical psychophysics were indirect. They failed to grasp the subject's sensory judgement directly, but were always founded on a judgement relative to only one aspect: the value of just noticeable intensity, in absolute or by difference. It was instead possible to utilize methods of direct detection (we shall see those suggested by Stevens) which show that the function that best describes the relationship between sensory judgement (Δ) and intensity of stimulation is a power function, and precisely:

$$\Psi = kI^n$$

where k and n are constants which depend on the kind of stimulus utilised.

Of the methods proposed by Stevens, most widely used is that of *magnitude estimation*. The subject is first presented with the so-called *module*, which as a rule consists of two stimuli, the first of very low intensity, the other of very high intensity. The subject must assign the arbitrary value of 1 to the first stimulus, and an arbitrary value of 10, or 100, or another value at any rate greater than the first, to the second stimulus. The subject is then presented, in random order, with the stimuli to evaluate. These have an intensity intermediate between the first two, and the subject is invited to attribute a numerical value proportional to the perceived intensity which falls within the range 1-10 or 1-100, according to the initial module.

An alternative method to *magnitude estimation* exists: namely the *magnitude production* method. In this case, the subject is given a device with which to produce stimuli of different intensities. The experimenter reads out the numbers and the subject's task is to produce stimuli with intensities corresponding to the numbers said.

Another frequently used method is so-called *cross-modality matching*. The subject is given apparatus which produces stimuli of the intensity desired by him/her for a certain sensory modality (for example, sound intensity). Once again, a module is presented – that is, a maximum and a minimum value comprising all the intensities of the stimuli that the subject will evaluate, that will be presented in random order, with stimuli of different intensity belonging to an other sensory modality (for instance, brightness intensity). The subject will evaluate the stimuli producing for each stimulus of the second modality A a stimulus of corresponding intensity in the first modality B.

Some concluding remarks are in order. According to Stevens, the power function (which Fechner did not discard a priori, however) is much more plausible than the logarithmic one. Stevens[18] called this "Fechner's compression," because we can also detect activities that follow the power function in the activity of the nervous system. Secondly, the power function reflects a more general tendency of the human mind, given that in 1714 the Swiss mathematician Gabriel Cramer pointed out in a letter to Nicole Bernoulli that the power function was the best solution to the above-mentioned 'paradox of St. Petersburg.'[19] Lastly, numerous authors[20] have pointed out that the difference between the logarithmic law and the power function is more apparent than real, and that it is largely determined by measurement procedures.

Helson's ALT instead states that one cannot analyse the sensation in abstract, independently of the (recent) history of the stimulation experienced by the subject in the same sensory modality. Hence if before the experiment (or in the initial part of it) the subject has been exposed to powerful stimuli, he/she will tend to evaluate the stimuli with a mean value as less intense. Conversely, if the subject has experienced stimuli of low intensity, he/she will tend to evaluate the stimuli of intermediate value as more intense. In other words, the exposure creates a certain level of adaptation, and the judgement (the sensation) is given, not in abstract but with reference to this level.

The SDT is instead an attempt to apply the theory of statistical decision-making to psychophysics. A short example follows. Let us imagine that we must detect the presence or the absence of a stimulus (the signal) when there is a background noise ('noise' is any background disturbance in each sensory modality, not just in the auditory domain). The task of the subject is detection and consists in judging whether the signal exists or not. There are four possibilities: saying 'yes' the signal exists when it really exists (*Hit,* H); saying 'yes' the signal exists when there is actually only background noise (*False Alarm*, FA); saying 'no' the signal does not exist when it actually exists (*Omission*, O); saying 'no' the signal does not exist when in actual fact there is only background noise (*Correct Rejection,* RC).

Now on the basis of the relationship between these four kinds of responses, the SDT (Luce's so-called choice theory, 1959, has a similar meaning) distinguishes two factors that intervene in detection process: the first one is linked to the sensitivity of the system (and that is expressed with the index d'), the second is linked to the subjective criterion used in the task, characterised by an attitude more or less conservative (and it is expressed with the index Δ). The gambler, in fact, will like better the H's than the RC's, but in this way he will increase also the number of FA's. The conservative, on the contrary, will privilege overall the CR's, but in this way we will have an increase of the O's.

A convincing example is the radar operator who must distinguish on a screen whether the dots that he sees are background noise or a ship of the enemy. Of course, his/her decision will be more risky if he/she thinks that the base can dispose of an unlimited quantity of ammunition, while he/she will be very cautious if only one bullet is left, and the risk of an error is consequently very high. According to the theoreticians of SDT, classical psychophysics has missed that sensitivity and the criterion are always deeply interconnected; this is the reason why they, having failed to distinguish these two factors, have always attributed to sensitivity alone results that need to be interpreted in a more complex way.

However, numerous criticisms have been brought against SDT. From the Gestaltist perspective,[21] it has been pointed out that the concept of criterion, when the observations are close to the threshold values, implies an Helmholtzian activity of 'unconscious inference.' In any case, with ALT, but especially with SDT, it become very clear that psychophysics implies in a true sense a real representational activity.

Dipartimento di Psicologia
Università di Firenze
Via San Niccolò, 93
I- 50125 Firenze
Email: luccio@cesit1.unifi.it

NOTES

[1] Cf. Donders 1868-69. Cf. Luccio 1993.
[2] Fechner 1840.
[3] Cf. Turner 1988.
[4] On the problem of the materialistic question in Germany at the birth of scientific psychology, cf. Romano 1990.
[5] Cf. Conti 1975.
[6] Cf. Fechner 1860, vol. 1, 237; vol. 2, 549.
[7] Cf. Bouguer 1760, 51.
[8] Cf. Pliskoff 1977.
[9] Cf. Bessel 1822; Luccio 1993.
[10] Cf. Buttman 1970.
[11] See Fechner 1824, 1875. Cf. Marshall 1988; Arnheim 1980.
[12] Fechner, 1838, and 1840. Cf. Turner, 1988. For a complete bibliography of the works of Fechner, see Gundlach 1988a, and for an analysis of his place in the history of ideas see Gundlach 1988b.
[13] Cf. Bergson 1889. Cf. Gomes Penna 1988.
[14] Cf. Bergson 1889, 15–16.
[15] Cf. Potrc and Vospernik 1996.

[16] Cf. Green and Sweets 1963; Beretta, 1969.
[17] On more recent developments in psychophysics see Luccio 1996, and Purghé 1996.
[18] Stevens 1970, 1043.
[19] Bernoulli 1733. Cf. Conti 1975, 19n.
[20] Cf. Krueger 1989, and the relative 'peer discussion.'
[21] Cf. Metzger 1974.

REFERENCES

Albertazzi, L.: 1996, *The Philosophy of Alexius Meinong*, special issue of *Axiomathes*, **7**.

Arnheim, R.: 1980, 'The other Gustav Theodor Fechner,' *Gestalt Theory*, **2**, 133–140.

Beretta, A.: 1969, *TDS*. Florence, OS.

Bergson, H.: 1889, *Essai sur les données immédiates de la conscience*. Paris, Alcan.

Bernoulli, D.: 1733, 'Specimen theoriae novae de mensura sortis,' *Commentarium Academiae Scientiarum Petropolitanae*, **5**, 175–192.

Bessel, F.W.: 1822, 'Persönliche Gleichung bei Durchgangsbeobachtung,' *Astronomische Beobachtungen der Sternwarte zu Königsberg*, **8**, 39–72.

Bouguer, P.: 1760, *Traité d'optique sur la gradation de la lumière*. Paris, Val.

Brozek, J. and Gundlach, H. eds.: 1988, *G.T. Fechner and Psychology*. Passau, Passavia Universitätsverlag.

Buttman, G.: 1973, *The Shadow of the Telescope*. New York, Scribner.

Carterette, E.C. and Friedman, M. eds.: 1974, *Handbook of Perception. Vol. 1. Roots of Perception*. New York, Academic Press.

Clark, L.H. ed.: 1954, *Consumer Behaviour*. New York, New York University Press.

Conti, C.: 1975, *Probabilità e valore nelle scienze sociali*. Milan, Mazzotta.

Dawes, W.R.: 1851, 'On a Photometrical Method of Determining the Magnitudes of Telescopic Stars,' *Monthly Notices of the Royal Astronomic Society*, **11**, 187–198.

Delboeuf, J.: 1872, *Examen critique de la loi psychophysique: sa base et sa signification*. Paris, Alcan.

Delezenne, C.E.J.: 1827, 'Sur les valeurs numériques des notes de la gamma,' *Recueil des travaux de la Societé des Sciences, de l'Agriculture et des Arts de Lille*, **58**, 1–56.

Donders, F.C.: 1868/69, 'Over de snelheid van psychische processen,' *Onderzoekingen gedaan in het Physiologisch Laboratorium der Utrechtsche Hoogeschool*, **2**, 92–120.

Fechner, G.T.: 1824, *Stapelia Mixta*. Leipzig, Voss.

Fechner, G.T.: 1838, 'Über die subjektiven Complementarfarben,' *Annalen der Physik und Chemie*, **44**, 221–245, 513–535.

Fechner, G.T.: 1840, 'Über die subjektiven Nachbilder und Nebenbilder,' *Annalen der Physik und Chemie*, **44**, 193–221, 427–470.

Fechner, G.T.: 1851, *Zend-Avesta oder über die Dinge des Himmels und des Jenseit*. Leipzig, Voss.

Fechner, G.T.: 1860, *Elemente der Psychophysik*. Leipzig, Breitkopf und Härtel.

Fechner, G.T.: 1875, *Kleine Schriften*. Leipzig, Breitkopf und Härtel.

Gomes Penna, A.: 1988, 'Bergson's Critique of Fechner's Psychophysics,' in Brozek and Gundlach, eds., 1988, 151–155.

Green, D.M. and Sweets, J.A.: 1966, *Signal Detection Theory and Psychophysics*. New York, Wiley.

Gundlach, H.: 1988a, *Index Psychophysicus*. Passau, Passavia Universitätsverlag.

Gundlach, H.: 1988b, 'Fechner Scholarship – Aspects of its History,' in Brozek and Gundlach, eds., 1988, 17–30.

Helson, M.: 1964, *Adaptation Level Theory: An Experimental and Systematic Approach in Behaviour*. New York, Harper and Row.

Krueger, L.E.: 1989, 'Reconciling Fechner and Stevens: Towards a Unified Psychophysical Law,' *Behavioural and Brain Sciences*, **12**, 251–320.

Legrenzi, P. and Luccio, R.: 1993, *Immagini della psicologia*. Bologna, il Mulino.

Luccio, R.: 1993, 'Il tempo della mente: Frans Cornelis Donders,' in Legrenzi and Luccio, 1993, *Immagini della psicologia*. Bologna, il Mulino, 43–79.

Luccio, R.: 1996, *Tecniche di ricerca e analisi dei dati in psicologia*. Bologna, il Mulino.

Luce, R.D. 1959, *Individual Choice Theory*. New York, Wiley.

Meinong, A.: 1896, 'Über die Bedeutung des Weber'schen Gesetzes. Beiträge zur Psychologie des Vergleichens und Messens,' *Zeitschrift für Psychologie und Physiologie des Sinnesorgane*, **11**, Sonder-Abdruck.

Metzger, W.: 1974, 'Consciousness, Perception and Action,' in Carterette and Friedman, eds., 1974, 109–122.

Plateau, J.: 1872, 'Sur la mesure des sensations physiques et sur la loi qui lie l'intensité de ces sensations à l'intensité de la cause excitante,' *Bulletin de l'Académie Royale des Sciences, des Lettres et des Beaux-Arts de Belgique*, **33**, 376–385.

Pliskoff, S.S.: 1977, 'Antecedents to Fechner's Law: The Astronomers J. Herschel, W.R. Dawes and N.R. Pogson,' *Journal of the Experimental Analysis of Behaviour*, **28**, 185–187.

Pogson, N.R.: 1856, 'Magnitude of Thirty-six of the Minor Planets for the First day of Each Month of the Year 1857,' *Monthly Notices of the Royal Astronomic Society*, **17**, 12–15.

Potrc, M. and Vospernik, M.: 1996, 'Meinong on Psychophysical Measurement,' in Albertazzi, ed., 1996, 187–202.

Purghé, F.: 1996, *Metodi psicofisici e scaling multidimensionale*. Torino, Bollati Boringhieri.

Romano, D.: 1990, *Paradigmi*. Milan, Cortina.

Stevens, S.S.: 1951, 'Mathematics, Measurement, and Psychophysics,' in Stevens, ed., 1951, 1–49.

Stevens, S.S. ed.: 1951, *Handbook of Experimental Psychology*. New York, Wiley.

Stevens, S.S.: 1970, 'Neural Events and the Psychophysical Law,' *Science*, **170**, 1043–1050.

Thurstone, L.L.: 1927, 'A Law of Comparative Judgement,' *Psychological review*, **34**, 273–286.

Torgerson, W.S.: 1954, 'A Law of Categorical Judgement,' in Clark, ed., 1954, 92–93.

Turner, R.S.: 1988, 'Fechner, Helmholtz and Hering on the Interpretation of Simultaneous Contrast,' in Brozek and Gundlach, eds., 1988, 137–150.

Weber, E.H.: 1834, *De pulsu, resorptione, auditu et tactu*. Leipzig, Köhler. En. tr. 1978 *De Tactu*. New York, Academic Press.

ROBIN D. ROLLINGER

LOTZE ON THE SENSORY REPRESENTATION OF SPACE

1. INTRODUCTION

While there is nowadays some appreciation for the vast and significant influence of the philosophical and psychological writings of Rudolf Hermann Lotze,[1] there is at the same time an ever growing need for more detailed considerations of various aspects of his work. For there can be little doubt that he was by far one of the most important philosophers in the post-idealist phase of nineteenth century German philosophy. In the following I shall attempt to make a contribution that will to some extent meet the need just indicated. Here I shall discuss his efforts to explain the sensory representation of space by means of the theory of local signs, which is vividly and succinctly described by William James as:

> The theory which denies that there can be in a sensation any element of actual locality, of inherent spatial order, any tone as it were which cries to us immediately and without further ado, 'I am *here*,' or 'I am *there*.'[2]

This theory was first presented by Lotze in 1846 in 'Seele und Seelenleben,'[3] stated in greater detail in 1852 in his *Medicinische Psychologie oder Physiologie der Seele*,[4] and later defended in subsequent works.[5]

Though the theory of local signs was one of the many theories which had been 'pillaged' from Lotze,[6] it was also not without its critics. Even in cases where it was rejected, it nonetheless received considerable attention throughout the second half of the nineteenth century and well into the twentieth century and is therefore not to be neglected in a detailed account of both the psychology and the philosophy of that time (which were indeed hardly separable).[7] As far as its place in philosophy is concerned, we must keep in mind, after all, that it is presented in Lotze's multi-volume popular work[8] and finally in the "system of philosophy,"[9] which was written for a less popular readership.

In the following I shall examine Lotze's views on sensory representation by first considering the motivation behind this theory and some of its alternatives which he rejects. Then I shall briefly present his theory of local signs and also

L. Albertazzi (ed.), *The Dawn of Cognitive Science. Early European Contributors*, 103–122.

consider it in application to some problems. Finally, some of the criticisms of this theory, namely the ones put forward by Franz Brentano, Carl Stumpf, and Wilhelm Dilthey, will be cited and discussed.

2. THE NEED FOR A THEORY OF THE SENSORY REPRESENTATION OF SPACE

As already mentioned, the theory of local signs received its first detailed formulation from Lotze in his *Medicinische Psychologie*. This work is a milestone in the history of psychology, for it is "the first book claiming to be a physiological psychology."[10] Such a discipline had indeed already been identified by Lotze years beforehand as central to the development of scientific psychology.[11] The intention of this "physiology of the soul,"[12] as Lotze also calls the discipline in question, is to develop "applicable views concerning the relations of mental life to the corporeal activities."[13] While the theory of local signs was later presented in different and sometimes broader contexts by Lotze, most notably in *Metaphysik* (1879), physiological considerations in conjunction with psychological ones nonetheless remain central to his wide-ranging philosophical concerns and particularly to the theory under discussion.[14]

For those of us who have been accustomed to philosophical inquiries conducted from our arm chairs the approach of Lotze might at first strike us as a little peculiar. We may initially be repelled by his theory, for it seems obvious that the immediately given content of sensations is spatial in character and nothing more needs to be said. There are however two considerations which might prompt us to give Lotze's theory of local signs a closer look.

First of all, some of the contents of our sensations are spatial in only a marginal and perhaps derivative sense. When we hear a melody, all of the tones are given together without being located as next to each other. Moreover, while we may in some sense speak of the vastness of a sound,[15] it is questionable whether we are speaking here in a strict and proper sense. The spatiality of odours and tastes is even less obvious than that of sounds. Such instances at least indicate that we need not attribute spatiality to the contents of sensations *as such*. Accordingly, the need arises to explain why some sensations, particularly visual and tactile ones, are spatial and others are not. It of course remains to be seen how Lotze deals with the apparently undeniable spatiality of the data of sight and touch, which have indeed been held as the primary source of our ideas of space since John Locke.[16]

The second consideration which should prompt us to sympathise with Lotze can be stated thus. Even if it is granted that we to some extent immediately perceive spatial properties, these data, as immediately perceived, are far from the full-fledged spatiality which we confront not only through sensation, but also through memory, the co-ordination of heterogeneous sensory data (e.g., between those of sight and touch), conceptualisation, and perhaps other cognitive processes. Hence, there is clearly a need for a theory concerning how it comes about that we represent not merely piecemeal spatial moments, but space as such and all the relations which it involves.[17] Whether Lotze himself successfully worked out such a theory is another matter. The point here is only that his theory of local signs is not to be dismissed without further ado, for it is an attempt to develop a theory of a matter which is most certainly in serious need of theoretical treatment.

There is in addition an important thesis which Lotze puts forward in his theory of sensation and also serves to motivate the theory of local signs. Sensations, he says, are not merely the products of certain stimuli, such as light waves or air vibrations, but also certain physiological processes.[18] Not only is the old Aristotelian view, according to which the sensory object actually in part (namely, as regards its form) enters the soul, thus flatly rejected by Lotze, but he also denies some of the modern views because they fail to allow for the production of sensations through the sensory mechanism. Above all, he stresses that this production is not to be conceived in terms of a transmission of *pictures* from the perceived objects to the perceiving soul or consciousness, as if the mere presence of these objects were a sufficient condition for the perception of them.[19] Once we forego thinking of perception in this way and adopt the notion of stimuli which need not in the least be matched in quality by perceptual images, it becomes plausible that spatial relations as well need not be pictured. While Lotze concedes that none of his contemporaries explicitly adopts this outmoded view, he nonetheless maintains: "we find all the more frequently that it repeats itself in a somewhat finer form."[20] In the consideration of sight, for instance, there are those who take the image on the retina to be the key to understanding what goes on in such perception. On this view the picture is seen, as it were, by a second eye, namely the soul, and this settles the matter. The great physiologist, Johannes Müller, and many of his disciples were inclined to this more refined, but still erroneous view.

Even if one takes a further step in such refinement and declares these images to be 'subjective,' Lotze emphatically denies that appealing to such images is of any explanatory value. Granting the terms of such a view for the sake of argument, he insists:

> Inevitably for them [the subjective images] too, in their transmission to the soul, the point occurs where their entire geometrical form is destroyed without a trace, and where they are replaced in the soul by a *sum of intensive stimulations* which, like a multiplicity of tones, contain no longer an indication of spatial extension or position. If we are therefore to gain an intuition of the actual position of external objects, this can be done not by *apprehension,* but

rather by *re-production of spatiality*. The extensive is everywhere transformed into the intensive, and only from the latter must the soul reconstruct a new internal world of space in which the images of external objects find their corresponding places.[21]

In order to see the plausibility of this claim, we must keep in mind what occurs in visual perception. The image on the retina is, after all, not the final result of the entire physiological process under consideration. Neural impulses must travel from the retina to the brain. In this phase it is by no means far-fetched to say, as Lotze does, that the geometrical form of the image on the retina is destroyed without a trace. As this is destroyed, the spatial properties inherent in it are likewise destroyed. Accordingly, there is a need for a theory to account for how we nonetheless come to perceive things spatially.

3. THEORIES WHICH LOTZE REJECTS

Let us now consider some of the theories which are meant to explain the sensory representation of space and why Lotze rejects them. It is Lotze's habit not to identify his opponents. Here we shall likewise not be troubled with this task, though it could well prove an interesting exercise elsewhere.

Lotze points out that it will not help to conceive of vision as a kind of touch as a means to explaining the sensory representation of space. "To some people," he says:

> [...] everything seems especially clear when they allow the soul to be immediately present in the eye and allow it, like a groping hand, to grasp immediately there with a thousand nerve endings the single points of colour precisely in the position which they simply have there; they forget that it would be equally difficult to show how tactile feelings, which the touching hand receives, would provide the right to relate the various points seized upon to fixed spatial locations; one would have to presuppose that the given place of the hand in space is already an object of perception, which would be left to be explained [...].[22]

In addition to this argument Lotze may insist once more that in *touch* the extensive is transformed into the intensive. Accordingly, the geometrical form, even if regarded as a pattern of tactile data, must be reconstructed in consciousness.

Instead of clutching at straws with such dubious metaphors as the one just considered, the more natural view is of course to say that the spatial organisation of visual data is determined by separate neural fibres which lead from the retina to different parts of the brain. In reply to this view, however, Lotze raises the question: "How does the soul come to an awareness of the fact that in this moment three central neural points are stimulated which lie at the corners of a triangle or in a straight line?"[23] The differentiation of the neural fibres can *at best* explain the differentiation of the visual data, but not how they are located in relation to each other. Indeed, Lotze points out that it is even doubtful that the differentiation of neural fibres alone can explain the differentiation of sensory contents within a given sense, for the same neural structure is to be found in the olfactory nerve "and yet it shows itself hardly able to arouse a manifold of separate sensations which correspond to the number of its single fibres."[24]

Besides explaining visual localisation in terms of tactile sensation, which itself needs to be explained, or in terms of the differentiation of neural fibres from the retina to the brain, one may finally attempt to deal with the issue by scrapping the old Cartesian concept of the soul as unextended. If the soul is extended, Lotze's opponent may contend, a spatial duplicate of the retinal image can be produced in the soul. This notion of a literal spatial duplicate in the soul, however, will not solve the problem of the sensory representation of space, according to Lotze, for he insists that the existence of such a duplicate does not in any way guarantee that its existence is actually known by the soul.[25] If, for example, there is a triangle in the soul, it does not follow from this that the soul knows of the existence of this triangle in itself. To be sure, Lotze finds this notion of an extended soul to be totally unacceptable. But even if it were accepted, he insists that it would be to no avail in solving the problem at hand.

Now at this juncture it is easy to imagine an objection which Lotze's opponent might raise against the view that the soul and its contents are unextended. According to this objection, the only content of sensation which would be left over in absence of any spatiality whatsoever would consist of *points* of the relevant quality. If, for instance, we see both blue and red while the geometrical form of the visual image has, as it were, been destroyed, then the only content left over is one or more points of blue and one or more points of red. Even if we grant the intelligibility of talking about such points, it is clearly far-fetched to say that we ever see them; it is all the more far-fetched to say that these are colours as originally given in visual sensations. Thus, according to the objection under consideration, Lotze must be wrong in thinking that sensory contents are inherently non-spatial and some theory for their inherent spatiality must be formulated.

In reply to this objection it can be said that the nonspatiality of colours and other contents of sensation does not entail the conception of them in terms of *points* any more than the combination of various notes in a chord entails the conception of them in these terms. I may hear a chord as consisting of various notes without regarding these notes as points. Likewise, a taste may be a combination of sweetness and sourness, but the sweetness and sourness which make up the entire content are by no means construed as points. Given these obvious examples of complex sensory contents whose parts are no more punctual than they are extended or located, it is to be stressed from Lotze's standpoint that punctuality is itself a spatial property which the contents of sensation cannot originally have. The objection just considered therefore begs the question by assuming that the contents of sensation must have some sort of spatiality.

4. THE THEORY OF LOCAL SIGNS

The theory of local signs is Lotze's alternative to the theories considered above. Yet, it may be better to speak of theories, rather than *the* theory, of local signs, for Lotze actually considers theree theories under this heading and, only after painstaking reflection, opts for the third.

Given that the visual and tactile contents of sensation are therefore devoid of any inherent spatiality, Lotze at first sees no other recourse but to say: "every excitation must by virtue of the point in the nervous system in which it occurs receive a peculiar colouring to which we want to give the name of its *local sign*."[26] This is Lotze's own preliminary and succinct statement of the theory of local signs. These are thus conceived of as sensations which accompany sensations of various visual and tactile data.[27] Let us look at this theory more closely in application to the visual representation of space, without entering into all of its technical details.

If we designate three distinct stimuli as A B C, three distinct places of the sensory organ as p q r, and the local signs which occur when these places are stimulated as π κ ρ, the following possibilities arise.[28] If a stimulus of the same kind A affects three different places of the sensory organ (p q r), the following combinations will result: Aπ Aκ Aρ. If A is, for instance, a sensation of red, there will accordingly arise a sensory representation of red in three distinct places. If, however, three different stimuli (A B C) affect the same sensory organ in three different places (p q r), the following combinations will arise: Aπ Bκ Cρ. Thus, there will be the sensory representation of three different data in three different places. Finally, Aπ Bπ Cπ would mean that three distinct types of data are represented in the same place (p) and are "therefore probably only conceivable as appearing successively in the same point of space intuition."[29] Here one may point out, however, that this last remark on Lotze's part is somewhat hasty, for there are cases where distinct data, or perhaps it is better to say distinguishable features of a single datum, can be perceived in the same place. The colour red, for instance, can be seen in a given place with a certain brightness which is distinguishable from the quality of redness as such. Yet, it is indeed inconceivable that one could perceive red and blue unmixed and in precisely the same place. Examples of this sort are no doubt what Lotze has in mind.

A much more serious problem for Lotze arises when we consider one of his criticisms of a competing theory and ask whether the same criticism does not apply to his own. Here we recall that he insists that the spatial relations of the stimulus or those which occur anywhere in the nervous system, e.g., those in the retinal image, cannot be used to explain the sensory representation of spatial relations. This is to be seen by keeping in mind the fact that the transition from physiological process to consciousness involves the destruction of all spatial relations in the former. Accordingly, the fact that the physiological process which causes a given visual or tactile sensation and the one which causes its local sign are transmitted through one and the same neural fibre will not in the least explain the resulting connection in consciousness.

In light of this difficulty Lotze proposes the following:

> For this reason it seems to me to be more natural to make the other assumption, that from the main impression [or stimulus] and the co-stimulation [i.e., the one which causes the local sign] there indeed arises in each fibre a total state which is as such conducted and also causes only a total sensation. This [sensation], if it remained alone, would not produce a motivation to distinguish anything in itself, as violet, for instance, would not challenge us to separate red and blue in it if we knew it alone; but since many different stimuli become more and more connected with the same neural process, there could arise from the comparison of these cases the separating activity which divided the total impressions into their components, but then learned at the same time to relate the thus separated local sign in each case to that qualitative impression from which it was separated only in thought.[30]

An occurrence of this kind, Lotze explains further, is to be found in our perception of sounds. It is a comparison between various sounds which allows us to distinguish the pitch from the timbre and again to relate a given pitch to the timbre from which it was distinguished.[31] Likewise, it is suggested, we can distinguish the place of a given tactile or visual datum only through such a comparison, though the relevant sensation and its local sign are one total state, just as the auditory sensation of both a certain timbre and a certain pitch is one total state. While Lotze maintains that this view of local signs is preferable to the one which had at first been discussed, he nonetheless maintains that a third one, which is to be considered presently and relies heavily on the notion of movement, has additional advantages.

To be sure, it seems to be artificial to think of the sensory representation of space in terms of comparisons of the sort just discussed. Artificiality is however hardly avoidable here because it can always be asked how the soul manages to keep together its impressions. Ultimately it is association which is appealed to here, but Lotze regards association as an unexplainable feature of the soul.[32] Thus, if one regards the visual or tactile datum which is transmitted by a given neural fibre as distinct from the local sign which is allegedly transmitted through the same neural fibre, one must make an appeal to association, which is no less artificial than considering the visual or tactile sensation and the local sign together as one entire state that arises from one single neural process and appealing to the activity of comparison in the manner suggested in order to explain how one can distinguish the visual or tactile datum from the local sign. This preference, however, still assumes that the local sign is caused in the stimulated nerve itself. It will now be seen that Lotze actually prefers a third way of conceiving local signs.

As the theory has been stated thus far, it is still in need of further development in order to explain the nature and origin of the sensory representation of space. There is still a need to conceive of local signs as quantitatively related, e.g., in terms of such formulae: $\kappa = \pi + \Delta$, $\rho = \pi + 2\Delta$, $\kappa = \rho - \Delta$.[33] Accordingly it be will possible to grasp how B, for instance, is not only located in between A and C, but there is also a continuity of space as it is represented. In order to account for this quantitative and continuous aspect of visual space, Lotze finds it necessary to appeal to the impressions resulting from eye movement as essential to the system of local signs.[34]

If we wish to see optimally, Lotze points out, the relevant stimulus must affect the centre of the retina, whereas those nerves further from the centre are less sensitive.[35] When the eye moves in such a way that the stimulus in question strikes the centre of the retina, there occurs a certain feeling of motion which varies quantitatively, according as the motion varies in magnitude and direction. The continuity which is conveyed through this feeling of motion is accordingly that aspect of the local signs which indicates the continuity of the visual field. To be sure, Lotze is well aware that we localise certain impressions within the visual field without making the eye movements in question. In these cases, however, the peculiar intensity transmitted from the particular point on the retina is associated with the feeling of movement in question and therefore suffices as a local sign. The feeling of intensity in question, after all, had on numerous occasions before been followed by the feeling of movement and thus an association is set up in accordance with an old familiar law. If, however, there had never been any feeling of movement before, the various intensities which are produced by the stimulation of points of the retina could not function as local signs.

Now there are some who might think that eye movement cannot be given such a crucial role in the visual presentation of space, for in order to move the eyes towards this or that one would first have to know where to move them. Lotze maintains, however, that the eye movements towards a given stimulus are for the most part involuntary and at the beginning of life entirely so. By the time we learn to move our eyes towards something which interests us, we have already experienced this movement many times before. Moreover, he points out that his theory of local signs cannot be empirically tested by considering what occurs in the visual experience of those who have gained sight for the first time through operations. While such people may no doubt have much to learn about visual space, they have already developed their tactile presentation of space and accordingly cannot be compared with the new-born who has to learn about space through both sight and touch.

There are a number of special problems which Lotze attempts to solve by appealing to his theory of local signs in vision. One of these concerns the localisation of the visual field itself rather than that of various impressions within it. "At first," says Lotze:

> [...] this image [which makes up the visual field] obtains a place with respect to the eye whose openings and closings, familiar to us in another way, are conditional for its existence or non-existence. That is to say, the visible world is *in front of us*. What is *behind* us does not yet merely not exist at all for us, but rather we do not even know that there is something that could be called 'behind us.'[36]

In order to localise the visual field further, more is required than the blinking of one's eyes. By movements of the body, such as turning one's head, various visual fields are connected, one consisting of a b c with one consisting of b c d, which is turn linked with c d e and so forth. The resulting spatial presentation is one of a spherical plane which has a 'back' and a 'front.' It is however doubtful whether Lotze's view on this matter is particularly illuminating. It appears rather as a matter of common sense in contrast with the previously examined aspect of his theory. What is even more surprising is that Lotze fails to point out that his theory of local signs in vision from the very outset entails the connection of visual fields, i.e., by virtue of eye movements alone. The visual field, after all, changes whenever the eyes move.

Another problem which Lotze attempts to solve is the one concerning the presentation of depth. While we may get the impression that the presentation of spherical space by means of movements of the body already implies depth, Lotze maintains that if someone's presentation of space were limited to this, he would in fact have no idea at all of depth. The presentation of depth, on his view, "cannot arise automatically, but only by experience which we have when we move among the seen objects."[37] There is however little that is innovative in this idea. Moreover, since Lotze himself says that what he has in mind here is too boring to describe, I shall not dwell on it further.

More interesting is what he says regarding another problem, namely the determination of up and down in the visual field.[38] While some have observed that the image on the retina is upside down and have tried to explain how it is ultimately turned right side up in consciousness, their attempts have been based on the assumption that the retinal image is the object which is primarily given in visual perception. As we have seen, Lotze maintains that vision never occurs until neural impulses are transmitted from the retina to the brain. Since the geometrical form of the retinal image is destroyed in this process, it need not be turned right side up. Rather, up and down are to be explained strictly in terms of the eye movements, associated with certain muscle feelings, most notably the feeling of heaviness which functions as a local sign for down.

5. SOME CRITICISMS OF LOTZE'S THEORY OF LOCAL SIGNS

Now let us look at some of the criticisms of the theory of local signs. Here I shall restrict the discussion to ones which were made shortly after Lotze's death by three significant nineteenth century philosopher-psychologists: Franz Brentano, Carl Stumpf, and Wilhelm Dilthey.[39] After briefly presenting these criticisms, I shall consider whether a certain common approach to the investigation of mental life determines these three to reject the theory of local signs.

In these criticisms the theory of local signs is called an 'empiricist' theory of space-representation, whereas the view that space is already inherent in visual and tactile is referred to as nativism.[40] This terminology may, to be sure, be somewhat confusing. After all, the same terms are often used to describe different positions regarding concept-formation. While the empiricist maintains that concepts are derived from experience, the nativist takes the view that some concepts are innate. What is under discussion at present is however a completely different dispute. Someone may be an empiricist in the theory of concept-formation and a nativist in the theory of space-representation.[41] The converse is also true. With this clarification in mind let us consider the criticisms of the theory of local signs.

The first criticism I wish to look at is to be found in a lecture which Franz Brentano gave in Vienna in 1883, entitled *Ausgewählte psychologische Fragen*.[42] There was, to be sure, great mutual respect between Brentano and Lotze, as indicated by the fact that Brentano sent two of his pupils, Anton Marty and Carl Stumpf, to study with Lotze and also by the fact that Lotze played an important role in Brentano's call to Vienna in 1874. Nonetheless, there were significant disagreements between Brentano and Lotze. One of disagreements was about the theory of local signs.[43] In the above-mentioned lecture Brentano says the following about the attempt to solve the problem of sensory representation of space by appealing to muscle sensations which result from movement:

> But this view is false. Let one make an observation with electrical sparks. Here I cannot execute any eye movements; there are no muscle sensations here. The basis for association is gone, hence also its systems. We would thus have to see only colour here without shape and extension, which is however not the case. Consequently this solution is impossible. Also the

difficulty would have to obtain continually of directing the eye to an object which has stimulated the eye laterally.[44]

Carl Stumpf was a pupil of both Lotze and Brentano. While it seems that the influence of Brentano on him was not only prior to Lotze's, but also the more definitive one, his great respect for Lotze is to be found throughout his writings and also in the article he wrote in memory of Lotze.[45] Moreover, Lotze had high regard for Stumpf as well.[46] In any case, as regards the theory of sensory representation Stumpf was on the side of Brentano rather than Lotze.[47] Here we may cite the succinct criticism which Stumpf states in his 1886 lecture on psychology:

Against the empiricist thesis it can however be objected a) that through the combination of several presentations, as experience testifies, a new presentation which is completely different from them never arises in consciousness, b) that even specific muscle sensations can be excluded (e.g., through instantaneous illumination of an object with the electrical flash), and still even then spatial perceptions are possible. We must therefore in principle return to the nativist view. This is also the only way of grasping the absolute impossibility of presenting a colour without any extension. Just like intensity, extensity is an integral moment of the sensory impression.[48]

Wilhelm Dilthey is of course widely known for his concern with hermeneutics, but in the late nineteenth century he was very much occupied with psychology both in Breslau and then in Berlin, where he had been called to the very chair which Lotze had just accepted before his sudden death in 1881.[49] In his 1883/84 lecture on psychology Dilthey says:

But the requirement of these local signs is unjustified: if an objective space exists, the difficulties are not solved; if space is a phenomenon, the problem does not at all obtain. If space is actual, those stimuli which have qualitative sensations as their cause also consist in processes of motion. If now the difference of position cannot have any effect on our consciousness, it is also impossible for us to distinguish red and green, for the motion consists, after all, in a change of position. Therefore: The same difficulty which Lotze finds in presenting space obtains also for qualitative sensation. If, however, the presentation of space is replaced by a nonspatial occurrence, the whole problem raised by Lotze disappears. In both cases only the requirement remains that in the final point of the physiological process the arrangement, which was the foundation of the spatiality of the object, was represented by a corresponding arrangement of elements of excitation. Furthermore, the local signs which are hypothetically introduced by Lotze, the muscle sensations, do not appear to be sufficient to explain the fineness of our distinguishing in space.[50]

Now the question arises whether Brentano, Stumpf, and Dilthey are in some way led to reject the theory of local signs because they share a common approach in psychology. It is of course well known that Brentano had insisted on a distinction between descriptive and genetic psychology.[51] Stumpf was also inclined to follow him in this regard.[52] While genetic psychology was meant to identify the causes of psychical phenomena and to involve a good deal of physiology, descriptive psychology was to characterise these phenomena universally, e.g., by the criterion of intentional inexistence which distinguishes them from physical phenomena, but also to classify them. Moreover, Brentano and Stumpf thought of descriptive psychology as a discipline of great philosophical significance. When we consider that Dilthey also defended a notion of descriptive psychology,[53] it seems plausible that this common approach between Dilthey and the school of Brentano was an important motivating factor in the preference for a nativist theory of the sensory representation of space over Lotze's theory of local signs. Lotze's psychology, by contrast, is for the most part genetically and physiologically oriented from the outset. The same goes for two other important psychologist-philosophers, Wilhelm Wundt and Hermann von Helmholtz, who developed their own particular versions of the theory of local signs.

The suggestion that the rejection of local signs is determined by a descriptive-psychological orientation, however, turns out to be rather dubious when we consider that Dilthey insists in the above-cited lecture, more particularly in a remark explicitly directed against Brentano, that a psychology devoid of physiological considerations is not realisable.[54] Furthermore, when we take into account that Stumpf's ultimate reason for preferring nativism lies in his observation that colour and extension are inseparable, it comes to light that it is not descriptive psychology as such that is ultimately at work in this preference. The phenomena under consideration here are coloured patches, hence physical rather than psychical phenomena. The discipline which Stumpf later came to identify as the one concerned with physical phenomena, or simply phenomena (*Erscheinungen*) as opposed to psychical functions (*psychische Funktionen*), is phenomenology, not descriptive psychology.[55] It thus becomes dubious to say that his practice of descriptive psychology is what prompts him to reject the theory of local signs.

Finally, in Lotze's own lectures on psychology he says:

> Our scientific needs would be fully satisfied when we 1) could present all single constituent parts of this life [of the soul] and the general forms of their connection (descriptive or empirical psychology); when we 2) could identify the nature of the subject this whole life, as well as the effective forces and conditions through which the whole of this life is produced and compelled to retain those experientially known forms of its course (explanatory, mechanical or metaphysical psychology).[56]

Yet, in the very same lecture Lotze puts forward the theory of signs. Accordingly Lotze did not see descriptive psychology as an obstacle to his theory of space-representation.

6. CONCLUSION

While Lotze's theory of local signs is perhaps one of the most counter-intuitive theories which have arisen in the history of psychology and philosophy, it has hopefully been shown that it is motivated by physiological considerations which are not easily dealt with by any alternative theory. Here we have focused on vision. It has also been shown that in addition to these physiological considerations there are some problems which can only be solved, according to Lotze, by denying that visual data are intrinsically spatial and accordingly that they require other sensations which function as local signs. While it is very difficult to deny that Stumpf is correct in his observation that a colour cannot be seen in separation from extension, only the advancement of neurology and more in-depth cognitive investigations could decide what theory would be able to replace the Lotzean one in order to accommodate the phenomena. By no means have I tried to provide such a theory in the foregoing, but the upshot of these considerations seems to be that there should indeed be one. Apparently this need can only be met by a genius as penetrating as Lotze's was.

Robin D. Rollinger
Husserl-Archiv, Universität Freiburg
Werthmannplatz, Kollegiengebäude I
D-79098 Freiburg i.Br.
Email: rollinge@ruf.uni-freiburg.de

NOTES

[1] Recently an intellectual biography of Lotze has been published by Pester 1997. This work goes far beyond the earlier biographies, the incomplete one in Falckenberg 1901 and also Wentscher 1924, 73–198, though these can still be read with profit. As regards literature by and about Lotze, the bibliography of Pester's book is very thorough-going.

[2] James 1890, 157 ff.

[3] This text was published as an article in Wagner 1846 and re-printed in Lotze 1885, 1–204. Localisation of sensations is treated in pp. 50–80.

[4] Passages cited from Lotze 1852 will be translated into English in the text, but also given in the German in the relevant footnote. Though the spelling of the title is not changed in accordance with modern German spelling, the spelling in passages quoted from this work will be modernised. The same practice will be adopted regarding other works by Lotze.

In order to insure consistency in translations, all passages which are quoted in this study from the writings of Lotze or other texts originally in German are translated by me. Many of Lotze's writings have however been translated long ago and by such notable thinkers as Bosanquet, as can be seen from the bibliographies of both Pester's biography and the edition of George Santayana's dissertation in Santayany 1971. These translations are accordingly difficult to obtain. Moreover, by providing my own, I have hopefully cast the cited passages into an

English which is somewhat more accessible to the twenty-first century reader.

[5] Included among these writings is an article in French (Lotze 1877, reprinted in Lotze Lotze 1891, 372–396. Concerning some of the anomalies in this particular exposition, see Stumpf 1893.

[6] Passmore 1968, 51: "Few philosophers have been so pillaged [...]." Cf. Pester 1997, 341 ff. In Santayana 1971,which was written in 1889 (thus not long after the death of Lotze in 1881), the theory of local signs is said to be "generally accepted" (p. 109). Concerning developments of this theory, see Woodward 1978.

[7] This period is, after all, aptly called 'the Lotzean period' by Kuntz in his introduction to Santayana 1971, 49.

[8] Lotze 1856/58/65.

[9] This two-volume work, consisting of Lotze 1874 and 1880, as well as Lotze 1879, was actually a second attempt on Lotze's part to put forward a system of philosophy. Early in his career he had already published a metaphysics (Leipzig, Weidmann, 1841) and a logic (Leipzig, Weidmann, 1843). In the later volume on metaphysics Lotze is concerned not only with the exposition of his theory of local signs (§§ 275–290), but also with issues which are more metaphysical in the narrower sense of the word, namely the demonstration of the thesis that space is purely subjective (§§ 99–117) and criticisms of the so-called 'deductions of space' (§§ 118–137). These issues, however, lie outside of the scope of the present paper. Whenever *Metaphysik* is referred to here, it is the work of 1879 which is meant.
Though the theory of local signs accordingly has its place in Lotze's system of philosophy, Santayana is right in regarding this as one of the theories which "have an interest and value in themselves, quite apart from Lotze's theory of the universe" (Santayana 1971, 111).The presentation of the theory of local signs in *Metaphysik* was, to be sure, the last version thereof which he published. The syllabus of his lectures on psychology (1888/81), which also includes a summary of the theory (§§ 27–41), was however posthumously published in Lotze 1912b.

[10] Boring 1950, 262.

[11] Pester 1997, 190.

[12] In the first chapter of the first book of Lotze 1852, he argues that the notion of the soul is needed in order to account for the unity of consciousness. Cf. Lotze's later treatment of this topic in *Metaphysik* (Book III, Chapter I). Throughout Lotze's defence of his theory of local signs he accordingly works with the concept of soul, but it is arguable that this concept is not essential to the theory under discussion.

[13] Lotze 1852: "anwendbare Anschauungen über die Beziehungen des geistigen Lebens zu den körperlichen Tätigkeiten."

[14] Here it may be mentioned that Lotze's background in physiology was by no means a product of dilettantism. Rather, in Leipzig he took doctorate degrees (1838) and later habilitated (1840) in both philosophy and medicine. Also, in the intervening year Lotze actually practised medicine and accordingly ranks together with John Locke and William James as one of the outstanding physician-philosophers.

[15] James 1890, 134 ff. James' attempt to explain the sensory representation of space as originally given in the datum of vastness, which he ascribes to all sensations, is of course opposed to Lotze's theory.

[16] Locke 1987, II.xiii.2. Locke's distinction between secondary and primary properties, however, is one which is unacceptable to Lotze, who maintains that our sensations cannot in any way be pictorial representations of the external world. This point will be elaborated on presently in the text.

[17] Cornelius 1896, 280 ff. Hans Cornelius correctly pointed out that 'localisation' can be used in two senses: "first, as a term for the immediate perception of spatial relations of a constituent content of a whole content of the optical or tactile realm through which the location of the former is determined within the latter; but also as a term for the judgements about spatial predicates which refer not to the relations of parts of a present perception, but rather to their

empirical connection with contents which are to be perceived at other times (not at present) [einmal zur Bezeichnung der unmittelbaren Wahrnehmung räumlicher Relationen eines Teilinhaltes innerhalb eines Gesamtinhaltes des optischen oder des haptischen Gebietes, durch welche der Ort des ersteren innerhalb des letzteren sich bestimmt; weiter aber zur Bezeichnung der Urteile über räumliche Prädikate, welche nicht auf die Relationen der Teile einer gegenwärtigen Wahrnehmung, sondern auf den empirischen Zusammenhang der letzteren mit anderweitig wahrzunehmenden (nicht gegenwärtigen) Inhalten Bezug haben]." The point here is that, even if it were insisted (as it is by Lotze's opponents) that the theory of local signs is not to be applied to localisation in the first sense, a theory of this kind is nonetheless needed for localisation in the second sense.

[18] Lotze 1886, 25 ff. See also Lotze 1852, 174 ff.

[19] The rejection of the picture theory of sensation was also taken up by Franz Brentano who prefers to call the contents of sensation 'signs' (*Zeichen*) rather than pictures or images of their stimuli and also describes the latter as 'space-like' (*raumähnlich*) and 'time-like' (*zeitähnlich*). See Brentano 1874, 128 ff. One of course is reminded of the role which Kant gives to 'thing-in-itself.' This Kantian concept was no doubt highly influential throughout the nineteenth century, even on such anti-Kantian philosophers as Brentano. Yet, the external world for both Brentano and Lotze still has a structure which is represented in sensation and is in this regard rather unlike the thing-in-itself.

[20] Lotze 1852, 327: "wir finden desto häufiger, daß er sich in einer etwas feineren Form wiederholt."

[21] Lotze 1852, 328: "Unfehlbar muß auch für sie in ihrer Überleitung zur Seele der Punkt eintreten, wo ihre ganze geometrische Gestalt spurlos zerstört wird, und wo sie in der Seele durch eine *Summe intensiver Erregungen* ersetzt werden, die gleich einer Vielheit von Tönen, keine Andeutung räumlicher Ausdehnung oder Lage mehr enthält. Sollen wir daher eine Anschauung der wirklichen Lage äußerer Objekte gewinnen, so kann es nicht auf dem Wege der *Auffassung*, sondern auf dem der *Wiedererzeugung der Räumlichkeit* sein. Überall wird das Extensive in ein Intensives verwandelt, und aus diesem erst muß die Seele eine neue innerliche Raumwelt rekonstruieren, in welcher die Bilder der äußeren Objekte ihre entsprechenden Stellen finden." This statement is highly reminiscent of the view presented in Herbart 1892b, § 110 ff., in Herbart 1892a, 88 ff. Regarding Herbart's influence on Lotze in the formulation of the theory of local signs, see what Lotze himself has to say in Lotze 1886, 56 ff. The philosophical views of Lotze and Herbart are moreover related in more ways than this, but a discussion of this topic certainly lies outside the scope of the present study.

[22] Lotze 1879, § 276: "Manchen erscheint alles besonders klar, wenn sie unmittelbar im Auge die Seele vorhanden sein und wie eine tastende Hand sie gleich dort mit tausend Nervenspitzen die einzelnen Farbenpunkte genau in der Lage erfassen lassen, die sie dort eben besitzen; sie vergessen, daß es gleich schwierig sein würde, zu zeigen, wie aus Tastgefühlen, welche die zufassende Hand empfängt, die Berechtigung flösse, die verschiedenen ergriffenen Punkte auf bestimmte Raumstellen zu beziehen; man müßte voraussetzen, daß der jedesmalige Ort der Hand im Raume schon Gegenstand der Wahrnhemung sei, die man eben noch erklären wollte [...]."

[23] Lotze 1879, § 276. "Wie erlangt die Seele eine Kenntnis der Tatsache, daß in diesem Augenblick drei zentrale Nervenpunkte gereizt sind, die in den Ecken eines Dreiecks oder in einer geraden Linie liegen?"

[24] Lotze 1879, § 278: "[...] und doch zeigt er sich sehr wenig fähig, eine Mannigfaltigkeit gesonderter Empfindungen zu erwecken, welche der Anzahl seiner einzelnen Fäden entspräche."

[25] Lotze 1879, § 276.

[26] Lotze 1852, 334: "[...] so muß jede Erregung vermöge des Punktes im Nervensystem, an welchem sie stattfindet, eine eigentümliche Färbung erhalten, die wir mit dem Namen ihres *Lokalzeichens* belegen wollen." It is to be seen that the local sign in the visual representation of space results from eye movement rather than the stimulation of a particular point as such.

[27] Concerning the view that the local signs are to be thought of only in purely physical terms or as unconscious, see Lotze 1879, § 282. It would take us too far afield to consider the difficulties in this connection, though it is significant that Lotze makes a concession to Herbart, who had great influence in the nineteenth century in promoting the notion of the unconscious.

[28] Lotze 1879, § 279.

[29] Lotze, 1879, § 279.

[30] Lotze 1879, § 280: "Es scheint mir daher die andere Annahme natürlicher, daß aus dem Haup-teindruck und der Nebenerregung in der Tat in jeder Faser ein Gesamtzustand entstehe, der als solcher fortgeleitet, auch nur eine Gesamtempfindung veranlaßt. Diese würde, wenn sie für sich allein bliebe, keine Veranlassung erzeugen, in ihr etwas zu unterscheiden, so wie etwa das Violet, wenn wir es allein kännten, uns nicht Rot und Blau in ihm zu sondern auffordern würde; aber da viele verschiedene Reize nach und nach sich mit demselben Nervenvorgange verknüpfen, so könnte aus der Vergleichung dieser Fälle die sondernde Tätigkeit entstehen, welche die Gesamteindrücke in ihre Komponente zerfällte, dann aber zugleich lernte, das so getrennte Lokalzeichen in jedem Falle auf denjenigen qualitativen Eindruck zu beziehen, von dem es eben nur in Gedanken abgesondert wurde."

[31] This proposal is highly reminiscent of the one which Hume had made regarding the 'distinctions of reason' and was taken up later in the nineteenth century by Hans Cornelius. See Rollinger 1993, 16 ff., 74 ff. Lotze, however, does not make any reference to Hume in the above-cited passage and in fact held British empiricism in low regard, as Stumpf points out in Stumpf 1918, 13.

[32] Lotze 1879, § 280.

[33] Lotze 1879, § 281.

[34] Lotze 1879, § 283.

[35] Lotze 1879, §§ 284 ff.

[36] Lotze 1912b, § 35: "Zuerst erlangt nun dieses Bild einen Ort mit Rücksicht auf das Auge, dessen Öffnungen und Schließungen, die uns auf andere Weise bekannt werden, bedingend für sein Dasein oder Nichtdasein sind. Nämlich die sichtbare Welt ist *vorn vor unsern Augen.* Was *hinter* uns ist, existiert nicht bloß für uns noch gar nicht, sondern wir wissen nicht einmal, daß es etwas gibt, was man 'hinten' nennen dürfte."

[37] Lotze 1912b, § 35: "Die Vorstellung nun, daß es so etwas wie eine dritte Dimension des Raumes gebe, kann nicht von selbst, sondern nur durch die Erfahrung entstehen, welche wir machen, wenn wir uns durch die gesehenen Objekte hindurch bewegen."

[38] Lotze 1912b, § 36.

[39] I do not wish to imply that the theory of local signs had been left uncriticised before Lotze's death. As will be seen, during Lotze's life Stumpf already subjected this theory to criticism. Below, however, I shall cite his succinct comments on the subject as these are found in his 1886/87 lectures on psychology.

[40] There is nothing peculiar about Brentano, Stumpf, and Dilthey in this regard. They simply used the received terminology of the time, which apparently comes from Hermann von Helmholtz. See the brief article under *Nativismus* in Eisler 1922, 419. The author of this article also acknowledges the distinction between nativism as a theory of space-representation and nativism in the epistemological sense.

[41] This can be said of both Brentano and Stumpf. Concerning Brentano's empiricist view of concept-formation, see Rollinger 1999, 35 ff. See also Stumpf 1939, 9–123, where a single-minded effort is made to argue for the empirical origin of a number of controversial concepts, though space is not one of them.

[42] A copy of this lecture is to be found in the Husserl Archives in Louvain, under the signature Q 9. It is written out in Husserl's Gabelsberger stenography. On the title page (Q 9/1) it is indicated that it was copied from the notes of Franz Hillebrand. Though Husserl himself did not begin attending the lectures of Brentano until 1884, in his formative years as a philosopher he made a special effort to acquire copies of notes of Brentano's lectures. On the title page of the lecture under consideration here Husserl also adds in parentheses to the title 'the space theories' (*die Raum-Theorien*).
I thank Professor Samuel IJsseling, former director of the Husserl Archives, for allowing me to quote from this lecture.
Brentano 1874, 112, only cautiously takes empiricist theories of space-representation, explicitly including Lotze's, into account as a possible reason why psychical phenomena cannot be characterised as those phenomena which lack spatial location or extension and thus contrasted with physical phenomena (i.e., the contents of sensations and corresponding phantasy or memory presentations).

[43] Another important disagreement between Brentano and Lotze is to be found in their classification of psychical phenomena. While Lotze adheres to the Kantian classification of them into thinking, feeling, and willing, Brentano classifies them into presentations, judgements, and acts of love and hate. See Brentano 1874, 244–247, where Lotze's view on this matter is cited and criticised.

[44] *Ausgewählte psychologische Fragen,* Q 9/29a: "Aber diese Auffassung ist falsch. Man mache eine Beobachtung mit elektrischen Funken. Hier kann ich keine Augenbewegungen ausführen; Muskelempfindungen sind keine da. Der Anhaltspunkt für Assoziation fällt weg, also auch seine Systeme. Wir müßten also nur Farbe hier sehen ohne Gestalt und Ausdehnung, was aber nicht der Fall ist. Somit ist diese Lösung unmöglich. Auch müßte fortwährend die Schwierigkeit bestehen, das Auge auf ein Objekt zu richten, welches das Auge seitlich gereizt hat." This is my transcription from the text in Gabelsberger stenography.

[45] Stumpf 1918. This article was published on the occasion of Lotze's one hundredth birth year.

[46] Pester 1997, 290 ff.

[47] This is already clear from Stumpf's early work (Stumpf 1873), which is dedicated to Lotze and also includes an appendix (pp. 315–24) where Lotze himself defends the theory of local signs. It should moreover be pointed out that, in spite of Stumpf's rejection of this theory, he found it important enough to test Husserl on it in 1887 for the sake of the approval of the latter's doctorate. Husserl had received his doctorate in Vienna in 1881 and needed it approved in Halle where he also habilitated in 1887. See Schuhmann 1977, 19.

[48] Rollinger 1999, 298 ff.

[49] Another interesting 'external' tie between Lotze and Dilthey is to be found in the fact that Georg Misch, a defender of Dilthey's philosophy against Husserl, was responsible for the new edition of Lotze's system of philosophy (Lotze 1911 and Lotze 1912a). In Lotze 1911 there is a lengthy introduction by Misch which is still well worth reading.

[50] Lotze 1997, 221: "Aber die Forderung dieser Lokalzeichen ist unberechtigt: existiert ein objektiver Raum, so lösen sich die Schwierigkeiten nicht; ist der Raum Phänomen, dann besteht das Problem gar nicht. Ist der Raum wirklich, dann bestehen auch die Reize, welche qualitative Empfindungen zur Folge haben, in Bewegungsvorgängen. Wenn nun die Verschiedenheit der Lage auf unser Bewußtsein nicht wirken kann, so kann auch rot und grün von uns nicht unterschieden werden, denn die Bewegung besteht ja in einer Verschiebung der Lage der Elemente. Also: Dieselbe Schwierigkeit, die Lotze im Raumvorstellen findet, besteht auch für die qualitative Empfindung. Ersetzt man aber die Raumvorstellung durch ein unräumliches Geschehen, dann verschwindet das ganze von Lotze hervorgehobene Problem. In

beiden Fällen bleibt nur die Anforderung, daß im Endpunkt des physiologischen Vorgangs die Anordnung, welche der Räumlichkeit des Gegenstandes zugrunde lag, durch eine entsprechende Anordnung von Erregungselementen repräsentiert wurde. Weiter erscheinen die von Lotze hypothetisch eingeführten Lokalzeichen des Auges, der Muskelempfindungen nicht ausreichend, die Feinheit unserer Unterscheidung im Raume zu erklären."

[51] Brentano 1982.

[52] Here one may also think of Husserl, who was a student of both Brentano and Stumpf, and also at one time, and indeed arguably so until the discovery of the phenomenological reduction in circa 1905, an adherent of descriptive psychology. See Rollinger 1999, 6–9. Though Husserl is known for his divergence from Brentano and Stumpf in that he claimed to be investigating consciousness from a phenomenological standpoint, as distinct from either descriptive or genetic psychology, it has been argued, in Rollinger 1993, 195–210, that some (if not all) of his early disagreements with Brentano were not at all methodological in character, but rather easily understood as alternative theories in descriptive psychology.

Given Husserl's descriptive-psychological orientation, it may be asked if he rejected the the theory of local signs. In his 1905 lecture on time-consciousness, which is still primarily an inquiry in descriptive psychology rather than transcendental phenomenology, Husserl does in fact mention them, but only to say that if one speaks of them one should speak of quality signs and of temporal signs as well. All of these 'signs' are conceived of by Husserl as sensory contents (called *hylé* in his later writings) which are, according to his theory, apprehended (or apperceived) in consciousness. Though the notion of temporal signs can be found in Stern 1897, 331 (which Husserl read in preparation of the lecture just cited and also criticised therein), it can be traced back to Lotze 1879, § 154. A discussion of Lotze's theory of temporal signs or its influence on Husserl would however take us much too far afield in the present discussion.

[53] Dilthey 1964, 139–240. Something else which both Brentano and Dilthey had in common is to be found in the fact that they both attended some of the same lectures given in Berlin by the Aristotelian and anti-Hegelian philosopher, Arnold Trendelenburg. The historical continuity from Trendelenburg to Brentano (and Dilthey) and then to Husserl (who, like Trendelenburg, published a work under the title *Logische Untersuchungen*) has unfortunately been neglected and awaits examination from historians of philosophy and psychology.

[54] Dilthey 1964, 204. It does not matter here what Dilthey came to think in his later discussions of descriptive and explanatory psychology, for the point is that in the 1883/84 lecture on psychology he rejected both the theory of local signs and insisted upon a physiological approach in psychology.

Anton Marty, like Stumpf a pupil of both Brentano and Lotze, also defended the Brentanian notion of a descriptive psychology and also rejected other attempts to present a descriptive psychology. See Anton Marty 1987, 49–66, especially 53 ff. While he does not name anyone in particular whose views are under attack on this point, it is highly likely that it is Dilthey's descriptive psychology which he thus dismisses. See Marek and Smith 1987, 33–47, especially 40–43.

[55] The distinction between phenomena and psychical functions is discussed at length in Stumpf 1907. The notion of phenomenology, as a neutral science and as a pre-science, is primarily worked out by Stumpf 1907, 26–32. See Rollinger forthcoming. It is most important to emphasise here that phenomenology in the Stumpfian sense has little in common with phenomenology in the Husserlian sense.

[56] Lotze 1912, § 1: "Unsere wissenschaftlichen Bedürfnisse würden vollkommen befriedigt sein, wenn wir 1. nach Anleitung der Beobachtung vollständig alle einzelnen Bestandteile dieses Lebens [der Seele] und die allgemeinen Formen ihrer Verknüpfungen darstellen könnten (*deskriptive* oder *empirische* Psychologie); wenn wir 2. die Natur des Subjekts dieses ganzen Lebens, sowie die wirksamen Kräfte und Bedingungen namhaft machen könnten, durch welche das Ganze dieses Lebens hervorgebracht und genötigt wird, jene erfahrungsgemäß bekannten

Formen seines Verlaufs inne zu halten (*erklärende*, *mechanische* oder *metaphysische* Psychologie." Lotze continues to mention a third endeavor which "could finally state the rational meaning for which all this exists, or the vocation that the life of the soul as such is to fulfil in the entire world (ideal or speculative psychology) [den vernünftigen Sinn angeben könnten, wozu dies alles vorhanden ist, oder den Beruf, den das Seelenleben überhaupt im ganzen der Welt zu erfüllen hat (*ideale* oder *spekulative* Psychologie)]," but he also adds that psychology in this sense is not strictly scientific.

REFERENCES

Boring, E.: 1950, *A History of Experimental Psychology*. Englewood Cliffs, N.J., Prentice Hall.

Brentano, F.: 1874, *Psychologie vom empirischen Standpunkte*. Leipzig, Duncker and Humblot. En. tr. 1995, *Psychology from an Empirical Standpoint*, by Linda McAlister et al., London, Routlege.

Brentano, F.: 1892, *Deskriptive Psychologie*, ed. by R.M Chisholm and W. Baumgartner, Hamburg, Felix Meiner. En. tr. 1995, *Descriptive Psychology*, by Benito Müller, London, Routledge.

Cornelius, H.: 1896, *Psychologie als Erfahrungswissenschaft*. Leipzig, B.G. Teubner.

Dilthey, W.: 1964, *Gesammelte Schriften*, vol. 5., ed. by Georg Misch. Göttingen, Vandenhoeck and Rueprecht.

Dilthey, W.: 1997, *Wilhelm Dilthey Gesammelte Schriften*, vol. 21, ed. by G. van Kerckhoven and H. Ulrich-Lessing, Göttingen, Vandenhoeck and Rueprecht.

Eisler, R.: 1922, *Handwörterbuch der Philosophie*. Berlin, E.S. Mittler and Sohn.

Falckenberg, R.: 1901, *Hermann Lotze*. Erster Teil: Das Leben und die Entstehung der Schriften nach Briefen. Stuttgart, Frommanns Verlag.

Herbart, J.F.: 1892, *Psychologie als Wissenschaft*, in Kehrbach, ed., J. F. Herbart's *Sämtliche Werke*, Langenzalza, Hermann Beyer and Söhne.

James, W.: 1890, *The Principles of Psychology*. New York, Holt and Co.

Locke, J.: 1987, *An Essay concerning Human Understanding*, ed. by Peter H. Nidditch, Oxford, Clarendon Press.

Lotze, R.H.: 1852, *Medicinische Psychologie oder Physiologie der Seele*. Leipzig, Weidmann.

Lotze, R.H.: 1874, 1880, *Logik. Drei Bücher vom Denken, vom Untersuchen und vom Erkennen*, 2 vols. Leipzig, Hirzel.

Lotze, R.H: 1877, 'De la formation de la notion d'espace. La théorie des signes locaux,' *Revue philosophique*, **4**, 345–355.

Lotze, R.H.: 1879, *System der Philosophie. Zweiter Theil: Drei Bücher der Metaphysik*. Leipzig, Hirzel.

Lotze, R.H.: 1885, *Kleinere Schriften* II, ed. by D. Peipers, Leipzig, Hirzel.

Lotze, R.H.: 1891, *Kleinere Schriften* III/1 and 2, ed. by D. Peipers, Leipzig, Hirzel.

Lotze, R.H.: 1911, *System der Philosophie. Erster Teil: Drei Bücher der Logik*, ed. by Georg Misch, Leipzig, Felix Meiner.

Lotze, R.H.: 1912a, *System der Philosophie. Zweiter Teil: Drei Bücher der Metaphysik*, ed. by Georg Misch, Leipzig, Felix Meiner.

Lotze, R.H.: 1912b, *Grundzüge der Psychologie. Diktate aus den Vorlesungen*. Leipzig, Hirzel.

Lotze, R.H.: 1856/58/65, *Mikrokosmus*. Leipzig, Hirzel.

Marek, J.C. and Smith, B.: 1987, 'Einleitung zu Anton Martys Elemente der deskriptiven Psychologie,' *Conceptus*, **21**, 33–47.

Marty, A.: 1987, 'Elemente der deskriptive Psychologie. Zwei Auszüge aus Nachschriften von Vorlesungen Anton Martys,' *Conceptus*, **21**, 49–66.

Passmore, J.: 1968, *A Hundred Years of Philosophy*. Harmondsworth, Penguin.

Pester, R.: 1997, *Hermann Lotze - Wege seines Denkens und Forschens. Ein Kapitel deutscher Philosophie - und Wissenschaftsgeschichte im 19 Jahrhundert*. Würzburg, Königshausen and Neumann.

Rollinger, R.D.: 1993a, 'Husserl and Brentano on Imagination,' *Archiv für Geschichte der Philosophie*, **75**, 195–210.

Rollinger, R.D.: 1993b, *Meinong and Husserl on Abstraction and Universals: From Hume Studies I to Logical Investigations II*. Amsterdam, Rodopi.

Rollinger, R.D.: 1999, *Husserl's Position in the School of Brentano*. Dordrecht, Kluwer.

Rollinger, R.D.: Forthcoming, 'Stumpf on Phenomena and Phenomenology,' *Brentano-Studien*.

Santayana, G.: 1971, *Lotze's System of Philosophy*, ed. by Paul Grimley Kuntz, London, Indiana University Press. Dissertation.

Schuhmann, K.: 1977, *Husserl-Chronik. Denk und Lebensweg Edmund Husserls*. The Hague, Martinus Nijhoff.

Stern, L.W.: 1897 'Psychische Präsenzzeit,' *Zeitschrift für Psychologie und Physiologie der Sinnesorgane*, **13**, 325–349.

Stumpf, C.: 1873, *Über den psychologischen Ursprung der Raumvorstellung*. Leipzig, Hirzel.

Stumpf, C.: 1893, "Zum Begriff der Lokalzeichen," *Zeitschrift für Psychologie und Physiologie der Sinnesorgane*, **4**, 70–73.

Stumpf, C.: 1907, *Zur Einteilung der Wissenschaften*. Berlin, Königliche preußsische Akademie der Wissenschaften.

Stumpf, C.: 1907, *Erscheinungen und psychische Funktionen*. Berlin, Königliche preußische Akademie der Wissenschaften.

Stumpf, C.: 1918, 'Zum Gedächtnis Lotzes,' *Kant-Studien*, **22**, 1–26.

Stumpf, C.: 1939, *Erkenntnislehre I*. Leipzig, Johann Ambrosius Barth.

Wagner, R. ed.: 1846, *Handwörterbuch der Physiologie* III. Braunschweig, Wieweg.

Wentscher, M.: 1924, *Fechner und Lotze*. Munich, Ernst Reinhardt.

Woodward, W.: 1978, 'From Association to Gestalt: The Fate of Hermann Lotze's Theory of Spatial Perception,' *Isis*, **69**, 572–582.

JAN SEBESTIK

ERNST MACH'S EVOLUTIONARY THEORY OF REPRESENTATION

1. INTRODUCING THE PROBLEM

Mach's thinking could be characterised in terms of an *epistemological turn* in the sense of empirical epistemology, like that of the British empiricists. With Mach, Austrian philosophy takes a new orientation, by leaving aside Bolzano's logical and mathematical interests as well as Brentanos's introspective study of the human mind.

Mach came to philosophy from science. Like Bolzano, he revived the tradition of the great philosophers of the past who were also scientists. But each of them worked in different, complementary fields. Mach knew practically nothing about recent advances in logic and his mathematical knowledge did not go beyond what is necessary for the working physicist.

His goal consisted in unifying human knowledge, in closing the chasm between mind and matter, in establishing a common ground for such different disciplines as psychology, physiology and physics in order to construct a unique world picture based on science, and one compatible with common sense and explicable in evolutionary terms. Mach undertook the task of analysing the formation and the structure of the world as it is perceived and of explaining the evolution of human knowledge. He asked: how did human thinking evolve from animal behaviour and reactions? How was science formed from elementary knowledge contained in perception? What were ultimate elements of knowledge and how could they be combined in order to obtain conceptual structures which constituted our science? For Mach, we do not perceive colours or forms of objects; what we really see are bodies, corporeal, material bodies in space. It is only by analysing our perception that we arrive at its components, namely sensations.

2. ONTOLOGY

Mach saw a profound split between our psychical activity on the one hand, and the behaviour of unanimated objects as reflected in the difference between

L. Albertazzi (ed.), The Dawn of Cognitive Science. Early European Contributors, 123–134.

psychology and physics. "Psychical life seems to be a world for itself, with laws of another order," (*mit Gesetzen anderer Ordnung*). On the other hand, we experience physical events as being foreign; they could be different. The material bodies could obey different laws without them disturbing us.

Would it be possible to reduce one of these apparently fundamentally different domains and sciences to the other? For example, can we conceive of cultural objects in terms of universal physics? Mach resists the temptation of radical physicalism. Even a purely technical object such as Watt's steam-engine, cannot be explained in purely physical terms. We can understand a particular, individual engine in terms of physics and engineering, but in order to explain kinds of steam-engines, we must also consider their place in industrial production and in the world economy, as well as their cultural history. Even less acceptable to Mach is the physicalist explanation of the thought of a physicist. One should take into account his previous thoughts, his perceptions, his personal history including his education and eventually also the history of physics interwoven into the web of universal history – and all that is an impossible task.[1]

Physics and psychology investigate two complementary aspects of reality. Both are legitimate and independent sciences. They both offer responses to ontological problems.

In his youth, Mach was not far from the monadological viewpoint. He considered monads (atoms) to be endowed with inner life. Nature was thought of as having two sides: a physical side and a psychological one.

> If psychical life is to be harmonised at all with the theories of physics, we are obliged, I reasoned, to conceive atoms as *feelings* (ensouled). The various dynamic phenomena of the atoms would then represent the physical processes, whilst the internal states *connected there with* would be the phenomena of psychic life.[2]

If we accept both the atomic hypothesis and thesis of the unity of the soul, we can arrive "at a tenable monistic conception."

To destroy this kind of prejudice is the goal of Mach's mature work. There are two prejudices implied in the monadological position. One is the philosopher's realism operating with the traditional concept of substance (atom) as the permanent bearer of its properties, and with the substantial ego, the indestructible soul. The second prejudice involves the naïve attitude of every person (even that of the philosopher outside his study) who believes in a physical world independent of our minds and in the causal relationship between external objects and our sensations and perceptions. The originality of Mach's position consists in destroying the twofold prejudice and at the same time in preserving the naïve attitude of the ordinary man.

First, Mach asks, how do we know about substances? Have we the right to conclude the existence a permanent source of our perceptions from the fact that we can have different perceptions of the same object? Must there be an indestructible substratum behind sensible data? According to Mach, the permanence of a thing does not imply its eternity or indestructibility. We have no experience of absolute permanence; our experience does not go further than mere relative constancy. And above all, we have no access to any mysterious source behind our perceptions; they are all what we have.

If the philosophical prejudice about abiding substances has to be rightly rejected, the natural attitude has to be explained. The latter is for Mach the fruit of the biological evolution of the human species. The identification of recurring, more or less permanent complexes of properties had been of vital importance in the struggle for survival. The need to schematise the extremely complex world where man was both prey and hunter led to the idea of something permanent producing the manifold of sensory experiences. The realistic fiction of objects hidden behind perceptions can be useful. In *The Analysis of Sensations*, Mach considers a conceptual word such as 'matter' to be a "highly natural, unconsciously constructed mental symbol for a complex of sensuous elements." The same also holds for the thesis of "the artificial hypothetical atoms and molecules of chemistry:" it has only the value of "economical symbolisation of the world of experience."[3] Such a hypothesis may serve as a mathematical model to describe certain experiences. The philosopher, however, must explain the origin of such fictions and the limits of their use. Hence Mach's *phenomenism.* Although physics operates with independent external objects and psychology operates with inner mental states, both fields deal with one and the same kind of reality, the only one to which we have direct access and which is all what is really given, namely sensations (*Empfindungen*). "There is no rift between the psychical and the physical, no *within* and *without*, no *sensation* to which an outward thing corresponds. There is but *one kind of elements.*"[4]

The world is a web of sensations, or to use, with Mach, a more appropriate word (because not associated with subjectivist philosophy), a web of *elements*, i.e., colours, sounds, pressures, odours, pains etc., and also of spatial and temporal elements. External objects, living beings, other people, and also myself, my 'soul' or 'organ of consciousness' (*mein Bewu• tseinorgan)*, are nothing else but more or less stable functional complexes of such elements. Neither the naïve, non-philosophical realist, nor the scientist looses anything. The world remains as colourful and tasteful as before, just as painful with some pleasures. Only the superfluous colourless and tasteless fictitious entities are withdrawn from this picture.

Nevertheless, such an attitude demands an authentic conversion, a renunciation of the mental and linguistic habits which the human species has developed since its origins. The proud *ego* of modern rationalistic philosophy, the Archimedean point of all certainty, has became a provisional combination of sensations, volitions, thoughts and feelings which migrate and join other combinations of elements endowed with an organ of consciousness and an organ of representation (*Vorstellungsorgan*). Mach recalls Lichtenberg's aphorism according to which we should not say *I think*, but *it thinks*, and adds that the ego cannot be saved (*das Ich ist unrettbar*). Personal mortality is but a dream. Does not everyday experience teach that parts of ourselves die long before our death? On the other hand, however, being composed of the same elements as other complexes of elements, and having no sharp boundaries, "the ego can be so extended as to ultimately embrace the entire world."[5]

The equally fictitious idea of causality – understood as real flowing from one body to another one – has to be explained in terms of functional dependence open to formulation in quantitative terms. Reformulated in this way, understanding causality continues to provide the impetus of all scientific research. Physical laws express what is abiding in nature: not the elements themselves, but the links between them.

3. THE THEORY OF REPRESENTATION

Mach sets out to determine the origin and the cognitive function of our representations. "All science has for its aim the *representation of facts in thought*, either for *practical* ends, or for removing *intellectual* discomfort,"[6] he wrote. Upon other occasions, he insists on the adaptation of thought to facts, without further justifying transferring a concept borrowed from biology to epistemology.

As we have seen, sensations are the ultimate elements of our experience, the simplest elements of the world forming the basis of our physical concepts. Sensations alone, however, do not suffice to produce scientific knowledge. Without memory, our life would be a kaleidoscope, a sequence of unrelated mental states. It is memory that transforms sensations into representations (*Vorstellungen*), defined by Mach as "traces in the memory of earlier experiences which co-determine and weave further new complexes of sensations."[7] Mach does not acknowledge sharp boundaries between different sorts of mental acts; a continuous transition leads from sensations to intuitive representations, to ideas, ordinary concepts and eventually scientific concepts. Sensations are more vivid, stronger; representations appear and disappear rapidly, combining into larger complexes or varying according to each person's imagination.

While subjective representations respond to individual needs, concepts meet the intellectual needs of the human species. Like other representations, they have a physiological basis in sensory elements. They are produced by a similar configuration of sensations which create strong associations of representations directed towards biologically relevant behaviour. Even animals have "seeds of concepts," even if they do not have the corresponding linguistic term at their disposal. Concepts are that element of human experience under which objects to which we react in the same manner fall. The difference between the behaviour of animals or prehistoric humans on one hand, and civilised people on the other, consists only in the fact that the latter are able to perform a variety of testing and checking activities and to organise concepts into systems of hypotheses and theories.

The adaptation of new facts finds its expression in the formation of judgements. Judgements consist primarily in the broadening, "supplementing or amendment of the deficiencies of a sensuous percept"[8] by other sensuous percepts. Stored in the memory, they become, according to Locke's terminology, 'intuitive knowledge,' the spontaneous recollection of facts. Two principles are at work in the further refinement of judgements: that of the broad generalisation and continuity and that of sufficient differentiation.

Both contribute to selecting pertinent elements of perception. If the progressive mental adaptation embraces a great number of facts, the discovery of new facts, possibly incompatible with the earlier ones, can lead to new conscious and purposive adaptation. Such is the beginning of scientific investigation that involves comparison and differentiation of a number of instances falling into the same category. We examine the influence of varying factors in repeated experiences and, having formed abstract concepts, we are able to solve the given problem.

We name concepts and accompany them with images. A word yields nothing but an impulse to perform a sensory operation and this is the mark of the concept. (Mach also counts as sensory operations the enumeration of the number of angles in the case of the concept of 'heptagon,' or the identifications of factors when considering the concept 'square number.') But "the concept is never a *finished* percept" (Mach 1897,162), for the operation adds a new sensuous element not present earlier. The concept is "an instruction to *test* a given representation with respect to certain properties or to *produce* a representation having determinate properties."[9] As an example, let us take the conic sections. I cannot directly see that the ellipse, parabola, hyperbola are subsumed under that concept; but I can discover the fact by cutting a cone and by constructing the equation for conic sections. Is the latter operation also a sensory one? As the examples have shown, Mach considerably extends the notion of sensory operation; for him, it included any operation involving counting or manipulation of signs. Closer to sensory operations are of course those performed by physicists and chemists.

The decisive moment in concept formation is *abstraction*, i.e., the separation and selection of sensory elements and complexes (form, colour, material aspects, use and nature of the object) in accordance with our biological needs. Thus we learn to consider biologically relevant aspects of an experience separately. But the abstraction in question is not only a process of taking away, a negative operation, a refraining from attending to the sensuous elements which accompany the abstracted entity in the given complex; it is at the same time an adding process: "on the other hand, it is turned *toward* other and *new* sensuous elements."[10]

The step from ordinary concepts to scientific ones consists in a particular intention. It is the intentional formation of concepts and their combinations that marks the beginning of scientific concepts.

In scientific concepts the intellectual domination of nature reaches its peak. According to Mach, a scientific concept is "a precise and definite *reaction-activity*, which enriches the fact with new sensuous elements," more precisely "the consciousness of reactions that we expect from the class of objects designated by the concept-word."[11] Hence, a scientific concept is a mental complex more or less permanently connecting memories and expectations related to selected aspects of the behaviour of an object. Let us take the concept of a chemical element, for example of sodium. It is a sum of expectations concerning atomic weight obtained by measurement, colour, solubility in water etc., and similarly with mathematical, physical or biological concepts like 'circle,' 'intensity of current' or 'whale.' To have a concept means to be able to submit it at any time to testing in order to obtain the expected reactions. A concept can always be traced back to intuitive elements, but such a link can be indirect or even only potential. The result of a physical investigation "is based upon an almost unending series of simple observations (sensations)," because we must also take adjusting the experimental apparatus into account. In this way, the concept appears simultaneously as a condensation of previous experiences and as an instruction to test and to produce specific representations or sensations, a sequence of operations. Eventually, language contributes to stabilise the concepts and to form a conceptual system, a scientific theory.

The following schema sums up Mach's theory of mental objects (*psychologische Gebilde*):

1. sensory experiences, sensations;
2. intuitive representations obtained by recollecting past experiences;
3. typical representations (even animals may have them; they are the "seeds of concepts");

4. everyday empirical concepts obtained by abstraction from intuitive and typical representations;
5. scientific concepts.

4. SCIENTIFIC KNOWLEDGE

For Mach, unity of the physical and psychological dimensions is fundamental.

Mathematics plays a secondary role in his theory of science which aims principally to connect sensory experiences to theoretical concepts. An excellent physicist and historian of physics, Mach was not a very good mathematician. Speaking of mathematics, he considers it simply an auxiliary instrument, not as a means to express the structure of physical theory. "All auxiliary conceptions, laws and formulas, are but quantitative norms, regulating my sensory representation of the facts. The latter is the *end*, the former are the *means*."[12] As we have seen, theoretical concepts transcend immediate experience. Mach explains the progressive distancing of theoretical concepts by a natural tendency inherent in our thoughts which

> are spontaneously impelled to *complete* all incompletely observed facts [...] The impulse in a certain measure *enriches* the single fact. Through it the latter is *more* to us. By this impulse we have always a *larger* portion of nature in our field of vision [13].

Although the world is constructed from sensory elements, our knowledge often complements them by formulating hypotheses which go far beyond our experience, extending the field of knowledge by extrapolation. We can mentally add elements which are not only absent from the sensory field, but which cannot even come into it. We can e.g., imagine the moon as an inert heavy mass without any possibility of touching it. Another famous example is Mach's principle which appeals to the action of *all* celestial bodies in order to explain the law of inertia.

It is therefore necessary to make a sharp distinction between what we see and what we mentally supply. Only this distinction enables us to see old theories as obstacles for new discoveries. Mach cites many examples: the phenomena of conduction and exchange of heat, which led Black to the discovery of specific heat. But the same idea – the constant quantity of heat-substance – kept Black from realising that heat can also be produced, as everybody knows, by friction. Huygens' fundamental discovery of the undulatory theory of light prevented him at the same time from rightly grasping the phenomenon of polarisation, which he himself discovered. The preconceived idea of fluids acting at distance on conductors stood in the way of the discovery of specific inductive capacity; only Faraday, a non-academic, could overcome the theoretical prejudices of other scientists. Mach insists that theories are only auxiliary instruments for definite purposes. They have no absolute value.

Constancy plays a major role in Mach's conception. Our expectations are based on it. Things (complexes of elements) are relatively constant and we have a natural tendency to think that they are always present, whether we perceive them or not. Chemical elements also appear unconditionally constant. We expect constant replies to our questioning in the entire realms of facts covering, for example, electricity, magnetism, light, heat. The notion of 'electric body,' for instance, means for us expectations of definite groups of facts. On the other hand, the unity of nature does not permit us to isolate one specific fact or group of facts: "there is no such thing as a specific *electrical* fact [...] all physical facts are made up, in ultimate analysis, of the same sensuous elements (colours, pressures, spaces, times)" (p. 168). Completing our observation, we speak of the constancy of celestial bodies, and even of the whole past and future, the entire passage of time.

Such projections are not founded, for unconditioned constancy of things does not exist. Here, Mach steps out of his world of ever-changing elements-sensation to reach a *structural* point of view: "There is but one sort of constancy, which embraces all forms, namely, constancy of connection (or of relation)" (p. 169). This constancy of relations is precisely what physical laws express. If the constancy of thought consists in the impulse to complete observed facts, physics formulates quantitative norms regulating our spontaneously flowing thoughts. History of physics can be characterised by the increasing degrees of constancy:

> Whenever we have a special interest in the representation of facts, we endeavour to support and corroborate ideas of lesser constancy by ideas of greater constancy or to replace them by the latter [14].

Analogy with already obtained results in mechanics guided Newton to conceive of the planets as projectiles. Similarly, Huygens (and many others) were led by the analogy of sound in optics. Because since the 17th century mechanics has become the paradigm for all natural science, the scientist have sought "to conceive electrical, optical, and thermal processes as *mechanical* processes."

We naturally prefer, as the foundation of this process, the strongest and most thoroughly tested thoughts, and these are given to us by our much exercised mechanical functions, which we may test anew at any moment without many or cumbersome appliances. Hence the authority of mechanical explanations, especially those by pressure and impact.[15]

Only one science can yield greater certainty, namely mathematics, because mathematical thought must conform its own norms even if it grows from extraneous impulses. Mathematics invariably carries "most of the material for experimenting about with it," as if internal mathematical experience imposed its norms to elements received from outside. A mathematical formula in physics, for example the sine law of refraction, "is a kind of geometrical model which simply *imitates in form* the refraction of light and *takes its place* in our mind"[16].

For Mach, geometrical space is a structure both of space-elements and of physical elements:

> it is by no means made up wholly of the system of space-*sensations* (of the senses of sight and touch), but consists rather of a large body of *physical* observations, having the space-sensations as their point of departure[17].

The most important of such observations concerns the behaviour of rigid bodies and the notion of superposition yielding congruence for measuring lengths and angles. "When we are *compelled* to imagine an isosceles triangle as having equal angles at its base, our compulsion is due to the remembrance of powerful past experiences." Thus the geometer goes far beyond the space given by sight and touch which is not homogeneous. The Euclidean method, too, presupposes "abundant geometrical experiences." "It serves to protect us from the possible errors which we have acquired"[18]. Mach does not attribute particular importance to proofs and to the deductive structure of mathematics.

"The *memory* of a given experience can reveal to the mind features which in the original observation escaped unnoticed." Such is the power of geometrical imagination, which holds also for pure mathematics, even for arithmetic's. "Even the theory of numbers must be looked at in some such manner; its fundamental propositions can hardly be viewed as entirely independent of physical experience"[19]. And Mach repeats almost literally Bolzano's remark, according to which the certainty of mathematics comes from the fact that its results "can very easily be tested on intuitions and experiences."[20]

The cogency of geometry (and of all mathematics) is due," not to "some select and special kind of cognition, but only to the fact that the empirical material which is at its base is particularly convenient and handy, has been put to the test an untold number of times, and can be subjected again at any moment to the same tests"[21].

The same also holds for time:

> When a physicist wishes to determine a period of time, he applies, as his instrument of measurement, *identical* processes or processes *assumed to be identical,* such as vibrations of a pendulum, the rotation of the earth etc.[22].

The result of his measurement is a number.

5. CONCLUSION

Mach wants to reconstruct the "natural world" of human experience,[23] purified of theological and philosophical prejudices. At the same time, he tries to explain common sense realism in terms of permanence of sensory complexes and the biological response of organisms to the challenges presented by their environment. But, unlike the phenomenologists, Mach's "natural world," is not in opposition to the abstract world of modern science whose paradigm is mathematical physics. For Mach, the world of science, though organised according to specific human needs, grows spontaneously out of original experiences.

Mach's thinking to a large extent anticipates further developments in the empiricistic, pragmatist and evolutionist trends of the 20th century. His legacy, together with that of Bolzano, merged into the powerful current of analytical philosophy, represented in the first half of the 20th century by the Vienna Circle and, as regards Bolzano, also by the Polish school.

Jan Sebestik
Université de Paris I and C.N.R.S
Institut d'Histoire des Science
13, Rue de Four
F-75006 Paris
Email: jan.sebestik@libertysurf.fr

NOTES

[1] Nevertheless, according to Martin Kusch, Wundt and almost all contemporaries of Mach took him for a reductionist, a physicalist.
[2] Mach 1897, 183.
[3] Mach 1897, 152-153. Mach continues protesting against all materialistic and physicalistic attempts to explain mental phenomena by material processes: "Still less, therefore, should the monstrous idea ever enter our heads of employing atoms to explain psychical processes; seeing that atoms are but the symbols of certain peculiar complexes of sensuous elements which we meet with the narrow domain of *physics*." Mach 1897, 153.
[4] Mach 1897, 153.
[5] Mach 1897, 10.
[6] Mach 1897, 153. By "intellectual discomfort," Mach means logical shortcomings such as incompleteness, contradictions and similar.
[7] "Erinnerungsspuren älterer Empfindungserlebnisse, die neue Empfindungskomplexe mitbestimmen und weiterspinnen," Mach 1895, 20.
[8] Mach 1897, 157.
[9] Mach 1900, 419.
[10] Mach 1900, 165.
[11] Mach 1900, 162; "[...] das an das Wort gebundene Bewusstsein von den Reaktionen, die

man von der bezeichneten Klasse von Objekten erwartet," Mach 1900, 131.
[12] Mach 1897, 156. If not states otherwise, all following Mach's quotations are from this work.
[13] Mach 1897, 171.
[14] Mach 1897, 172.
[15] But Mach reminds us that "pressure and impact are by no means simpler phenomena than are for example the phenomena of gravitation. The contention that in physics everything can be reduced to the motion of smallest particles is, at best, but an improper draft on the future." Mach 1900, 184n.
[16] Mach 1897, 187.
[17] Mach 1897, 177.
[18] Mach 1897, 177n.
[19] Mach 1897, 178.
[20] Bolzano 1810, 150.
[21] Mach 1897, 178.
[22] Mach 1897, 179.
[23] The natural world of human experience (Husserl's *Lebenswelt*) should not be confused with the world of nature.

REFERENCES

Blackmore, J.T.: 1972, *Ernst Mach, his Work, Life and Influence*. Berkeley, University of California Press.

Blackmore, J.T. ed.: 1992, *Ernst Mach – a Deeper Look. Documents and New Perspectives*. Dordrecht, Reidel.

Bolzano, B.: 1810, *Beiträge zu einer begründeteren Darstellung der Mathematik*. Prague, Caspar Widtmann.

Cohen, R. S. and Seeger, R. J. eds.: 1970, *Ernst Mach, Physicist and Philosopher*. Dordrecht, Reidel.

Mach, E.: 1872, *Die Geschichte und die Wurzel des Satzes der Erhaltung der Arbeit*. Prag, Calve.

Mach, E.: 1883, *Die Mechanik in ihrer Entwicklung, historisch-kritisch dargestellt*. Leipzig, F.A. Brockhaus ; 9th ed., 1933. En. tr. by J. McCormac, *The Science of Mechanics. A Critical and Historical Account of its Development*. New Introduction by Karl Menger, 6th ed. 1960, La Salle Ill., Open Court.

Mach, E.: 1897, *Contributions to the Analysis of Sensations*. La Salle, Open Court (German original 1886, *Beiträge zur Analyse der Empfindungen*. Jena, Fischer).

Mach, E.: 1900, *Die Prinzipien der Wärmelehre, historisch-kritisch entwickelt*. Leipzig, J.A. Barth, 2nd ed. (1st ed. 1896). En. tr. 1896, *Principles of the Theory of Heat, Historically and Critically Elucidates*. Dordrecht, Reidel.

Mach, E.: 1896, *Populär-wissenschaftliche Vorlesungen*. Leipzig, J.A. Barth. En. tr. 1895, *Popular Scientific Lectures*, La Salle, Open Court.

Mach, E.: 1905, *Erkenntnis und Irrtum. Skizzen zur Psychologie der Forschung*. Leipzig, J.A. Barth. En. tr. 1976, *Knowledge and Error. Sketches on the Psychology of Inquiry*, by P. Foulkes, Dordrecht, Reidel.

Mach, E.: 1921, *Principien der physikalischen Optik. Historisch und erkenntnis-psychologisch entwickelt*. Leipzig, J.A. Barth. En. tr. 1926, *The Principles of Physical Optics*. New York, E.P. Dutton.

Sebestik, J.: 1998-1999, 'Mach et Duhem: épistémologie et histoire des sciences,' *Philosophia Scientiae*, **3**, n° 2, 121–40.

Stadler, F.: 1982, *Vom Positivismus zur "Wissenschaftlichen Weltauffassung"*. Vienna, Löcker.

GIANNI ZANARINI

HERMANN VON HELMHOLTZ AND ERNST MACH ON MUSICAL CONSONANCE

1. A YOUNG PROFESSOR AND AN OLD THEME

In 1855, a young professor of anatomy and physiology in the University of Bonn (he was then only 34 years old) wrote a letter to a friend, who was also an academic colleague: indeed, he taught physiology in the University of Berlin. In that letter, the sender expressed his intention to devote himself to the study of acoustics. One year after, his scientific project was already well defined, and he could proudly write: "I have gradually accumulated a considerable amount of material for the reform of physiological acoustics."[1]

The young professor was Hermann von Helmholtz, and his Berlin friend was Emil du Bois Reymond, who was to become a world famous physiologist. They had met in the forties, when they both, together with Ernst Brücke, were pupils of Johannes Müller.

The master had on these exceptionally clever pupils a very deep influence. It was Müller, in fact, who wrote the famous *Handbook of Human Physiology* where he discarded any explanation of life processes which was not based on the grounds of physics and chemistry. And few years after, in 1847, his pupil Hermann von Helmholtz wrote that paper *On the Conservation of Energy*, which would alone allow its author (who at that time was no more than a young post graduate) to rank among the greatest scientists of nineteenth century.[2]

Already in this paper, the heritage of Müller shows itself very clearly: the concepts of physics, and more specifically matter and forces, are assumed as unifying tools to describe the whole world, both the living and the inanimate one.

Helmholtz, however, did not pretend to reduce completely physiology to physics: he rather required physiology not to invoke any sort of process (and, in the first place, any sort of *perpetuum mobile*) which was incompatible with physics. As we shall see, this was also the approach he adopted in his work on acoustics towards physiology, psychology and aesthetics.

L. Albertazzi (ed.), The Dawn of Cognitive Science. Early European Contributors, 135–150.

When he sent to du Bois Reymond the letter previously quoted, Helmholtz was writing the first volume of his *Handbook of Physiological Optics*, which was to be completed only in 1867, i.e., 12 years later. Perhaps, acoustics appeared to him more challenging, and he decided to postpone the conclusion of his *Optics*, devoting himself to the mastering of this still ill-defined discipline.

With the word 'acoustics', Helmholtz meant in the first place musical acoustics. As many middle class well brought-up young people of his time, Helmholtz had a sound musical training. Since his youth, he was a good piano player and a passionate concert-goer, as his letters from college to his father inform us.[3]

His masterpiece in acoustics is the *Treatise on the Sensations of Tone as a Physiological Basis for the Theory of Music.*[4] This long and detailed title informs us about the weight he gave to a scientific theory of acoustics for the understanding of music.

Also in this case, his epistemological position was very cautious: he was deeply aware of the impossibility of completely reducing music to physical acoustics. "I should consider a theory which claimed to have shown that all the laws of modern Thorough Bass were natural necessities to stand condemned as having proven too much."[5]

In the whole of his work, he thoroughly explored the boundaries between different sciences: "I always have worked around the borders of science," he wrote in 1871 to Benno Erdmann.[6] Obviously, this intellectual attitude was particularly well fitted for the study of musical acoustics.

In the present work an attempt will be made to connect the boundaries of two sciences, which, although drawn towards each other by many natural affinities, have hitherto remained practically distinct – I mean the boundary of physical and physiological acoustics on the one side, and of musical science and aesthetics on the other.[7]

Such a scientific choice, so difficult and (as we shall see) so easily misunderstood, was based on a deep epistemological belief: science doesn't deal with the true nature of the world, but is an intellectual construction built on sensations. And sensations, in turn, are:

> nothing but signs, which inform us, by the equality or the inequality of their appearance, whether we are dealing with the same or with different objects and properties of objects [...] Beyond this, they tell us nothing.[8]

Hence comes Helmholtz's interest towards connecting the one with the other different models of the world (studied in the perspective of physics, physiology, aesthetics), longing not for reduction, but for compatibility and for mutual enrichment.

In 1857, Helmholtz gave a lecture in Bonn under the title *On the Physiological Grounds of Musical Harmony*.[8] At that time, his ideas on musical acoustics were already completely defined, and he lacked only the time to complete the detailed analyses and calculations for the *Treatise*. He was then able to expose his scientific program on this topic:

When I considered physical acoustics from a physiological point of view, and thus more closely followed up the part which the ear plays in the perception of musical sounds, much became clear of which the connection had not been previously evident. I will attempt to inspire you with some of the interest which these questions have awakened in my own mind, by endeavouring to exhibit a few of the results of physical and physiological acoustics.[9]

Obviously, when Helmholtz quotes 'the ear' he refers to the inner ear and to the most up to date knowledge about the neurophysiology of hearing.

The approach here proposed had never been pursued before Helmholtz. It is true that every scientist who, since 1600, had devoted himself to acoustics had acknowledged, in principle, the rôle of the ear in the construction of auditory sensations. In particular, the French scientist Joseph Sauveur (who coined the word 'acoustique') considered that this science had to deal not only with the production of sound and with its transmission, but also with its reception by the ear.[10] But a research program like this one was very difficult to pursue, not only because of the slower development of physiology with respect to physics, but also because of the lack of scientists with sufficient multidisciplinary knowledge and interest.

Helmholtz was a physiologist and a physicist, and had an excellent musical training; moreover, he was a passionate explorer of 'boundaries.' So, the research programme put forth by Sauveur found in Helmholtz the scientist who could carry it out.

The study of acoustics pursued by Helmholtz has a central and specific goal: and he already points to it in this lecture.

[...] I shall select the most important [point] of all, which will best show you the significance and results of scientific investigation in this field: I mean the foundation of concord.[11]

2. CONSONANCE AND DISSONANCE

The theme of 'concord' here quoted by Helmholtz refers to the ancient problem of consonance and dissonance in music. Helmholtz himself goes back to Pythagoras in his quotations.[12]

Broadly speaking, consonance may be defined as some sort of harmony of two sounds. But for philosophers as Pythagoras, Plato, Saint Augustine, Kepler, consonance is not in the first place a sensory experience, i.e., an agreeableness. It is rather 'a sense of agreement, a judgement of being well adjusted, of being in accord,'[13] 'a relatedness between sounds.'[14]

This accord is the source of a an intellectual pleasure, with deep cosmological implications. In fact, the oldest explanation of musical consonance may be qualified as a metaphysical one, which after Pythagoras was reformulated by philosophers and scientist, from Plato (mainly in *Timaeus*) down to Johannes Kepler. This line of explanation asserts that consonance is the way in which music reflects the harmony of the heavens, the 'harmony of the spheres.'[15]

In this purely intellectual perspective, Pythagoras highlighted the consonance of unison (i.e., the harmony of the couple of sounds obtained from two identical strings), of the octave (the harmony of the musical interval obtained from a string and its half), of the fifth (obtained from a string and the $^2/_3$ of it) and of the fourth (obtained from a string and the ¾ of it).[16]

Only during the scientific revolution, in seventeenth century, we can find a scientific interest in consonance as a sensory experience and we meet an effort to explain it in scientific terms. In particular, Galileo managed to explain from a scientific point of view the traditional ranking of consonances (the unison first, then the octave, then the fifth and the fourth), enlarging it to thirds and sixths (recently admitted by music theoreticians like Zerlino).

He explained this ranking in terms of more or less regular and simple oscillations of the ear-drum under the compound vibration due to the two simultaneous sounds.[17]

Galileo was the first to investigate musical consonance from a physical point of view. However, there is then a sort of contradiction between Galileo as a scientist and Galileo as a musician (he was in fact a good lute player): the first one rejected the approximations accepted in the practical music of the second one. In fact, according to his theory, the almost perfect (but never absolutely pure) consonances of practical music ought to sound as the worst dissonances.

Joseph Sauveur was perhaps less prone to accept philosophical tradition, and consequently was more free in linking together scientific explanation and musical practice. He based his scientific explanation of consonance on the absence of beats: an acoustic phenomenon which was very well known to players of musical instruments, who exploited it for tuning their instruments.[18] The explicit reference to musicians marks the difference with respect to previous centuries, showing the new interest of scientists in practical music.

The explanation of consonance given by Sauveur is free from the difficulty of Galileo's theory. In fact, the frequency of beats goes smoothly to zero when the musical interval approaches a consonance.

But two new difficulties affect Sauveur's theory. First, his approach is of no use in order to establish a ranking of consonances, since (by his definition) all consonances are equally free from beats. Second, it cannot explain why slightly mistuned octaves or fifths give rise to a dissonance, in spite of the large value of the corresponding frequency interval.

We can then conclude that Sauveur's theory has to do mainly with consonance as a measure of tuning rather than with consonance as a ranking of agreeable sensory experience.

Differently from Pythagoras, who discarded sensory experience, but also differently from Sauveur, who proposed nothing more than a particular theory, Helmholtz tried to give a sound physical and physiological basis to the ranking of sensory consonance. As we shall see, he was able to confirm the ranking codified in music theory: first of all unison, then octave, fifth, fourth, and so on.

3. A SCIENTIFIC EXPLANATION

In order to obtain this result, Helmholtz started from Sauveur's beat theory, but exploited also another research by Sauveur, namely his thorough study of overtones.

In the middle of the nineteenth century, when Helmholtz started his investigation, the fact that complex sounds are made of simple sounds was already universally accepted. This was due not only to empirical studies (like those pursued by Sauveur), but also to the fundamental results obtained by mathematical physics: the wave equation and, more recently, the Fourier theorem, which stated the possibility of decomposing any periodical waveform into sine functions with harmonic frequencies.[19]

The French mathematician Fourier has proved the correctness of a mathematical law, which in reference to our present subject may be thus enunciated: any given regular periodic form of vibration can always be produced by the addition of simple vibrations, having pitch numbers which are once, twice, thrice, four times, &c., as great as the pitch numbers of the given motion.[20]

Some years before the beginning of Helmholtz's research, Ohm had supposed that the ear was a sort of Fourier analyser of sounds;[21] but he lacked the physiological knowledge needed to explore further on this hypothetical statement. Helmholtz, on the contrary, had at his disposal the most recent achievements of inner ear physiology.

The essential result of our description of the ear may [...] be said to consist in having found the terminations of the auditory nerves everywhere connected with a peculiar auxiliary apparatus, partly elastic, partly firm, which may be put in sympathetic vibration under the influence of external vibration, and will then probably agitate and excite the mass of nerves [...] There must be different parts of the ear which are set in vibration by tones of different pitch and which receive the sensation of these tones [...] Hence every simple tone of determinate pitch will be felt only by certain nerve fibres, and simple tones of different pitch will excite different fibres. When a compound musical tone or chord is presented to the ear, [...] the chord must be resolved into its individual compound tones, and the compound tone into its individual harmonic partial tones.[22]

This neuro-physiological model of the inner ear shows itself clearly as an application of the specific energy theory proposed by Johannes Müller.[23] In this respect, hearing appears to Helmholtz much more complex than sight, because the number of nerves with different specificities is much greater. This may be one reason for Helmholtz's deep interest in acoustics.

During the years, Helmholtz considered different neuro-physiological mechanisms for the inner ear: but he never changed his mind about the capacity of the ear to analyse complex sounds into their constituents.

The proposition enunciated and defended by G. S. Ohm must be regarded as proved, viz. that the human ear perceives pendular [harmonic] vibrations alone as simple tones, and resolves all other periodic motions of the air into a series of pendular vibrations, hearing the series of simple tones which correspond with these simple vibrations.[24]

Let us now briefly summarise the development of Helmholtz's arguments. First of all, he observes that beats between simple sounds can have different sensory effects, depending on their frequency.

When a slight difference in pitch [is] produced, the beats are heard at first as long drawn out fluctuations alternately swelling and vanishing. Slow beats of this kind are by no means disagreeable to the ear.[25]

It must be noted that these beats are the ones studied by Sauveur, which are here considered by Helmholtz of minor importance for explaining the ranking of consonances.

But when the interval between the two tones increases to about a Semitone, the number of beats becomes 20 or 30 in a second, and [...] the effect of the collective impression of such a rapid beat is quite different [...] Such rapidly beating tones are jarring and rough.[26]

At this point, after having highlighted roughness as the characteristic sensory experience of dissonance, Helmholtz looks for an answer to the second difficulty of Sauveur's theory: how can roughness be experienced also when a large mistuned interval (e. g., a mistuned octave, or fifth) is presented to the ear?

The beats hitherto considered were produced by two simple tones, without any intervention of upper partial [...] tones. Such beats could only arise when the two given tones made a comparatively small interval with each other. As soon as this interval increased even to a minor Third, the beats became indistinct. Now it is well known that beats can also arise from two tones which make a much greater interval with each other, and we shall see hereafter that these beats play a principal part in setting the consonant intervals of our musical scales.[27]

The answer is then found taking into account also overtones (upper partial tones), and beats between them.

Even when the fundamental tones have such widely different pitches that they cannot produce audible beats, the upper partial tones may beat and make the [resulting complex] tone rough. Thus, if two tones form a fifth (that is, one makes two vibrations in the same time as the other makes three), there is one upper partial in both tones which makes six vibrations in the same time. Now, if the ratio of the pitches of the fundamental tones is exactly as 2 to 3, the two upper partial tones of six vibrations are precisely alike, and do not destroy the harmony of the fundamental tones. But if the ratio is only approximately as 2 to 3, then these two upper partials are not exactly alike, and hence will beat and roughen the tone [...].[28]

The ranking of consonances can now be simply obtained by examining how many upper partial tones of the two sounds are so near to each other to produce beats, and computing the importance and the degree of roughness of these beats. Obviously, beats between upper partials can take place also when the fundamental tones are perfectly in tune (e.g. if they are at a perfect major third interval).

Helmholtz is well aware of the fact that the degree of consonance also depends on the composition of the two complex tones, and therefore on the characteristics of the instruments which play them. But the ranking is a sufficiently general result to be exposed without reference to the composition of the complex sounds.

The following is the order of the consonant intervals, beginning with those distinctly characterised, and then proceeding to those which have their limits somewhat blurred, so to speak, by the weaker beats of the higher upper partial tones:

1. Octave 1:2	2. Twelfth 1:3	3. Fifth 2:3
4. Fourth 3:4	5. Major Sixth ... 3:5	6. Major Third... 4:5
7. Minor Third .. 5:6 [29]		

In order to obtain a more precise ranking of consonance and dissonance, Helmholtz takes into account also the combination tones discovered by Tartini. Of these tones he gives a physico-mathematical explanation based on the non linear response of the ear.[30]

The explanation of consonance given by Helmholtz requires, at this point, only a final passage: a neurophysiological explanation of roughness.

Beats produce intermittent excitement of certain auditory nerve fibres. The reason why such an intermittent excitement acts so much more unpleasantly than an equally strong or even a stronger continuous excitement may be gathered from the analogous action of other human nerves. Any powerful excitement of a nerve deadens its excitability, and consequently renders it less sensitive to fresh irritants. But after the excitement ceases, and the nerve is left to itself, irritability is speedily re-established in a living body by the influence of arterial blood [...] Every one knows how unpleasant and annoying is any flickering light, even it is relatively very weak, coming, for example, from a little flickering taper or rushlight [...] A jarring intermittent tone is for the nerves of hearing what a flickering light is to the nerves of sight, or scratching to the nerves of touch.[31]

Towards the end of the exposition of consonance theory, Helmholtz summarises his findings in this way.

Consonance is a continuous, dissonance an intermittent sensation of tone. Two consonant tones flow on quietly side by side in an undisturbed stream; dissonant tones cut one another up into separate pulses of tone. The description of the distinction at which we have arrived agrees precisely with Euclid's old definition, 'Consonance is the blending of a higher with a lower tone.' Dissonance is incapacity to mix, when two tones cannot blend, but appear rough to the ear.[32]

Such a deep research in the physico-physiological roots of consonance and dissonance could not, obviously, be satisfied with tempered tuning. And, in fact, Helmholtz deals at length, in his *Treatise*, with the different sensory experiences which are produced by the same intervals but with different tunings.

This line of research was perhaps too old fashioned with respect to the musical and harmonic taste of his time: for instance, Helmholtz tried to explain his findings to Johannes Brahms, but without any success.[33]

4. CONSONANCE AND MUSIC

In the third part of his *Treatise*, Helmholtz studies the historical evolution of musical scale and of the laws of harmony. He does this in a very peculiar way, always looking for the physical and physiological constraints to this development, which however are by no means its sole determinants.

When we spoke previously, in the theory of consonance, of agreeable and disagreeable, we referred solely to the immediate impression made on the senses when an isolated combination of sounds strikes the ear, and paid no attention to artistic contrasts and means of expression; we thought only of sensuous pleasure, not of aesthetic beauty. The two must be kept strictly apart, although the first is an important means for attaining the second.[34]

In a very classical perspective, Helmholtz underlines the convergence between scientific research, music theory and composition practice of great masters.[35]

These phenomena of agreeableness of tone, as determined solely by the senses, are of course merely the first step towards the beautiful in music. For the attainment of that higher beauty which appeals to the intellect, harmony and disharmony are only means, although essential and powerful means. In disharmony the auditory nerve feels hurt by the beats of incompatible tones. It longs for the pure efflux of the tones into harmony. It hastens towards that harmony for satisfaction and rest. Thus both harmony and disharmony alternately urge and moderate the flow of tones, while the mind sees in their immaterial motion an image of its own perpetually streaming thoughts and moods.[36]

In a sense, in the *Treatise* we meet a rather unexpected synthesis between advanced scientific research and very classical aesthetic views.[37] However, Helmholtz's classical aesthetic preferences did not imply a closure to new musical perspectives: in his last years, according to Koenigsberger, he devoted himself passionately to the study of piano reductions of Wagner's scores.[38] Nowhere in the *Treatise*, in fact, do we find limitations imposed to aesthetic research.

The system of Scales, Modes and Harmonic Tissues does not rest solely upon inalterable natural laws, but is also, at least partly, the result of aesthetic principles, which have already changed, and will still further change, with the progressive development of humanity.[39]

Helmholtz's stance is never a scientistic one: rather, it is deeply scientific, and in this perspective he longs for a scientific development also for aesthetics. But, after having shown the fundamental rôle of physics and physiology in the historical development of musical systems, he leaves this task to others.

Here I close my work. It appears to me that I have carried it as far as the physiological properties of the sensation of hearing exercise a direct influence on the construction of a musical system, that is, as far as the work especially belongs to natural philosophy. For even if I could not avoid mixing up aesthetic problems with physical, the former were comparatively simple, and the latter much more complicated [...] The real difficulty would lie in the development of the psychical motives which here assert themselves [...] But however alluring such an aim may be, I prefer leaving others to carry out such investigations, in which I should feel myself too much an amateur, while I myself remain on the safe ground of natural philosophy.[40]

5. AN ENTHUSIASTIC SUPPORTER

In 1865, only two years after the publication of the *Treatise*, a young physicist gave a lecture on *The Causes of Harmony*, in which he explained the new and deeply appreciated theory by Helmholtz. At the end of his speech, he triumphantly exclaimed:

Beats are the only sin, the sole evil of music. Consonance is the coalescence of sounds without appreciable beats! This principle is competent to introduce wonderful order and logic into the doctrines of the fundamental bass. The compendiums of the Theory of Harmony which

(Heaven be witness!) have stood hitherto little behind the cookbooks in subtlety of logic, are rendered extraordinarily clear and simple.[41]

The young scientist was Ernst Mach. In order to reach this conclusion, he started with a psycho-physiological argument: how can we convince ourselves that complex sounds are made of simple sounds? Since Sauveur to Helmholtz, this problem had been carefully studied, and Helmholtz himself had given the answer adopted by Mach: the identification of a complex sounds as a unitary object is not to be considered as a pure sensation, but an intellectual construction (a perception) due to habit.

We have, as it were, to distinguish between the material ear of the body and the spiritual ear of the mind [...] As you are aware, no perceptions obtained by the senses are merely sensations impressed on our nervous systems. A peculiar intellectual activity is required to pass from a nervous sensation to the conception of an external object, which the sensation has aroused [...] Now it is an universal law of the perceptions obtained through the senses that we pay only so much attention to the sensations actually experienced as is sufficient for us to recognise external objects [...] Hence my former distinction of a material and a spiritual ear.[42]

A deep attention is then required in order to be able to decompose sounds into their constituents, but this reversal of habit is indeed possible. "[...] a certain amount of undisturbed concentration is always necessary for analysing musical tones by ear alone."[43] Attention, of course, can be refined by practice. Mach, in his lecture, gives a number of examples in which the pianoforte is used as an help: first he shows how to decompose chords into their constituents, and then how to decompose complex sounds.

If I strike a harmony, or chord, on the piano, by a mere effort of attention you can fix every tone of that harmony [...] With a prolonged exercise you can go further, becoming able to decompose into a series of sounds a musical sound which previously was supposed to be simple.[44]

In this way, the importance of harmonic overtones is proved. Then Mach devotes himself to the demonstration of the existence and of the importance of beats. Also in this case, he suggests simple and convincing experiments to be done by means of two metronomes. But dissonance depends also on beats between upper partial tones: to explain this fact, Mach employs a visual model.

To make this perfectly clear to you, I have constructed a model [...] It represents a claviatur [keyboard]. At its top a movable strip of wood with the marks 1,2,...,6 [corresponding to the position of harmonic overtones] is placed. By setting the strip in any position, for example in that where the mark 1 is over the note c of the claviatur, the marks 2,3,...,6, as you see, stand over the overtones of c. The same happens when the strip is placed in any other position. A second, exactly similar strip possesses the same properties [...] Moving one of the two strips along the other is equivalent to a departure from unison. All the overtones of the one sound now fall alongside those of the other; beats are at once produced; the combination of the tones becomes unpleasant: we obtain a dissonance [...] Only in a few, quite definite positions do the overtones partially coincide. Such positions, therefore, signify higher degrees of euphony – they point out the consonant intervals.[45]

With the help of this model, he is able to give a visual idea of the ranking of consonances. For instance, it is visually clear that a perfect fifth is much more consonant than a minor second (because in the first case the importance of beats between overtones is much lower). After a number of examples, he can say:

[...] The beauty of the theory is that it bears upon its face the stamp of the truth. It is no phantom of the brain. Every musician can hear for himself the beats which the overtones of his musical sounds produce [...] This is the answer which Helmholtz gave to Pythagoras [...].[46]

6. A CRITICAL REVISION

Twenty years after, Mach developed a deep critical revision of Helmholtz's beat theory that he had so enthusiastically accepted in the past. This revision was made in the perspective of harmony theory.

The fact that a musician will never confound a more perfectly consonant chord on a poorly tuned piano with a less perfectly consonant chord on a well tuned piano, although the roughnesses in the two cases may be the same, is a sufficient indication that the degree of roughness is not the only characteristic of Harmony [...] The positive physiologico-psychological characteristics which distinguishes one harmony from another is not given by beats.[47]

And in *The Analysis of Sensations* he further developed this criticism.

The theory of Helmholtz as to tone sensations seemed, when it was first promulgated, to be an admirably complete and classical achievement. Yet, on a fundamental examination, it has not been able to stand against criticism [...] In thinking that a task, which provides ample work for psychologists and physiologists as well as for physicists, could be mastered, in all its main features, from physical points of view, Helmholtz was doubtless under a delusion.[48]

At this time, Mach denies that a theory of musical consonance can be based on beats. This is not a rejection of the absolute validity of the beat theory. On the contrary, Mach admits that in part the agreeableness of a superposition of musical sounds can be explained by means of this theory. But he questions the musical relevance of this physico-physiological approach. He points then to the distinction between psycho-acoustical and harmonic consonance.

Every third, every fourth, every major or minor triad has its characteristic colour, by which it is recognised independently of the pitch of the fundamental and independently of the number of beats [...].[49]

How is it possible to explain this feature, which, according to Mach himself, is "the result of a purely physical and intellectual analysis, and has nothing to do with sensations?"[50] Mach gives no clear cut response, and rather looks for the weak points of Helmholtz's theory: in particular, he criticises his resonance theory, and questions the very production of beats inside the ear. In so doing, he puts into evidence a high-level property of harmony which cannot be reduced neither to physics nor to physiology.

7. THE COMPLEXITY OF PERCEPTION

In the same years, Helmholtz was deeply criticised also by Carl Stumpf, a psychologist and a musicologist who was also a former pupil of Franz Brentano. He stressed the autonomy of psychology from physiology and from physics, and the irreducibility of musical experience to objective measurements. In particular, he proposed a new version of an old conception of consonance: i.e., the definition given by Euclid of consonance as fusion of sounds.[51] But also Helmholtz considered this definition as very near his own: we must then try to understand the deep reasons for this criticism.

Still more radical than Stumpf, the musicologist Hugo Riemann stated that consonance has a musical meaning only if it is studied inside the context of harmony. In his view, musical listening must be considered as a very complex activity of mind.

As we can see, the positions of Ernst Mach previously quoted are not far from these, and are probably inspired by them. We could say that the acknowledgement of the complexity of musical world is the common ground on which these authors rejected the physico-physiological theory by Helmholtz. In other words, they all stressed the creative rôle of perception, which in a few years was to be fully recognised by Gestalt psychology.

Significantly, Ernst Mach was the first to stress the *Tongestalt* (i.e., the 'tonal figure') as the property of a melody which is conserved after transposition.[52] This global approach to music as a complex psychic experience was further developed by Christian von Ehrenfels and Alexius Meinong.[53] Also the influence of Carl Stumpf on the rise of Gestalt theory from music psychology was very important. Wertheimer, Koffka and Köhler were all his pupils, and concepts as *Intervallfarbe* ('colour' of a musical interval) and *Vokalität* (global character of vowels, which cannot be reduced to their frequency composition) are well in the mainstream of Gestalt psychology, and in sharp opposition to the physical-physiological theory of Helmholtz: 'a sort of illusion', according to Köhler, nothing more than a 'summative theory.'[54]

It must be said, however, that some hints to this complexity of music and music perception can be found also in the work of Helmholtz.[55] But they are not so fully developed as the physico-physiological arguments, and it can be understood that the image of Helmholtz among music psychologists was, *malgré lui*, that of a reductionist.

In 20th century, a lot of research on these themes has been carried on. And all this research points to the creative rôle of sound perception and music perception, which builds its objects, its streams, its fusions and its segregations. The paradox of the musical world is that it is built on acoustic perception, but can shape, on its turn, this same perception.[56] Harmony, in particular, although being based on physical sounds, appears to be a world in itself, where consonance cannot be reduced neither to absence of beats nor to exact coincidence of overtones.

For instance, inside harmony the dissonance of a chord can assume much more the meaning of a tension which demands resolution than the meaning of a non agreeable sensation. Or else, a dominant chord needs to be resolved into a root chord, and this need is much stronger than any sense of agreeableness or disagreeableness.

Moreover, as Mach himself noted, harmonic consonance can be observed also between mistuned musical notes. In this higher level world of meanings, the mistuning of a chord can well become a second order effect. In other words, inside a musical context consonance must be considered a function, more than a property, of dyads and triads.[57]

But, at the end of 20th century, also the image of a definite and forever fixed harmonic world has been the object of a deep critical assessment. In a sense, only now, after the dissolution of traditional harmonic theory, is it possible to appreciate fully the most important contribution by Helmholtz to acoustics, which is the deep study of timbre of music sounds. In other words, Helmholtz's work can be revisited in terms of *Tonfarben*, of – so to say – 'colours' of sounds.

To Mach and to *Gestalt* psychology we must acknowledge the merit of stressing the impossibility to reduce to sensations the world of music. But this is only one of the first steps in a long history of research in progress, which can be subsumed under the heading of *Klangfarbenmelodie* (timbre melody) that great intuition by Schoenberg[58] which only at this new turn of century is being exploited, through a rediscovery of Helmholtz's seminal work on timbre.[59]

Gianni Zanarini
Università di Bologna
Dipartimento di Fisica
Via Irnerio, 46
I-40126 Bologna
Email: zanarini@df.unibo.it

NOTES

[1] Cf. Koenisberger 1902, 157.
[2] See for instance Bevilacqua 1993, 291–333.
[3] Cf. Cahan 1993.
[4] Cf. Helmholtz 1863.
[5] Cf. Helmholtz 1863, vii.
[6] Cf. Cahan 1993, 2.
[7] Cf. Helmholtz 1863, 1.
[8] Cf. Koenigsberger 1902, 49.
[9] Cf. Helmholtz 1857, 47.
[10] See for instance Cohen 1984.
[11] Cf. Helmholtz 1857, 47.
[12] For a history of consonance and dissonance, see for instance Cohen 1984. See also Gozza 1989.
[13] Cf. Cazden 1980, 127.
[14] Cf. Tenney 1988, 9.
[15] See for instance Cohen 1984.
[16] For simplicity, the intervals are given here their modern names.
[17] Cf. Galilei 1638, 672.
[18] Cf. Fontenelle 1716, 143.
[19] Cf. Beyer 1999.
[20] Cf. Helmholtz 1863, 34.
[21] Cf. Turner 1977.
[22] Cf. Helmholtz 1863, 143–148.
[23] Cf. Hatfield 1993, 12. On Müller's specific energy theory, see also the paper by S. D'Agostino in this volume.
[24] Cf. Helmholtz 1863, 56.
[25] Cf. Helmholtz 1863, 167.
[26] Cf. Helmholtz 1863, 168.
[27] Cf. Helmholtz 1863, 179.
[28] Cf. Helmholtz 1857, 72.
[29] Cf. Helmholtz 1863, 183.
[30] Cf. Helmholtz 1863, 197 ff.
[31] Cf. Helmholtz 1863, 169–170.
[32] Cf. Helmholtz 1863, 226.
[33] See for instance Serravezza 1996, 80.
[34] Cf. Helmholtz 1863, 234.
[35] Cf. Serravezza 1990, 351.
[36] Cf. Helmholtz 1857, 75.
[37] Cf. Martinelli 1996, 527.
[38] Cf. Koenigsberger 1902, 410.
[39] Cf. Helmholtz 1863, 235.
[40] Cf. Helmholtz 1863, 371.
[41] Cf. Mach 1896, 44.
[42] Cf. Helmholtz 1857, 64–67.
[43] Cf. Helmholtz 1863, 51.

[44] Cf. Mach 1896, 36–37.
[45] Cf. Mach 1896, 42–43.
[46] Cf. Mach 1896, 44–45.
[47] Cf. Mach 1896, 384–385.
[48] Cf. Mach 1886, 305–306.
[49] Cf. Mach 1886, 287.
[50] Cf. Mach 1886, 290.
[51] For the criticisms of Stumpf to Helmholtz, see Serravezza 1996, 45 ff.
[52] Cf. Mach 1886.
[53] For a thorough discussion of psychology of music as the origin of Gestalt Theory see Martinelli 1998.
[54] Cf. Martinelli 1998, 106.
[55] See for instance Helmholtz 1863, 213–214.
[56] Cf. Leman 1996, 4.
[57] Cf. Cazden 1980, 124.
[58] Cf. Schoenberg 1911.
[59] See for instance Barrière 1991.

REFERENCES

Barrière, J.B. ed.: 1991, *Le timbre, métaphore pour la composition*. Paris, Bourgois-Ircam.

Bevilacqua, F: 1993, 'Helmholtz's "Über die Erhaltung der Kraft": The Emergence of a Theoretical Physicist,' in Cahan, ed., 1993a, 291-333.

Beyer, R.T.: 1999, *Sounds of Our Time*. New York, Springer Verlag.

Bregman, A.: 1994, *Auditory Scene Analysis*. Cambridge, Mass., MIT Press.

Cahan, D.: 1993, 'Helmholtz at the Borders of Science,' in Cahan, ed., 1993a, 1-16.

Cahan D. ed.: 1993a, *Hermann von Helmholtz and the Foundations of Nineteenth-Century Science*. Berkeley, University of California Press.

Cahan D. ed.: 1993b, *Letters of Hermann von Helmholtz to his Parents: The Medical Education of a German Scientist (1837–1846)*. Stuttgart, Franz Steiner Verlag.

Cahan D. ed.: 1995, *Hermann v. Helmholtz, Science and Culture: Popular and Philosophical Essays*. Chicago, University of Chicago Press.

Cazden, N.: 1980, 'The Definition of Consonance and Dissonance,' *International Review of Aesthetics and Sociology of Music*, **11**, 123-168.

Cohen, A.: 1984, *Quantifying Music*. Reidel, Dordrecht.

Fontenelle, B.: 1716, 'Sur la determination d'un son fixe,' in *Histoire de l'Académie Royale des Sciences*, **143**, Paris, Académie des Sciences.

Galilei, G.: 1638, *Discorsi intorno a due nuove scienze*. Roma, Edizione Nazionale.

Gozza, P. ed.: 1989, *La musica nella rivoluzione scientifica del seicento*. Bologna, Il Mulino.

Hatfield, G.: 1993, 'Helmholtz and Classicism,' in Cahan, ed., 1993a, 522-558.

Helmholtz, H.: 1857, 'Über die physiologischen Ursachen der musikalischen Harmonie,' En. tr. in Cahan, ed., 1995, 47-75.

Helmholtz, H.: 1877, *Die Lehre von den Tonempfindungen als physiologische Grundlage für die Theorie der Musik*, Brauschweig, Vieweg. En. tr. 1954, *On the Sensations of Tone as a Physiological Basis for the Theory of Music*, New York, Dover.

Koenigsberger, L.: 1902, *Hermann von Helmholtz*. Braunschweig, Vieweg. En. tr. 1965, New York, Dover.

Leman, M.: 1996, *Music and Schema Theory*. Heidelberg, Springer Verlag.

Mach, E.: 1886, *Die Analyse der Empfindungen und das Verhältnis der Physischen zum Psychischen*. Jena, Fischer. En. tr. 1952, *Analysis of Sensations*, New York, Dover.
Mach, E.: 1896, *Populär-wissenschaftliche Vorlesungen*. Leipzig, Barth. En. tr. 1943, *Popular Scientific Lectures* , Open Court, La Salle.
Martinelli, R.: 1996, 'Fantasia musicale e scienza dei suoni,' *Intersezioni*, **16**, 517–529.
Martinelli, R.: 1998, 'Musica e teoria della Gestalt,' *Il Saggiatore Musicale*, **VI**, 1, 93–110.
Serravezza, A.: 1990, 'Helmholtz, Stumpf, Riemann. Un itinerario,' *Rivista Italiana di Musicologia*, **24**, 347–422.
Serravezza, A.: 1996, *Musica e scienza nell'età del positivismo*. Bologna, Il Mulino.
Tenney, J.: 1988, *A History of Consonance and Dissonance*. New York, Excelsior.
Turner, R.S.: 1977, 'The Ohm-Seebeck Dispute, Hermann von Helmholtz, and the Origin of Physiological Acoustics,' *British Journal for the History of Science*, **10**, 1–24.

SALVO D'AGOSTINO

THE BILD-CONCEPTION OF PHYSICAL THEORIES FROM HELMHOLTZ TO HERTZ

1. INTRODUCTION

This study will be restricted to a consideration of some aspects of the conceptual history of theoretical physics, as they appeared in the scientific literature in the decades around the turn of the century. The span of time under examination corresponds to the rise and the rapid growth of theoretical physics in Europe, approximately between 1870 and 1930.[1]

Of course, modern science was always divided into theory and experiments,[2] and that division was not unique to physics. However it must be acknowledged that it was only in physics that theoretical work developed into a major teaching and research speciality in its own right.[3]

If it is true that theoretical physics (TP) was mainly a creation of German physics, where it received full institutional recognition, it is also undeniable that other European countries had an important part in its creation.[4] From the beginning of the nineteenth century, in France, England and other European countries, mathematical physicists[5] prepared the analytical tools and, in part, the approaches for the new discipline. A fruitful school of mathematical physicists flourished in Italy around the turn of the century. Consequently, one can speak of TP, as a typically European discipline. At the turn of the century, the methods and conceptions of TP spread all over Europe and the world.

It is not an easy matter to define the nature and scope of TP and of its adept, the theoretical physicist. Ludwig Boltzmann, one of its major contributors, confessed in 1895 that, even the formulation of this concept [of a theoretical physicist] is not entirely without difficulty.[6] The fact is that TP is still a major part of today's physics and we are still immersed in its ways of approaching physical knowledge.[7]

It is well known that the development of TP implied new theories – relativity and quantum mechanics were major examples – and an expanded experimental activity in many European laboratories. However, the fact that

L. Albertazzi (ed.), The Dawn of Cognitive Science. Early European Contributors, 151–166.

this expanded activity was somehow related to the new form of theories and to a modified conception of the theory-experiment relationship, this fact has been scarcely noticed in the scientific literature.

According to Ernst Cassirer[8] a crisis in fundamentals characterised much of the second half of nineteenth century physics and still affects modern research. In fact, since Galileo's and Newton's times, theories were widely differing in content, but "their ontological significance was never seriously challenged."[9] It was this challenge at the end of the century which brought on the crisis, causing "the demand for *reflective criticism* in the natural sciences to become urgent with ever mounting emphasis."[10] (Italics mine).

The naive descriptive conception of physics, i.e., the idea that physical concepts and theories refer to entities existing in nature (i.e., refer to their ontological status) was first rejected by the great Helmholtz by denying that physics concepts are just images of things (see the following pages). The same descriptive conception was deeply criticised by such physicists as Hertz and Boltzmann.

Maxwell and Helmholtz were perhaps the first nineteenth century physicists to realise that the Kantian criticism had to be accounted for in meeting the problems inherent in the new form of physics. In his rectorial address at the Berlin University, *The Facts In Perception,* Helmholtz explicitly stated that the same philosophical problem was common to science and philosophy, although the two disciplines encountered it proceeding from two opposite directions.

What is true in our intuition and thought? In what sense do our representations correspond to actuality? Philosophy and natural science encounter this problem from two opposite sides, it is a task common to both.[11]

Hertz and Boltzmann adopted the Kantian term *Bild* to designate the new conception of theory, not a faithful image of nature but an intellectual construct whose relation to the phenomena was to be analysed in terms not very different from those pertaining to the philosophical tradition. Einstein and Heisenberg used the same term to designate their conception of theory. Schrödinger referred to the *Bild*-conception as an established tradition in the history of TP.

2. HELMHOLTZ'S SECULARISATION OF KANT'S *ANSCHAULICHKEIT*

It has been said of Helmholtz that, through his contributions, physics was reborn from the bosom of physiology.[12] In fact, his optical research and his theory of vision confirm this statement: one of the most relevant antefacts of Helmholtz's physiological researches and, in general, of his epistemology, was Johannes Müller's *theory of the specific energy of sensory nerves*:

The stimulation of the optic nerve produces only sensations of light whether that stimulation be caused by objective light (vibrations in the ether), by an electrical current through the eye, by pressure on the eyeball, or by rapid directional changes of the eye.[13]

This theory suited Helmholtz's optical researches. In 1878, starting from the premises above, Helmholtz denied that the eye, and our senses in general, are passive receptors of a supposedly faithful image of the world. His problem was to discriminate those stimulations which are caused by objective light (vibrations in the ether) from the various stimulations which produce sensations of light. However, due to *the specific energy of sensory nerves,* even those stimulations that are caused by objective light, and which thus represent the report of what is peculiar to the external influence, are not the image of this influence but its symbolic representation. This report is related to the external influence as a symbol is related to the symbolised object.

Here is the whole passage:

Inasmuch as the quality of our sensation gives us a report of what is peculiar to the external influence by which it is excited, it may count as a symbol of it, but not as an image. For from an image one requires some kind of alikeness with the object of which it is an image.[14]

In the case of vision, the only criteria which allows to connect in the eye the external influence to the subjective 'lux' is the regularity by which concepts, such as that of vibrations in the ether, are regularly related in permanent and determinable laws and the various sensations of lux [representations] are correspondingly regularly related in the eye. This parallelism of regularities [parallelism of laws] holds in general:

Every law of nature asserts that upon preconditions alike in a certain respect, there always follow consequences which are alike in a certain other respect. Since like thinks are indicated in our world of sensations by like signs, an equally regular sequence will also correspond in the domain of our sensations to the sequence of like effects by law of nature upon like causes.[15]

Reality is thus reached only through the help of scientific laws. But perceptions are given to us in their intuitive immediacy while concepts are grasped only by the intellect.[16]

It is known that Hermann Helmholtz devoted a remarkable part of his contributions to an analysis of the Kantian problem of the a-priori in physics. Kant did not exercise his influence on Helmholtz through the mediation of the romantic philosophy of Shelling and Schopenhauer, less then ever through the Hegelian *Naturphilosophie*, which Helmholtz vehemently rejected. Helmholtz met Kant right in the middle of his physiological researches[17] and he originally contributed to a refinement of the role of the a-priori in science.

His researches on the physiology of vision led Helmholtz' interests towards the general epistemological problems of perception and to Kant's doctrine of forms of intuiting and thinking given prior to all experience:

> Johannes Müller's investigations into the physiology of the senses [...] summarised in his law *of specific energy of sensory nerves* has now brought the fullest confirmation [to Kant's doctrine]; one can almost say to an unexpected degree.[18]

However, according to Helmholtz, this confirmation somehow modified Kant's positions, by adding a distinction which was foreign to Kant. Helmholtz resolved the Kantian a-priori forms of intuition into two species: a *general form* and a *narrower specification*. According to him, what was truly a-priori in every sensory perception is a *general form devoid of any content*, as exemplified in the spatial perception of place, or, in the visual perception of an aggregate of coloured surfaces, which always compose our visual field.

Quite differently, the *narrower specification* concerns the various types of spaces described by the axioms of geometry or, in visual perception, it's content itself, the "particular colours which appear on this or that occasion, their arrangement or sequence."

In an attempt to better explain how the resolution above operates, he thus exemplified its inhering in optical and in spacial perceptions:

> Everything our eye sees, it is an aggregate of coloured surfaces in the visual field – that is our [general] form of intuition. The particular colours which appear on this or that occasion – their arrangement and sequence – this is the result of external influences and is not determined by any law of our makeup.
>
> Similarly, from the fact that space is a form of intuiting, nothing whatever follows about the facts expressed by the axioms. If such propositions are taken to be non empirical ones, but to belong instead to the necessary form of intuition, then this is a further particular [narrower] specification of the general form of space; and *those grounds which allowed the conclusion that the form of intuition of space is transcendental, do not necessarily for that reason already suffice to prove, at the same time, that the axioms too are of transcendental origin.*[19] [Italics and parentheses, S.D.]

Applying to a study of sensations his distinction of the Kantian a-priori forms, Helmholtz found that among the various sensations there exist differences in form of two neatly distinguished types: one form concerned sensations of different type such as those belonging to different senses, like sight and sound; he called this a *difference in the modality (of sensations).*[20] The other form characterises sensations of the same type, and he called it a *difference in quality*.

The *modality* form belonged to the *general form* of intuitions which, by transcending the narrower specifications, is the truly Kantian a-priori and is thus compatible with various systems of axioms. For instance, optical modalities are distinguished from the acoustical ones by their characteristic features and thus their modality belongs to the *general form* of intuitions. A truly a-priori form "must be devoid of content and free to an extent sufficient for adsorbing any content whatsoever that can enter the relevant form of perception."[21]

Quite differently, the spatial determinations of geometry which characterise an Euclidean geometry, as distinguished from a non-Euclidean, concern just differences in quality among the same spatial perceptions and not their more fundamental differences in modality.

According to Helmholtz, the specification that Kant introduced into our spatial intuition (i.e., it's Euclidean three-dimensionality) contradicted the *general form*, because it "limits the form of intuition of space in such a way that it can no longer absorb every thinkable concept, if geometry is at all supposed to be applicable to the actual world."

At difference with Kant, the axioms of Euclidean geometry represent the intuition's *narrower specification* which is not a-priori, because this form of intuition limits our perception of space to its three-dimensional formulation and is too narrow (too full of a particular content)[22] for representing all possible contents of our experience. For this reason, the space three-dimensionality can be adopted to represent the empirical situation.

Helmholtz thought that through his distinctions above he had advanced Kant on his own ground, here Kant was not critical enough in his critique.[23]

He considered his contribution to be the result of his own physiological investigations, and, as such, an important advance with respect to Kant, because:

> the processes related to the modality features of perceptions, had to remain still unformulable in words, and unknown and inaccessible to philosophy, as long as [Kant] investigated only cognitions finding their expression in language.[24]

It can be argued that, in his approach to the study of cognitions as distinguished from a merely linguistic approach, he had introduced a new theme in the Kantian philosophy: by considering that the processes above are accessible to science only via an operational approach, Helmholtz opened a new field of inquiry which is still fruitfully continued in our modern psychophysical research.

Helmholtz was convinced that in his research he always studied the correlation between the presumed external agent and the observable effects on our sensation (perceptions) with the aim of describing these effects. In his view, an *ideal description* should thus achieve a perfect adequacy between concepts and the corresponding perceptions.[25] At the close of his 1878 lecture (his rectorial Address), he emphasised that his favourite approach, had no less than Goethe[26] and Kirchhoff as its supporters. He also confirmed Kirchhoff's statement that "the task of the most abstract amongst the natural sciences, namely mechanics [is] to describe completely and in the simplest manner the motions occurring in nature."[27]

It is clear that the term 'description' in Helmholtz's and Kirchhoff's usage merely referred to the phenomena and not to the presumed underlying agents of our perceptions. They are usually considered the representatives of the German phenomenological school in physics.

It can be also argued that by limiting the theory's scope merely to a description of phenomena, Helmholtz's phenomenologism somehow diminished the cognitive import of the Kantian transcendental forms of intuition and of the Kantian *Kategorien*. He thus contributed, so-to-speak, to their 'secularisation.'

In their perspicuous introduction to the Hertz-Schlick translation of Helmholtz's papers, R. S. Cohen and J. Helkana referring especially to Helmholtz's 1886 "magisterial paper," consider his thought on the a-priori as situated on the:

> watershed between a regulative principle [in the Kantian sense] and an heuristic principle [as Hertz saw it], i.e., a guideline in the effort to formulate laws for the new classes of phenomena, destined either to show its power or to fade away.[28]

M. Schlick expressed his agreement on the point that Helmholtz supported a physiological and psychological interpretation[29] of those form of intuition which Kant had credited to a transcendental aspect of knowledge.

As regards the historical perspective, Helmholtz's thought stands on the watershed between two great traditions in the development of modern physics: on one hand, through the attention given to perceptions *per se* and psychology, he opened a way to Mach's phenomenology and, perhaps, to Copenhagenism in Quantum Mechanics, but, on the other hand, through his special upholding of Kant's a-priorism, he favoured Hertz's and Einstein's philosophies, thus contributing to the rise and development of TP.[30]

3. HERTZ REINSTATES A KANTIAN ORTODOXY AND THE *BILD*-CONCEPTION OF PHYSICAL THEORY

It has been said that the intellectual heir of the Kantian philosophy of Helmholtz was Hertz. Of course, some aspects of agreement and continuity between theories and conceptions of Helmholtz and those of his favourite pupil Hertz can be certainly found. However, the historian also finds many elements of dissent between their epistemologies.

To begin with, let us explore Hertz's *conception of a binarial aspect of physical theory.* In his Introduction to *Electric Waves*, Hertz examined different modes of representation of Maxwell's electromagnetic theory, i.e., the representation of the theory in Maxwell's own work, its representation as a limiting case of Helmholtz's theory, and its representation in his own work.

He concluded:

the representation of the theory in Maxwell's own work, its representation as a limiting case of Helmholtz's theory, and its representation in the present dissertation – however different in form – have substantially the same inner significance. This common significance of the different modes of representation [and others can be certainly be found] appears to me to be the undying part of Maxwell's work. This, and not Maxwell's peculiar conceptions or methods, would I designate as 'Maxwell's theory.' To the question, What is Maxwell's theory? I know of no shorter nor more definite answer than the following: Maxwell's theory is Maxwell's system of equations.[31]

In this often quoted statement, Maxwell's theory is identified with his system of equations. However, the system of equations is also qualified as the common and inner significance of a plurality of modes of representation of the same theory. This second qualification was very often neglected by the past commentators.

The statement that "Maxwell's theory is Maxwell's system of equations," does not preclude that "a mode of its representation, i.e., Maxwell's peculiar conceptions or methods," is to be taken into account in a more general acceptation of the meaning of physical theory (henceforth I will write a capitalised *Theory* in order to designate this more general acception of a physical theory). Here is, in its entirety, the continuation of Hertz's passage:

Maxwell's theory is Maxwell's system of equations. Every theory which leads to the same system of equations, and therefore comprises the same possible phenomena, I would consider to be a form or special case of Maxwell's theory [Theory]. Hence in this sense and in this sense only, the two theoretical dissertations in the present volume can be regarded as representation of Maxwell's theory. In no sense can they [i.e., the two theoretical dissertations in the present volume] claim to be a precise rendering of Maxwell's ideas. On the contrary, it is doubtful whether Maxwell, were he alive, would acknowledge them as representing his own views in all respect.[32] [Parentheses are mine.]

Thereafter Hertz affirms his binarial conception of Theory. Any Theory comprises two aspects: a mode of representation and the fundamental equations. Different modes of representation of the same Theory may have in common the same equations which thus symbolise their common significance.

The common significance of the different modes of representation (and others can certainly be found) appears to me to be the undying part of Maxwell's work.[33]

Notice that only the fundamental equations are undying, i.e., possess a perennial validity and not the modes of representation; the latter are historically mutable conceptions, as is evident from Hertz's above exemplification.

In Hertz's view, the equations are to be accepted as the minimum denominator of the various modes of representation, i.e., of the various interpretations of the theory's physical content. The various modes of representation have in common the equations, i. e., the mathematical structure of the Theory. Hertz thus promotes Mathematics, the undying part of the theory, to a highly significant role in physical theory. The mathematical side of the theory, similarly to Kant's '*Kategorien*,' is certain and perennial but purely formal i.e., empty of a physical content; this is the price mathematics has to pay in order to live its perennial life in the sky of the pure forms.

Notice that by introducing into physical theories the binarial aspect of modes of representation and of mathematics, the traditional conception of theory as a indistinguishable unity of a mathematical apparatus and a physical content was deeply modified.

I argue that Hertz experienced the necessity of this separation between the mathematical form and the physical content of theory since his early approach to Maxwell's in his 1884 theoretical paper. The mathematical form[34] of Maxwell's equations which Hertz derived in 1844 from the physically empty[35] conception of an independent existence of forces in space, represented for Hertz a pure logical form[36] conforming to the requirement of a logical permissibility (*Zulässigkeit*), still to be filled up with the physical content that was later presented to him by a reversal of Helmholtz's polarisation theory.[37]

As a further analysis of the Hertzian ideas, let us now explore how they developed in the concluding work of his scientific life,[38] *Prinzipien der Mechanik in neuen Zusammenhänge Dargestellt* (In the following respectively referred to as: *Prinzipien, The Principles*).

We have seen that, in Hertz's thought, representations play the role of intermediaries between mathematical symbols and the empirical content of theories, i.e., they correlate mathematical symbols to observables. However, we have seen that the correlation is not univocal, i.e., is not univocally determined either on the side of the equations, or on the observations' side. There is a certain amount of freedom in connecting the two, due to the possibility of different modes of representation. This is a feature of Hertz's theorising that contradicts Helmholtz's parallelism of regularities above.

Other contradictory features between the two German scientists' epistemologies are to be found in Hertz's *Prinzipien der Mechanik.* In his Introduction, Hertz criticised Energetism, i.e., the Energetist representation of physics that he considered somewhat related to the Helmholtzian principle of energy conservation.[39] In his view, the Energetist representation of physics had defects such as lack of exactness and lack of logical permissivity, which, in the end, suggested to Hertz its rejection as a general conception of physical theory.[40]

He adopted instead a mass-time-space representation of Mechanics, including hidden masses and rigid constraints as to their connections, as substitutes for the absence of forces which were not included among the theory's primitives.

I think that Hertz's rejection of Energetism has to be justified by more fundamental reasons than the defects listed in the passage above. At bottom, every defect in a theory might be adjustable,[41] but I think that there existed more fundamental reasons for Hertz's rejection of Energetism and for his preference for a mass-time-space representation. One reason was that this last representation was consistent with his model-like *Bild*-conception of theory because it offered a conceptual model, that was absent in Energetism.[42] At bottom, Hertz conception of theory contrasted with Helmholtz's phenomenology on the matter of role of theoretical representations (*Bilder*).

In fact, it is interesting to notice that in a passage[43] almost totally ignored by the historical literature, Helmholtz claimed that the presence in Hertz's conception of this model of a 'mechanical representation,' was the point of distinction between his and Hertz's conceptions of theory: in his *Vorwort* to Hertz's *Prinzipien der Mechanik,* Helmholtz remarked that Hertz, his favourite student, adopted 'mechanical representations' instead of the "simple representation of physical facts by systems of differential equations," which Helmholtz privileged. For this reason, in the same passage, he ranked Hertz with Kelvin and Maxwell as representative of the school of mechanical explanations:

> English physicists – e.g., Lord Kelvin in his theory of vortex atoms, and Maxwell, in his hypothesis of systems of cells with rotating contents, on which he bases his attempt at a mechanical explanation of electromagnetic processes – have evidently derived a fuller satisfaction from such explanations than from the simple representation of physical facts and laws in the most general form, as given in systems of differential equations. For my own part, I must admit that I have adhered to the latter mode of representation [i.e., system of differential equations] and I have felt safer in so doing; yet I have no essential objections to raise against a method which has been adopted by three physicists of such eminence.[44]

The mechanical explanations Helmholtz credited to Hertz, as the characterising feature of his philosophy, are to be related to Hertz's *Bild* conception of theory as a theoretical model, a representation, where concepts do not necessarily correspond to observables because hidden quantities are included within the theory.

4. *BILD*- CONCEPTION AND HIDDEN MASSES

The fact is that, according to Hertz, Helmholtz's requirement above of a parallelism between concepts and perceptions does not unambiguously determine the choice of an appropriate Theory. A *multiplicity of representations* is thus consistent with the Helmholtzian requirement:

> The images [*Bilder*] which we form of things are not determined without ambiguity by the requirement that the consequents of the images must be the images of the consequents. Various images of the same objects are possible, and these images may differ in various aspects.[45]

For Hertz, theories are not completely determined on the empirical level i.e., various images may correspond to one perception (object). Reciprocally, there exist concepts to which no perception corresponds (Hidden quantities).[46] The related problem has been named[47] the underdeterminationism problem. It can be affirmed that underdeterminationism (UD) represented an important element of distinction between Helmholtz's and Hertz's epistemologies, for Hertz takes it as the basis for the further development of internal (non-empirical) criteria of validation in theory choice.

In Hertz's system hidden quantities, such as Ether, have a fundamental role, because Helmholtz's parallelism is not only underdeterminate, but also impossible in general, if theory is to be limited to observable quantities.[48]

In order to eliminate UD, Hertz introduced other requirements with the role of new selective criteria for the choice of an appropriate theory. These requirements, that he discussed at length in his Introduction to *Prinzipien*, are then to be considered higher level formal criteria, consistent with Hertz's ideal of an adequate theory. Simplicity[49] is one and the most relevant requirement[50] for the choice of a representation among various representations. But one can require simplicity in theories in as much as theories are *Bilder*, i.e., they concern "images of our own creation not nature:"

> It is true we cannot a priori demand from nature simplicity, nor can we judge what in her opinion is simple. But with regard to images of our own creation we can lay down requirements. We are justified in deciding that if our images are well adapted to the things, the actual relations of the things must be represented by simple relations between the images.[51]

Simplicity consists in the requirement of a minimum number of axioms compatible with the deduction of physical laws. In his Prefatory Note to Book 1 of *Prinzipien*, Hertz showed how, by shaping his theory through the requirements above, he was inspired by the Kantian conception of a-priori judgements:

> The subject matter of the first book is completely independent of experience. *All the assertions made are a-priori judgements in Kant's sense.* They are based upon the laws of internal intuition of, and upon the logical forms followed by, the person who makes the assertions; with his external experience they have no other connection than these intuitions and forms may have.[52] [My Italics.]

He continued, in an orthodox Kantian fashion:

> The time of the first book [of *Prinzipien*] is the time of our internal intuition [...] in itself it is always an independent variable. The space of the first book is space as we conceive it. It is therefore the space of Euclid's geometry, with all the properties which this geometry ascribes it.[53] [My Italics.]

Let us notice that, in the passage above, Hertz limited our conception of space to the Euclidean geometry. In so doing, he forgot Helmholtz' statement that the Euclidean system of axioms pertains to the qualitative form of our perception of space, which is not truly a-priori in Helmholtz's view. One can thus argue that Hertz reinstated a Kantian orthodoxy at the cost of neglecting or even contradicting his master's criticism of the Kantian a-priori. However, one should also admit that Hertz's orthodoxy was a necessary component of his logic of research which led him to the discovery of electric waves.

In fact, the strength of the axiomatic apparatus in guiding his experiments is significantly exemplified by his discovery of electromagnetic waves. He was primarily guided by the principle of *uniqueness and independent existence of electric forces.*[54]

This principle was also consistent with his assumption of contiguous action. In as much as waves of electric force could also be explained by Helmholtz's theory of a dielectric polarisation acting at a close distance, experience alone was unable to suggest and support Hertz's conception of contiguous action. A striking confirmation thereof is the fact that, more than ten years after Hertz's celebrated experiment, Boltzmann still considered contiguous action as "completely beyond the facts." According to him, contiguous action, "however a-priori it may seem to some, still goes completely beyond the facts and to date remains well beyond what can be elaborated in detail."[55]

Boltzmann was right because no crude fact could directly prove contiguous action. Hertz would have contended that only an a-priori assumption, in the form of a principle, can give support for contiguous action to be shown by an experiment. But, once accepted, this principle revealed a surprising power in directing the experiment towards Hertz's goals.[56]

In his often quoted dictum, Hertz hinted to this unity between philosophical speculation on principles and practical results:

> By the experiment above sketched the propagation in time of a supposed action at-a-distance is for the first time proved. This fact forms the philosophical result of the experiments, and, indeed, in a certain sense the most important result.[57]

5. CONCLUSIONS

I think that the *Bild*-conception was central to the overall process through which classical physics led to modern physics. In absence of a *Bild*-conception, the different forms of generalised mechanics to which both Hertz and Boltzmann contributed, could not have been acceptable within the traditional conception of theory as a description of factual occurrences.

The explicit intervention of hidden quantities into physical theory and its justification through Hertz's requirements represent a turning point in the development of theoretical physics at the close of the Nineteenth Century. Since hidden quantities cannot be observed, they belong to a pure theoretical framework and that gives scientific acceptability to the idea that this latter framework has exigencies of its own, which are to be distinguished from those pertaining to the old concept of an empirical science or even to the Helmholtzian parallelism of laws. In his *Bild*-conception, Hertz substantially expressed the idea that theory should also meet standards of its own.

Once accepted, the *Bild*-conception loosened the bonds which still justified some requirements in generalised mechanics: reasons for keeping on constructing theories founded on Lagrange-type equations and mechanical magnitudes were no longer plausible. Hertz's *Prinzipien* and Boltzmann's *Mechanik* were perhaps the last attempts in that direction. Einstein's General Relativity definitely loosened the above bonds: in this author's judgement, it was a *Bild*-type theory.

Salvo D' Agostino
Centro della Storia Contemporanea e dei Quaranta "Edoardo Amaldi"
Accademia delle Scienze detta dei Quaranta
Villa Lontana, via Cassia Antica, 35
I- 00100 Roma
Email: s.dagostino@iol.it

NOTES

[1] In this study, I do not take into account the outstanding figure of Max Planck, the first German specialist in theoretical physics. I think I can be excused for this omission, due to the importance of his contributions which deserve a separate treatment.
[2] Cf. Jungnickel and Mc Cormmach 1986, vol 1, Preface, xix.
[3] Cf. Wien 1915, 241–59, 241. Quoted by Jungnickel and Mc Cormmach 1986, vol 1, xviii.
[4] Ibid. xvi.
[5] In France, Fourier and Ampere inspired a trend in Mathematical Physics, which had as its adepts Poisson and Chauchy, among the others. Comte's philosophy somehow expressed the foundational conceptions of this trend, which had also as its roots the teaching and research of the Ecole Polytechnique.
[6] Cf. Boltzmann 1905, 94. Quoted by Jungnickel and Mc Cormmach 1986, vol. 1, Preface, xv.
[7] Moreover, these ways have not been immutable, e.g. when one passed from Relativistic theories to Quantum Mechanics and contemporary Physics (Cf. Jammer 1979, 46).
[8] Cf. Cassirer 1950, 82.
[9] Ibid. 84.
[10] Ibid. 82.
[11] Helmholtz 1977, 117.
[12] Cf. Cappelletti 1976. Also, Cohen and Helkana 1977, Introduction, xix.
[13] Helmholtz 1977, 119.
[14] Ibid. 122.
[15] Ibid. 122.
[16] Ibid. 122.
[17] Ibid. 118.
[18] Ibid. 118, 119.
[19] Ibid. 162.
[20] One can argue that, in Helmholtz's conception, the eye's known lack of distinction between pure and mixed colours could not be considered any more as an eye's imperfection or incapability (as it was for Newton), because this lack of distinction was rather the eye's own modality, which differentiated vision from other senses, e. g., from hearing.
[21] Ibid. 162.
[22] M. Schlick, notes to *Hermann von Helmholtz Epistemological Writings*, in Helmholtz 1977, 172, note 33. For Helmholtz the general form is that schema devoid of any content which he declared to be the true form of intuition, in respect of which Kant's doctrine of the a-priori is to be upheld.
[23] Helmholtz 1977, Appendices, n. 3, 163.
[24] Helmholtz 1977, 121. Helmholtz compared the different perceptive characteristics of eye and ear and concluded that the formers incapability to detect beat frequencies represented its modality: "could the optical nerve at all follow in sensations the enormously rapid beats of light oscillations, then every mixed colour would act as a dissonance."
[25] Helmholtz believed that, within his science, he was faithful to Goethe's views, because his conception of a necessarily symmetrical correspondence between perceptions and concepts agreed with Goethe's request for a unity between perception and conception. He diverted Goethe's accusation that concepts obscured facts, to those abstract concepts which do not possess a corresponding perception.
[26] He concluded that his views was fully in accordance with Goethe's positions: "I consider it a propitious sign that Goethe, both here and in other matters, finds himself with me on the same path. His theory of colour can be considered an attempt to save the immediate truth of

sensory impressions from attacks by science" (Helmholtz 1977, 141). He agreed with the poet that one should simply demand from science that it should be only an artistic arrangement of the facts and form no abstract concepts going beyond this.

[27] Ibid. 141.

[28] Schlick, Helmholtz 1977, xxii.

[29] Helmholtz 1977, 172.

[30] Following the first tradition, Boltzmann introduced a hereditary Darwinian interpretation of the a-priori in physics. Cf. D'Agostino 1990, 381–398.

[31] E.W. cf. 21.

[32] Hertz 1962, 21.

[33] Hertz 1962, 21.

[34] Hertz's above quoted statement on the role of mathematics in Maxwell's theory have not as yet received an adequate interpretation. Concerning this aspect I like to consider Hertz's often quoted statement that Mathematics is at times more clever than its author himself as a consideration on the power of the logical-mathematical aspects of theories, in the sense of a neo-kantian emphasis on the relevance of logical consistency as an a-priori requirement on theory. In this connection, one is struck by the intrinsic fruitfulness and indispensability of Hertz's 1884 ideas, almost a confirmation of the Kantian tenet on behalf of the indispensability of transcendental ideas in physics when they are correctly employed, i.e., in their immanent role.

[35] In the absence of a polarisation theory, the independent existence of forces in space was an empty conception, because it could be filled with a variety of modes of representation – e.g. Faraday's physical lines of force, or Hertz's ethereal polarisation – in order to endow it with a physical meaning.

[36] Hertz 1895, 313 and note 1. Hertz insists on questions concerning the problem of the logical consistency of his new theory with the traditional theory.

[37] D'Agostino 1993, 51.

[38] Hertz 1956, 2.

[39] In the same passage, Hertz expressed the opinion that Energetism theory was influenced by the overpowering impression made by the discovery of the principle of conservation of energy. In his presentation, Hertz chose Hamilton's Principle of Least Action as the most suitable principle, among the integral principles of Mechanics, to represent mathematically the Energetist conception. One of the advantages of this choice is that one 'can deduce a whole series of relations, especially of mutual relations between every kind of possible force, which are wanting in the principles of the first image [i.e., that of customary mechanics].' These relations, he adds, were demonstrated by Helmholtz in his *Über die physicalische Bedeutung des Prinzips der kleinstens Wirkung*. Cf. Hertz 1956, 16.

[40] In fact, Hamilton's Principle cannot be applied to certain mechanical systems as, for instance, when bodies of three dimensions roll on one another without slipping. A second defect is that it calls for an infinite value of potential energy, when one assumes an infinite total mass for the Universe. Cf. Hertz 1956, 22.

[41] In effect, the third defect (presented as a difficulty against its acceptance as a general conception of physical theory) does not concern the energy theorem in itself but the selection of integral principles among other possible mechanical principles and it could then be overcome by other appropriate selections. In the use of 'extremum' principles, like Hamilton's, there is, however, another difficulty: their implicit finality. This difficulty is of a philosophical nature and, as such, Hertz attributes to it a major significance.

[42] Hertz 1956, 14. According to R. S. Cohen and J. Helkana, in his 1886 "magisterial paper on the physical meaning of the principle of least action," a continuation of his series on monociclic systems, Helmholtz "turned from the physics of masses in motion to a general physical science which was to be characterised by two principles, conservation of energy and minimal action." As remarked by the two authors above, it is doubtful that the quantity Helmholtz minimised in his essay was intended as energy rather than as 'force,' i.e., the quantity that featured in his 1847 seminal paper on the conservation of energy.
[43] Leo Königsberg reports Helmholtz's passage without any comment. On the other hand, he emphasises Hertz's application in *Prinzipien* of Helmholtz's mechanical theories (Helmholtz's 1886 essay).
[44] Helmholtz, 'Preface,' in Hertz 1956, 2. Helmholtz never ceased to adhere to this conception in his Elektrodynamics. In his Preface, he was still unwilling to renounce his double force view (Preface, viii).
[45] Hertz 1956, 2.
[46] Hertz's construction of a general theoretical system encompassing mechanics and electrodynamics, implied the concept of a *hidden substance*, i.e., Ether, as its unification element. This was Hertz's unfinished program in *Prinzipien*.
[47] Howard, 1996.
[48] Hertz 1956, 25: "If we try to understand the motion of bodies around us and to refer them to simple and clear rules, paying attention only to what can be *directly observed*, our attempt will in general fail. We soon became aware that *the totality of things visible and tangible* do not form an universe conformable to law, in which the same results always follow from the same conditions. We become convinced that the manifold of the actual universe must be grater than the manifold of the universe which is directly revealed to us by our senses." (Italics S.D.). In Hertz's view above, only the introduction of hidden quantities allowed parallelism to reach the status of a general principle.
[49] One argues that, for Hertz, the necessary relations in our mind, *qua-logical*, have also an epistemological import.
[50] Simplicity is consistent with the '*Zulässigkeit*' requirement.
[51] Hertz 1956, 23.
[52] Ibid. 45.
[53] Ibid. 45.
[54] Hertz, 'On the finite Velocity of Propagation of electric Action' in Hertz 1962, 105–123, 122. Cf. D'Agostino 1975, 261–323, 288 ff. Electric forces exist independently from their sources, from the charges and currents, in as much as they are properties of a hidden substance, Ether, in the form of its polarisation and motions.
[55] Boltzmann, 'On the Fundamental Principles,' (1899) in Boltzmann 1974, 119; Poincarè, too, was doubtful about the cruciality of Hertz's experiment: cf. Bevilacqua 1984, 239–256.
[56] D'Agostino 1990, 390.
[57] Hertz 1962, 21.

REFERENCES

Bitbol, M. and Darrigol, O. eds.: *Erwin Schrödinger, Philosophy and the Birth of Quantum Mechanics*. Paris, Editions Frontières.

Boltzmann, L.: 1974, *Theoretical Physics and Philosophical Problems*, ed. by B. McGuinness, Dordrecht, Reidel.

Cappelletti, V. ed.: 1976, *Opere di Hermann von Helmholtz*. Turin, Utet.

Cassirer, E.: 1950, *The Problem of Knowledge. Philosophy, Science and History since Hegel.* New Haven, Yale University Press.

Cohen R.S. and Elkana Y. eds.: 1977, *H. von Helmholtz, Epistemological Writings.* Boston Studies in the Philosophy of Science, Vol. xxxvii. Reidel, Dordrecht.

D'Agostino, S.: 1990, 'Boltzmann and Hertz on the *Bild*-Conception of Physical Theory,' *History of Science*, **28,** 380–398.

D'Agostino, S.: 1992 'Continuity and Completeness in Physical theory: Schrödinger's Return to the Wave Interpretation of Quantum Mechanics in the 1950's,' in Bitbol and Darrigol, eds., 1992, 339-362.

D'Agostino, S.: 1993, 'Hertz's Researches and Their Place in Nineteenth Century Theoretical Physics,' *Centaurus*, **36,** 46–82.

D'Agostino, S.: 1975, 'Hertz's Researches on Electromagnetic Waves,' *Historical Studies in the Physical Sciences*, **6**, 261–323.

Giere, R.N. and Richardson, A.W. eds.: 1996, *Origins of Logical Empiricism.* Minneapolis, University of Minnesota Press.

Helmholtz, H. von: 1876, 'Die Tatsachen in der Wahrnehmung,' in *Populäre wissenschaftliche Vorträge*, 2 vols. Braunschweig, Vieweg. En. tr. 'The Facts in Perception,' in Cohen and Elkana, eds., 1977, 115-185.

Hertz, H.: 1914, *Untersuchungen über die Ausbreitung der elektrischen Kraft.* Leipzig, Barth. En. tr. 1962, *Electric Waves, being researches on the propagation of electric action with finite velocity through space*, New York, Dover.

Hertz, H.: 1910 *Prinzipien der Mechanick in Neuen Zusammenhange dargestellt.* Leipzig, Barth. En. tr. 1956, *The Principles of Mechanics presented in a New Form*, New York, Dover.

Howard D.: 1996, 'Relativity, Eindeutigkeit, and Monomorphism: Rudolf Carnap and the Development of the Category Concept in Formal Semantics,' in Giere and Richardson, eds., 1996, 1-67.

Jammer, M.: 1979, 'A Consideration on the Philosophical Implication of the New Physics,' in Radnitsky and Anderson, eds., 1979, 41-62.

Jungnickel, C. and Mc Cormmach, R. eds: 1986, *Intellectual Mastery of Nature, Theoretical Physics from Ohm to Einstein*, 2 vols. Chicago, The University of Chicago Press.

Radnitsky, G. and Anderson, G. eds.: 1979, *The Structure and Development of Science.* Boston Studies in the Philosophy of Science, Dordrecht, Reidel.

Wilhel W.: 1915, 'Ziele und methoden der theoretischen Physik,' *Jahrbuch der Relativität und Elektronik*, **12,** 241–59, 241.

KARL SCHUHMANN

REPRESENTATION IN EARLY HUSSERL

1. CARL STUMPF

Husserl's early philosophy of cognition grew out of the investigations of the Brentano school, where he was especially influenced by Brentano's "unforgettable course of lectures" (as he was later to call them)[1] on "Selected Questions in Psychology and Aesthetics," a Vienna lecture course from Winter 1885/86, and the lectures on psychology given by Carl Stumpf in Halle a year later. Where Brentano had especially dealt with phantasy presentations in their relation to perceptual presentations, i.e., sensations, Stumpf had given a comprehensive survey of psychical life along the lines of the Brentanist classification of psychical acts into presentations, judgements, and emotional phenomena, thereby insisting especially on the various types of presentations such as sensations, phantasy presentations, the presentations of space and time, and abstract and symbolic presentations. Both thinkers considered sensations to be fundamental, while all other presentations are of a higher level and therefore to a degree depend on them.

Husserl's student notes from the afore-mentioned Brentano lectures are apparently no longer extant,[2] but those he took in Stumpf's course fortunately have survived.[3] According to Stumpf, phantasy presentations can be distinguished from sensations, firstly, in that the intensity of their object is extremely small, as compared to that of any, be it even the weakest sensation. There exists an unbridgeable gulf even between an actually sensed pianissimo and a merely presented fortissimo of the same harmony. The continuum of intensity is broken between these two classes of presentations. Secondly, phantasy presentations "possess a smaller fullness of immanent features in comparison with sensations with the same name."[4] I even cannot in all detail imagine a person whom I am accustomed to see every day. This incompleteness may go so far that there is little or even no content at all which the phantasy presentation would share with the corresponding sensation; sometimes it offers nothing but some relational determinations. This is to say that it may contain nothing but a set of relations to something we already

L. Albertazzi (ed.), The Dawn of Cognitive Science. Early European Contributors, 167–184.

know. To illustrate this, Stumpf mentions a person who was unable to produce tone imaginations in the proper sense of the term, but managed to do so indirectly by imagining the different muscle sensations one has in the larynx when singing the tones in question.[5] An additional criterion Stumpf offers is that a sensation will in most cases last as long as its external stimulus lasts. Phantasy presentations are on the contrary brought about only by what was at that time called psychical causality. They are triggered off by some other presentation which is actually present: my sensation of A reminds me of B which itself is not present, but merely imagined. In contrast to sensations phantasy presentations are however extremely volatile and unstable. Moreover we often are unable to change the content of our sensations, at least as long as we do not also change the situation of our body vis-à-vis the object perceived, but we are far more free in imagining *ad libitum* any of this object's sides.

Time presentation differs from other phantasy presentations, first, in that it is brought about involuntarily. It is moreover not a special content of its own, but something of a character attaching itself to any presented content by means of an original association. Space presentation in contrast is built on a certain amount of original space sensations, i.e., on contents of a given three-dimensional sort. True, these contents offer us only a limited part of space. But we can expand the amount of actually given spatial contents by means of imagination. We may imagine that we would walk around a material thing of which only the forefront is seen, and we may judge that the space we experience extends in fact beyond the horizon of what is actually given.

As to symbolic presentations, i.e., presentations which function as signs for other ones, we are to distinguish between two main classes. There are cases in which it is profitable for us to rely on signs, because we thereby save the energy necessary for turning directly to that which the sign in fact refers to. When addressing a letter, it is enough to mention the street name and the house number; we need not describe the house we want to send the letter to in all relevant details. The use of signs is here merely a matter of abbreviation and economy. But there are also cases where the sign is absolutely indispensable, because an authentic presentation of the object is not only difficult to obtain, but for reasons of principle excluded. Thus in sense intuition we will practically not get beyond a presentation of the number five,[6] and therefore "a concrete presentation of 1000 is impossible."[7] And yet any child may easily calculate with large numbers, precisely by using number signs which contain nothing but certain relations to what we can actually intuit: the number sign 1000 refers to a number which is ten times 100, and 100 refers to a number which is ten times ten – and ten is in each case two times five. "Thus there arises that remarkable way of running the household with surrogates," Stumpf remarks;[8] thus, this is to say, there arises mathematics. But also ordinary language, as we know and use it, has for most purposes the function to represent contents which we ordinarily do not or even cannot present in a proper way. "The names stand in for the presentations."[9] Here nominalism has a healthy function: we often call to mind only the name, instead of the concept it stands for, because this is the most practical way of doing things.

As a good Brentanist, Stumpf is more interested in the exact description of individual cases than in establishing an overall system of presentations. Thus much can however be derived from what he says. Sensations are of course at the base of all other types of presentations; they are that which is primarily given. Moreover all differences between the different kinds of presentations are differences which affect only the content of these presentations. For as far as the act moment is concerned, all presentations, after all, fall into one and the same class of psychical acts, namely presentations, and to present is just to present. Or, as Stumpf succinctly states: "Since the presentation of black is not black, neither is the contrast between black and white a contrast of presenting."[10] In the case of sensations versus phantasy presentations their difference was seen to lie in the different intensity, completeness, duration and steadfastness of their content. Symbolic presentations however differ from all the rest, in that in presentations of the usual kind the object itself is the direct target of the act, whereas in symbolic presentations it is only represented by some (directly given) sign. Since this sign 'saturates,' so to speak, the act, it is no longer necessary that its proper goal should be present to our mind, in order to allow for an adequate handling of the sign in question. Signs do most of the job. Such symbolic presentations can be distinguished into those, the target of which can in principle be directly presented, and others where it cannot. This distinction cuts across the earlier ones, since, as we have seen, both in mathematics and in phantasy presentations there do occur objects which cannot be directly given, but can be referred to only by means of certain relations between the sign and what it stands for.

2. INTENTIONS[11]

It is within this framework, and starting from these assumptions, that Husserl sets out to develop his own view of the phenomena of representation. While there is no special treatment of time consciousness in early Husserl, he repeats with regard to space for the most part what Stumpf had said.[12] As a trained mathematician, he insisted more than his mentor had done on the 'idealisation' which leads up to that special type of space on which (Euclidean) geometry is built.[13] His most substantial innovations concern however a careful specification of the global treatment of presentation in Brentano – Stumpf, including a more detailed study of intuition and phantasy presentation.

Although this re-elaboration of the notion of presentation is primarily intended as a refinement or at best a modification of the relevant Brentanist doctrines, it leads Husserl in due course rather unexpectedly to an almost complete reversal of these doctrines. This forced him in the end to give up a fundamental element of the Brentanist view of intentionality, namely the idea that each presentation would have an immanent content or object of its own, so that presentations could be distinguished from each other exclusively by means of their specific content.

As a matter of fact, Husserl sets out from an analysis of perception, instead of sensation as Brentano and Stumpf had done. Though he nowhere gives a reason for this deviation, one can surmise that to him the notion of sensation pertains rather to the reflective analysis of perception, and not so much to the phenomenological description or, to put it in Brentanist terms, to the descriptive psychology of perception. The element which, for Brentanism in its more orthodox version, turns sensations into perceptions, is a psychical act in which this presentation is accepted – it is, in other words, a judgement. Now this interpretation of perception as a judgement about some sense presentation is rather counter-intuitive; it is peculiar even in the Brentanist framework, insofar as judgements in principle always allow for a yes/no – distinction, where the idea of a perception rejecting the existence of what is perceived seems somewhat strange. So Husserl ranges perception at least implicitly, together with imagination, under the common heading of presentations and correspondingly distinguishes between perceptual presentations on the one hand and phantasy presentations on the other.[14]

Now the Brentanists had been particularly intrigued by what Meinong had called "temporally distributed objects" such as melodies, which are apprehended as unities, but of which only parts are successively given in sensation. In harmony with the Brentanist doctrine of time presentation, Stumpf had declared that each sensation receives an original time mark at the time we present it, a mark which continually changes from the now into the past and the further past, and so on. Thus the content of the perception is incessantly modified by receiving each time a different time mark; from something given and perceived it is turned into something merely imagined and remembered. A complete melody, then, is never given to us, but can only be imagined on the basis of its last tone, a tone which I perceive presently now. In his *Philosophy of Arithmetic* Husserl had however discovered that not only spatially distributed objects such as swarms of geese or lines of soldiers are held together by what he called "figural moments," moments which are grasped immediately, but that this applies also to temporal objects such as melodies. The configuration of the tones is one of the things directly intuited in the very same complex intuition which also gives me the tones it is made up from. The melody is however not intuited as an additional element supervening upon what I perceive, but is a structure that is fused with these very elements or contents.[15] Thus the important thing is not so much the heterogeneity of sensations versus phantasy presentations – after all, the founding contents of sensation underlying one and the same Gestalt may, as both Mach and again von Ehrenfels had shown, considerably vary. Rather, one must insist on the unifying relations given and therefore intuited together with what we sense. Husserl correspondingly is less interested in the problem of how sense presentations might become phantasy presentations, but rather in the possibilities and limits of our intuition. Since intuition often is not so much of complex objects of a higher order, such as swarms or clusters, but of simple objects such as bees or stars, Husserl moreover applies the principles that obtain in the case of seeing streets and hearing melodies, to the objects they comprise, such as houses and tones.[16]

Regarding intuition, Husserl distinguishes between a broader meaning of the term and a narrower one. In the strict sense, I intuit only what is given to me directly and at this very moment. Standing in front of a house, I see only the front of the house. Since to intuit is to have the object in question given, in the sense that nothing of it remains unintuited, the restricted object of the momentary intuition may be said to be contained or comprehended in that intuition; it is its content. The object of a momentary intuition, Husserl states in Brentanist terminology, is its immanent content. To this momentary intuition Husserl opposes the notion of a coherent succession of such intuitions, in which the different aspects of the thing intuited are given one after the other. A thing, in the usual sense of a material entity, is nothing but the correlate of such a closed continuum of intuitions in which each intuition is lawfully connected with both a set of preceding and of following ones. Each of these intuitions has one of the sides of the house as its actual immanent content, but the house itself is not given in any of these acts; it is rather an "ideal object,"[17] in the sense that I cannot grasp it in one single, i.e., in any possible act.

The important thing here is that a given momentary intuition contains in itself an indication as to how to connect the subsequent ones to it. In this sense it points beyond itself and predelineates the next step that is to occur in the flow of my intuitions. To state that an intuition is present, is to say that I am aware of another possible intuition which links up with the first one and at the same time continues it. Intuitions make up unitary processes and continuous systems of experiences, precisely because each of them has, so to speak, hooks and eyes connecting it with other intuitions that are no longer or not yet present. Their temporal character consists in this that they are rooted in earlier ones and head on towards later ones. An intuition always goes beyond itself, not only in the sense that it is about its object (and not about itself), but also in the sense that it comes from an earlier one and flows over into a following one, so that as a matter of fact there exists a unitary stream of consciousness in which the actually present intuition is only a unit marked off in a more or less vague and arbitrary way.[18] Though Husserl does not say so explicitly, one could state that all intuitive presentations are essentially representations, to the degree that they point towards some future intuitions which, notwithstanding the fact that they are not yet present, are experienced as real possibilities dovetailed into the given one. Husserl comes closest to this when he states that given intuitions always function as *signs* for new and other ones.[19] Although the notion of sign is used here in a somewhat loose sense, i.e., in a way which, as we shall see, is not completely in harmony with Husserl's own definition of the sign, it expresses at least implicitly the view that each presentation is at the same time a representation. A more careful consideration of this fact would, by the way, probably also have prevented Husserl from affirming that there could be an "ultimate fulfilment" of intuitions, where their content "bears no representative function whatsoever."[20] It would have shown him, in other words, that intuition is always a momentary act in a succession of acts, i.e., that consciousness is essentially temporal or, more precisely, that it is itself a temporally distributed object. True, later Husserl talked of an original constitution of time-consciousness, on the basis of which objective time becomes constituted in a derivative way. And also early Husserl already knew that "the flow of intuitions is a temporal flow." Moreover early Husserl is right (against later Husserl, I would suggest) in adding that "this temporal aspect plays no role" in intuition.[21] Yet he should have seen that intuitions can continue their flow only if the object continues to exist and to remain present too, and if it does so *in the same time* in which intuition unfolds. This temporal aspect need not be taken into account, precisely because it makes no difference, i.e., because the time in which consciousness develops and in which the thing exists, is one and the same.

Husserl's most spectacular move beyond Brentanism consists however in pointing out that the unitary Brentanist concept of presentation is in fact equivocal. Thus far we have talked about presentations in the sense of intuitions. According to Husserl, the term presentation receives however "a totally different meaning" in the case of representations proper, i.e., in the case of symbolic or inauthentic presentations.[22] They originate from intuition, whenever the flow of intuitions does not continue along the lines I am, on the basis of my given intuitions, entitled to expect. Here the predelineation contained in any given intuition is not fulfilled, but frustrated. Whenever the general tendency of the flow of intuitions is in this way disrupted, there arises inevitably a feeling of dissatisfaction with what is now given to me, and an intention towards the correct fulfilment of my original expectation. I take the present content no longer as the content of my actual intuition, but as an indication that the true fulfilment of my intention lies somewhere beyond it. I make the immanent object, this is to say, represent something absent. Whether a content is intuited or serves as a representative of something not intuited, does therefore depend not on this content itself, but on my cognitive interests and expectations. When I am directed only at this content, then I intuit it, but if my intention moves however beyond it, it acquires a representative function. Thus it is the act, and more precisely the interest directing the act, and not the content, as Brentanism usually has it, that makes all the difference.[23]

The intention has, properly speaking, no proper immanent content at all, and it is precisely this aspect which characterises it as an intention. Although it is motivated by some given intuitive content, it is not directed at it. To the degree intentions tend toward intuitive fulfilment of themselves, they can also be seen as temporary disharmonies and deviations from intuition. Thus there exists an essential link between intuition and representation, in the sense that intuition is both the origin and the goal of representation. Intuitions, we have said, allow for the genesis of representations, because they bear representative elements already in themselves. "Intuition is fused with representation," as Husserl has it.[24] In this sense it is therefore more adequate to understand representation proper or 'mere' representation not so much as a type of presentation of a totally different order, as Husserl usually does, but rather as one of the extremes of a continuum ranging from intuition to representation; while intuition itself, in the sense of an interest that comes to complete rest and satisfaction in what is given, is nothing but the other extreme of this very same continuum. And it is an open question, whether these extremes do really occur in pure and unadulterated version in our psychical life, or whether they are nothing but ideal limit concepts that help us to think the continuum of consciousness.

3. IMAGES

Representations as tied to intuitions play a role in the flow of perception itself and may not therefore be confounded with phantasy presentations which usually do not claim such a direct relation to the given perceptual field. According to Brentanism, the content of phantasy presentations can, as we have seen, be described as weak, incomplete and volatile. In Husserl's view, this is indeed correct. But it is no more than a first rough approach to the phenomenon in question.[25] For we should not overlook the fact that notwithstanding such fluctuations the object I imagine can be, and usually is, a fixed and complete entity. I may try to imagine an event that never happened (e.g., the death of Achilles), and sometimes succeed better, or at another time do not succeed at all in doing so. Yet all these imaginations are still about one and the same event, an event which is and remains completely fictitious. Thus one is to distinguish between the shaky phantasms which make up the actual content of my phantasy presentation, and the object of this presentation which may be relatively stable. It is often through a varying or sometimes even almost non-existent content that we are directed at the object of imagination. "The image changes with incredible unsteadiness, but invariably means the unchanged object."[26] So there are two things directly present in the act of imagining: a representing content and a represented object. Indeed, the object of imagination is always something not present, i.e., something that is only represented – no matter whether it is something fictitious or also something real. Exactly because of its absence it can manifest itself only in or through something different, namely by means of a direct content of the phantasy presentation, a content which as such is of course present. It is by its very presence that this often volatile content is in a position to represent the identical object I intend. Thus the content in question is the only thing that is there for me, and yet it is not what the act of imagination is about. A phantasy presentation therefore is a composite act in which "a second object is intentionally given to me, but not in the form of a second appearance,"[27] precisely because this second object, i.e., the object I am in fact imagining, does not appear at all.

In this respect imagination differs from perception, where the object is presented as something that offers and manifests itself in a direct way. True, when reflecting on my perception I will become aware of the fact that the object as perceived is not simply the object as it is in itself. Sometimes I perceive the table from a perspective or under light conditions that are far from ideal or satisfactory, and I know that as a matter of fact it looks different from what I now happen to see. But I also know that it is the very table itself which for the time being appears under unfavourable conditions. The real table and what I am perceiving now, are not two different tables, but one and the same thing that is given either under conditions that allow me to see it as it is, or under conditions which prevent me from grasping it adequately. So also in perception there occurs, at least in principle, a duality between the appearance and the object. But this duality has an immanent teleological structure. The case I am interested in and for the sake of which I usually perform sense perception, is the very case in which the appearance allows the object to appear as it is. When I feel forced to reflect on their non-coincidence, I feel embarrassed or uneasy, precisely because perception does not do the job it is meant to do. Correspondingly I will try to eliminate the interfering factors, in order to allow the object to present itself in optimal fashion. For this unimpeded presence of the object itself is what perception is after.

In the case of imagination, things are however completely different. It belongs to the very nature of imagination that the object does not appear as it is in itself, and in fact does not appear at all. In opposition to perception, phantasy presentation always has, as we pointed out, a double object. On the one hand there is the unstable, constantly changing object which in fact appears to me. However, what is meant in my presentation and what it aims at, is not this 'Protean' object, but rather the theme or subject this presentation directs me at. This theme as such remains however beyond my actual reach. Phantasy presentations therefore always are mediated presentations. It should be noted that the peculiar nature of the imagination's theme, including its existence or non-existence, makes no intrinsic difference as regards the given presentation. Whether I imagine a dragon or the desk I usually work at, they are both given through images in my mind, though only one of them can be identified in perceptive acts, i.e., can be rendered directly present. Insofar as the object is imagined, the question of its possible existence remains therefore irrelevant.[28] In phantasy presentation even a real thing is after all 'only' imagined. In a sense this must of course also be said of the imagination's immediate object. Though it is in fact present to me and appears itself, it also appears *as* itself, i.e., as something that is nothing but an image and has no existence of its own after the manner of a real thing. I take it as an image by limiting its role to the function of pointing beyond itself and of helping me to fix attention on the real and ultimate subject-matter of my imaginative act. The immediate object of the act therefore is not a special sort of entity, but rather a function of the phantasms that fill my mind. This is not to say that there are phantasms floating around, and that at a given moment I take a new and further step and apprehend or interpret them in a certain way. On the contrary, to imagine is to have phantasms, and these phantasms are phantasms (i.e., "contents that are imagined,"[29] and not building blocks of consciousness or even sensations), precisely because they are apprehended as elements allowing for a representation of the object I try to imagine. The only question that arises here, is how I succeed in preparing the phantasms that I need for getting at an absent object. But this is in first instance a genetic question; it should certainly be answered by pointing out, with Brentano and Stumpf, that there exists a difference between reproductive and productive phantasy. Where the former finds, or at least intends to find, its materials among the contents of memory, the latter is far more active in consciously varying and modifying such contents.[30]

4. PICTURES

Husserl pays also a good deal of attention to another type of phantasy consciousness in which the primary materials are indeed something already there, namely picture consciousness. This type of consciousness arises when a given thing, i.e., something we are aware of as being part of the real world, functions as a trigger that makes us conceive a phantasy presentation. In view of the fact that, as we have seen, phantasy presentations always have a double object, we must therefore say that a picture presentation has three objects at once: the thing which sets the mechanism of imagination in motion, the image of something, and this imagined something itself. In Husserl's terms: "The physical picture awakens the mental image, and this in turn presents something else: the subject."[31] This type of consciousness occurs not only in figurative art, but generally in all cases where we have objects which, just as in the case of photographic representation, picture something else. Since the primary object of picture consciousness is something that is part of the world, Husserl calls this type of presentation also "physical imagination."[32] Its prerequisite is, to repeat it, a physical object which as such is subject to all the laws physical objects are subject to. This object must in some way be similar to another object which it represents by means of the comparability and/or likeness that obtains between them. Since the object which is represented, is itself absent, there applies in this case, too, what had applied already in that other case of mere phantasy presentation, namely that any question about its existence or non-existence is irrelevant. A given onlooker may understand some painting as showing the Queen of the Amazons, where another might believe that it portrays the Queen of England. The important thing is only some sufficient similarity between the picture and its purported object, a similarity which must be strong enough to allow for a correct identification of what is pictured. Unfortunately Husserl does not enter into a closer analysis of the notions of likeness, analogy or similarity involved in this case. It is however clear that this likeness as such is insufficient to make a physical thing a picture or portrait of somebody or something. For in that case any egg would be the perfect picture or portrait of any other egg, since there is nothing that is as much like an egg as an egg. But, Husserl says, in the case of twins the first one is not a picture of the other.[33] Upon the perception of a thing there must therefore always be built another act in which this thing is apprehended *as a picture*. This apprehension is in most cases triggered off by the fact that the thing given in perception reminds me of a thing I already know, or with regard to which I have some indirect relational knowledge. Thus a painting may represent the god Zeus known to me from my earlier interest in Greek mythology, or about whom I am being told just now, when I for the first time see a statue of him, that he is the highest of the Olympic gods. What constitutes the object as a picture, is however in all cases a psychical act of a higher order, in which the perceived object with its physical features is taken or understood as a representative of something different. What matters in this case is not the physical nature of the object – its weight or colour or size, – but the degree to which I can interpret these features as referring to another object which now enters into the focus of my interest and attention.

With regard to phantasy presentations elicited by pictures, Husserl is clearer than in his discussion of phantasy presentation as such, that its appearing object is nothing but a content of the imagining act, and not an object in the sense of an entity which would exist in its own right. This primary object is rather a pseudo-object (*Scheinobjekt*),[34] a mere fiction, and therefore may be called object "only in a modified sense of the term."[35] This object which, as was said, is the only thing present to me, is therefore, properly speaking, not an object at all. It is rather a non-entity, and the appearances we are faced with, are, according to Husserl's paradoxical expression, "appearances of a non-entity."[36] And yet they are there and are an indispensable part of picture consciousness. In the case of a photo, for instance, the physical thing I perceive is the surface of a rectangular sheet of rather thick paper covered, at least in the case of black and white photos, with shades of grey. The subject of this image is a person wearing clothes in different colours. What appears to me is however, in contradistinction to the physical picture, a three-dimensional entity, but, in contradistinction to the person represented, a colourless shape. Where the real person has a certain size, moves around and talks, the person, insofar as it appears to me *in* the photo, is tiny, still and silent. The appearance, this is to say, usually deviates considerably from the picture's subject. And yet this appearance is the only thing given to us in picture consciousness.

How are we to understand this paradoxical situation? If we keep in mind Husserl's tenet in his *Logical Investigations* that "to appear" means "to be experienced,"[37] then a solution suggests itself. Among things in the world there are sheets of paper and living persons. A person however is, properly speaking, never *in* a photo. To say that this person is to be seen in it, means only that I apprehend this paper as a photo of the person in question. What is added to the paper, is, as Husserl rightly insists, not some new objective datum that would enrich the content of my perception, but a psychical act of mine, a certain experience of the imagining kind. This experience is characterised as representing a person, and as doing so on the basis of what I apprehend as a certain likeness of the person in the dots and shades on the paper. Consequently it is primarily this person, secondarily the paper, but under no circumstances my act that I am directed at. On the other hand, this person is not there; rather, the only thing that, in addition to the paper appears to me, albeit not as the object of my act, is this act itself. Insofar as I am indirectly aware of any act I perform, I may say that it appears to me; insofar as its object (the person I imagine) does not appear at all, this appearance is an appearance of nothing. The act may be called its own 'object' only in an improper sense, that is to say, in a modified way – and an improper object is no object. Yet also its proper object (i.e., its subject-matter) may be said to be present only in an improper way, which again is to say that it is not present at all. In sum, it is more correct to speak, in the case of mere imagination, of just one object, namely the subject-matter this imagination is about,[38] and correspondingly in the case of picture-consciousness of no more than two objects, each time leaving out the immediate and internal object of the act, the 'immanent' or 'internal' image-character, as Husserl sometimes calls it.[39] For it is not an object at all, as Husserl had shown already in his 1894 paper on "Intentional Objects" with its general criticism of this Brentanist notion. In stating such an intermediary object, Husserl had unwittingly fallen prey to the Brentanist doctrine – a doctrine he criticised in his 1904/05 lectures on imagination – that "imagining is said to be something undifferentiated," and that therefore all differentiation belongs to its content.[40] True, later Husserl gave up this view which in a somehow Meinongian way multiplies entities beyond necessity, in order to state that both sensations and phantasms, i.e., the contents of presentations, are not pregiven thinglike entities, but are already consciousness themselves, so that "consciousness is consciousness through and through."[41] This statement, which is completely in line with his later transcendental idealism, overlooks however the fact that any talk of contents of consciousness belongs to descriptive psychology and therefore does not in any way prejudice the genetic problem of the origin of these contents. Now Brentano had already insisted that, as far as this second question is concerned, physics and more specifically physiology comes into play, so that there is no need to have recourse to some transcendental subject.[42]

We have seen that an object of imagination may be firmly intended, no matter whether this appearing (pseudo-)object of imagination is constantly changing or kept fixed thanks to an underlying physical picture. The fact that the stable or unstable nature of the content of my imagination does not influence its very subject, can in view of our results be understood as a sequel to the fact that in both cases it is one and the same imaginative act that is directed at its object. Moreover we can now also understand the fact that the content of my phantasy presentation does not belong to the field, and therefore also not to the world of perception, so that there is no continuity between what is real and what is imagined (an issue insisted upon already by Stumpf). The field or world of imagination is not a second or different world at all, because there is nothing out there that would correspond to my imagination. It is after all the very sense of the imagining activity to deny all such independent existence of what I imagine. The reason for this discontinuity is simply that in imagining I do not in any way add to the objects which make up the world, but only perform a certain psychical act internal to me which has the characteristic of being a presentation, and more specifically a phantasy presentation.

5. SIGNS

Let us now turn to a third and more common type of representational consciousness, the consciousness of signs or symbols. In picture consciousness I am to actively perform the act of recognition in which I take the object I perceive as a picture of something different. I am helped and supported in this by the analogy between the two, so that in a sense this imaginative act can be performed automatically. In sign consciousness however there exists no such natural relation between the two objects. In order to remind me of what is signified by the sign, I have to have at my disposal some previous association between them. In most cases I am to consult my memory which tells me about their habitual relation as experienced before. Sign consciousness is therefore not a straightforward type of cognition, but rather involves recognition. Though Husserl, not unlike Stumpf, opposes perception and imagination as authentic types of presentation to symbolic presentations as inauthentic ones, he allows in some sense also for a transition between these two classes. For to the degree that pictures need not always be faithful renderings of what they are about, but may be more or less inadequate, one can understand sign consciousness as a zero type of picture consciousness.[43] This interpretation will work at least in those cases where we take something we actually intuit as a sign for some other concrete thing.[44] But a sign may also stand for something abstract, as in the case of words which usually signify general concepts. In all cases however the sign and what it signifies are in some sense heterogeneous. True, also the symbol can be similar to what it symbolises. But it is not this accidental similarity which constitutes an object as a sign or symbol, but rather the fact that, under the aspect one takes into account, it has nothing whatever to do with the thing signified.[45] Moreover the sign induces us to move beyond itself towards the object which this sign points at. This presupposes, as was said, some lasting associative link between the two and therefore a disposition or inclination to attach B to A, whenever I encounter A. This implies that of signs it must be said what we said about the notion of picture, namely that they arise only through some psychical activity. Where in the case of pictures it is the similarity between the original and the portrait that activates the mechanism which connects the second to the first, it is, in the case of signs, association which does this job. In both cases however a more or less automatic transition from the one to the other is guaranteed which requires no spontaneous or well-considered decision to be reinstated each time I come across the sign.

In "the widest conceivable sense" of the word 'sign,' "any character of something" may be called a sign of this thing.[46] Thus smoke is a sign of fire; the sign-post tells me that the town is nearby; fever indicates an illness; a brand proves who owns the cow; fossil bones show that in earlier times other species lived on earth; and the claw I see alarms me that the lion is around. Such signs, which are traditionally called natural signs, differ from phenomena of the kind already discussed, in that they do not deflect consciousness to something of necessity absent, but on the contrary point to something at least in principle more or less present. They are therefore, or at least have the tendency to become presentations themselves, instead of remaining forever mere representations. For this reason it would be better to call them indications, symptoms, marks, traces or clues, rather than signs, precisely because they do not represent something absent, but rather announce, be it only in a mediate way, the (hidden) presence of something. It is their task to lead us towards a thing which is intimately connected with them, in order to render that second entity manifest to us. They are nothing but bridges destined to carry us to the other side. In this sense, their presence makes us assume that there is more present than just these marks themselves. This assumption is not based on insight into some internal or necessary connection between the mark and what it hints at, but rather on earlier experience and therefore on memory and association.[47]

Signs in a stricter sense of the term, in which they can be identified (as we already did) with symbols, are however essentially representations. So there exists no direct way leading from these so-called artificial or conventional signs to the thing they signify. Rather, those things signified are and forever remain outside the realm of the sign. To this degree, signs are a special type of representations; they are representations to which the element of substitution or replacement is indispensable. Just as in the case of pictures, and in contradistinction to that of the mark, the existence of the thing signified is not in any relevant sense tied to that of the sign. There can be signs for existing as for non-existing things. Although we cannot perceive something non-existent, we can easily think about it by means of our concepts. Perceptions, this is to say, are never signs, but rather give us direct access to things. Concepts however are, precisely because they allow us only a mediate grasp of what they are about – and moreover do so without guaranteeing that these things in fact exist. An ontological proof for the existence of any entity is therefore, for reasons of principle, impossible; and if it were not, we might prove even the existence of that which does not or cannot exist at all, since we may conceive factually non-existent things as well as impossible ones.

For Husserl the most interesting thing about symbolic presentations is however not so much the impossibility to reach out, on their basis, to the things represented in them, but rather the fact that such a tentative grasp is completely unnecessary. It is the very idea of representation and replacement that it functions in the place of that which is replaced, without ever attempting to become or be that which it replaces. Acts performed in the name, by authority and general proxy of the one whom we represent, are for all practical purposes identical with acts done directly by the person who is represented. The acts of the deputy are as good as those of his authorising source. In the same sense and for the same reason it usually suffices for all practical purposes to employ symbolic presentations or surrogate presentations, i.e., mere representations, without ever having to take the pains to render truly present that which they represent. It is, in one word, often a matter of thought economy to work with symbols. Indeed, this is one of the main reasons why representational thought is so widely used in mental life. It is, as Stumpf already recognised, a matter of abbreviation and of saving time and energy.

Not unlike Stumpf, Husserl too distinguishes between symbolic presentations, for which "the authentic presentations which their surrogates replace, are themselves at any moment at our disposal,"[48] and others where no such authentic presentation is ever possible. To the representations of the second class there belong such complex ideas as the concepts of science and art, the Renaissance and the French revolution, Africa and the human race. In fact there is nobody who could possibly have an exhaustive idea of everything contained in these notions. Yet one might argue that this is due to the limitations of our intellect, and that at least in principle a powerful intellect might be conceived which could be aware of all these features. But what about the notion of an actually infinite God, or what about such contradictory concepts as wooden iron and round square?[49] However one might enlarge one's intellectual powers, a presentation which would give these objects directly and as they are in themselves is for all purposes excluded, whether this is because such objects are strictly inconceivable or because they simply cannot exist. Limiting ourselves to the handling of symbolic presentations is in these cases not only a matter of expediency, but a sheer necessity, for otherwise we could not handle them at all. Juggling with representations is therefore not only an inevitable necessity for a cognising being with finite intellectual powers; it is at the same time also a powerful tool which allows for an almost limitless expansion of our cognitive faculties. An excellent case in point is mathematics, as soon as we get beyond the most elementary numbers. Without the help of symbolic presentations we could not operate with irrational or transcendent, let alone with infinite or transfinite numbers.[50] In order to allow signs to do this job, Husserl adds to their definition a requirement that is not in Brentano or Stumpf: signs must not only characterise that which they stand for, they must moreover do so "in an unambiguous way."[51] If signs are in this way standardised, we can operate blindly with them without ever risking to run into error and misunderstanding due to some multivalence of the sign.

6. CONCLUSION

The originality of philosophers of the Anglo-American brand is often a function of their avowed disinterest in what their predecessors have said and done. To the informed reader this sometimes offers the rather bizarre spectacle of a huge waste of time and energy in their efforts to invent the wheel. Continental philosophers on the contrary generally know what they ought to know, better. But it is an all too human inclination to think up excuses which allow one to get around the detailed and difficult study, often in a foreign language and a foreign terminology, of what earlier thinkers have contributed to the progress of thought. One of the more recent manifestations of this type of *ignava ratio* is the sweeping affirmation that, from its very beginnings, Western philosophy was governed by a 'métaphysique de présence' which culminates in Husserlian phenomenology, for the very reason that a doctrine of phenomena is a doctrine about what is actually given and therefore present.

A more careful study of Husserl's early texts shows however that at least his early philosophy is, on the contrary, haunted by absence. Already intuition and perception contain very little that would be authentically present. As late as 1921 Husserl opened a lecture course with the remark that "perception is a pretence that cannot make itself true,"[52] precisely because it pretends to have things given to itself, where this is true only in a most restricted sense. Moreover all so-called higher cognitive functions are nothing but various types of representation, in most cases without any pretence to ever be in a position to render present in some way or other that which they are about. This Husserl sees as "the extremely noteworthy" nature of representation which is "of fundamental importance for the whole of psychology, and in particular for the psychology of cognition."[53] Our mental life is indeed in almost all its aspects a way of coming to grips with reality by means of representations. These are the rational, in the sense of economical, elements of mental life, for they yield incredibly rich results by means of a minimal and even negligible import. Consciousness is in fact a manner of processing very sketchy and imperfect data in a way that allows for such powerful tools of cognition as mathematics and logic, but also for our ability to move around in space, to make plans and work towards their realisation.

Karl Schuhmann
Fakulteit Wijsbegeerte
Universiteit Utrecht
Heidelberglaan, 8
NL - 3884 CS Utrecht

NOTES

[1] In his own 1904/05 lecture course on "Main Chapters from the Phenomenology and Theory of Cognition" (quoted in Hua X, XV ff – I will quote the Husserliana volumes by the usual abbreviation 'Hua,' adding volume numbers in Roman numerals and page numbers in Arabic ones).
[2] These notes must have been among the ones Husserl handed in 1935 to the then director of the Brentano Archives in Prague, Emil Utitz. But they were apparently destroyed in the wake of the German invasion of Czechoslovakia in 1938. Unfortunately Franziska Mayer-Hillebrand's edition of some relevant Brentano materials (Brentano 1959, 3–87) is almost useless for our purposes.
[3] Husserl's two note-books from Winter 1886/87 are kept at the Husserl Archives in Louvain under signature Q 11/I and II. I use Elisabeth Schuhmann's transcription.
[4] Stumpf's "Syllabus for Psychology" from Winter 1886/87 (kept at the Husserl Archives under signature Q 12), § 18, as translated in Rollinger 1996, 229.
[5] MS Q 11/I, 247ff.
[6] This is the source of Husserl's famous fifth habilitation thesis "Nobody can, properly speaking, count beyond three" (Hua XII, 339).
[7] Q 11/II, 506.
[8] Ibid.
[9] Q 11/II, 510.
[10] Stumpf's "Syllabus," § 19 (Rollinger 1996, 230).
[11] For a more detailed treatment of this topic cf. Schuhmann 1991, 46–75.
[12] Stumpf had worked out his theory in his book-length study of 1871.
[13] Cf. Hua XXI, esp. 262, 272ff, 275ff, 286–293.
[14] Hua XXII, 103.
[15] Cf. Hua XII, 204–210; Hua XXII, 269–272. On Husserl's concept of 'fusion' which he had inherited from Stumpf, cf. Rollinger 1996, 85 ff.
[16] Cf. Hua XXII, 110.
[17] Hua XXI, 281.
[18] Cf. Hua XXII, 273: "Ideally, the unified act of the enduring intuition might be analysed into instantaneous acts."
[19] Hua XXI, 282 ff.
[20] Hua XXII, 109.
[21] Hua XXII, 280.
[22] Hua XXII, 283.
[23] In this sense Münch 1993, 168 is no doubt right in affirming "that Husserl's concept of intention is new not only as compared with Brentano, but that it also contradicts his *Psychology from an Empirical Standpoint*."
[24] Hua XXII, 406.
[25] Hua XIX/1, 525.
[26] Hua XXIII, 134.
[27] Hua XXIII, 116.
[28] Cf. Hua XIX/2, 621.
[29] Hua XXIII, 126.
[30] Cf. Hua XXI, 306.
[31] Hua XXIII, 29.
[32] Hua XXIII, 21.
[33] Hua XXIII, 141.

[34] Hua XXIII, 48, 54, 279.
[35] Hua XXIII, 110.
[36] Hua XXIII, 49.
[37] Hua XIX/1, 359 ff.
[38] Cf. Hua XXIII, 85: "The phantasy appearance [...] refers to its object as simply as does perception."
[39] Hua XXIII, 50 and 52.
[40] Hua XXIII, 9.
[41] Hua XXIII, 265.
[42] This criticism can also be formulated as follows: In the necessary relation of all acts to the ego "by 'ego' is meant the empirical ego, the human personality. Its central core [...] is the body. All 'psychical,' 'internal' experiences appear as located in it [...] Now there comes, lagging very much behind, the philosopher and contrives, by accepting the body only as an external, physical object, the notion of a pure unity of consciousness or of the purely spiritual ego [...] I am to fight the philosophical fiction of a pure ego which has grown from a mere analysis of the word 'consciousness.'" (Hua XXIII, 200). This Husserlian criticism of Natorp can easily be leveled also against later Husserl.
[43] Cf. Hua XXIII, 142 ff.
[44] Cf. Hua XXI, 273.
[45] Cf. Hua XIX/2, 587 ff.
[46] Hua XII, 340 ff.
[47] Cf. Hua XIX/1, 32.
[48] Hua XII, 354.
[49] Cf. Hua XII, 356.
[50] In harmony with what we said this is not to imply that an infinite intellect could have the objects of higher mathematics given in direct presentations; it rather means that God does not calculate (Hua XII, 192 n. 1).
[51] Hua XII, 193. This element is apparently taken over from Ernst Schröder's *Vorlesungen* published in 1890 and reviewed by Husserl a year later. See Hua XXII, 11ff.
[52] Hua XI, 3.
[53] Hua XXII, 120.

REFERENCES

Brentano, F.: 1959, *Grundzüge der Ästhetik*. Bern, Francke. Repr. 1977, Hamburg, Meiner.

Münch, D.: 1993, *Intention und Zeichen*. Frankfurt a. Main, Suhrkamp.

Rollinger, R. D.: 1996, *Husserl's Position in the School of Brentano*. PhD Utrecht, Dordrecht, Kluwer.

Schuhmann, K.: 1991, 'Intentionalität und intentionaler Gegenstand beim frühen Husserl,' *Perspektiven und Probleme der Husserlschen Phänomenologie*, Freiburg, Karl Alber, 46–75.

Stumpf, C.: 1871, *Über den psychologischen Ursprung der Raumvorstellung*. Leipzig, S. Hirzel.

JENS CAVALLIN

CONTENTS, PSYCHO-PHYSICAL PRODUCTS AND REPRESENTATIONS. SOME NOTES ON THE THEORIES OF KAZIMIERZ TWARDOWSKI

1. TWARDOWSKI'S THEORIES OF CONTENT OF MENTAL PHENOMENA

1.1 Twardowski's early theory in 'Zur Lehre vom Inhalt und Gegenstand der Vorstellungen'[1]

The content of a presentation is, generally speaking, one particular kind of *object*. All general features, notably mereological features, belonging to objects in general therefore also apply to contents. As a mental phenomenon[2] we could direct a second presentation to it, in reflection: we have a presentation of a presentation. We could also submit judgements, have feelings, etc., about it.

Twardowski's basic ambition was to clear up an ambiguity in the use of 'object' by, among others, Brentano. He replaced Brentano's 'secondary' object of presentation (the 'immanent' or 'intentional' object referred to by e.g., Höfler and Meinong in their *Logic*)[3], by the notion of *content*. The most prominent logician of Twardowski's time, Sigwart, was satisfied to distinguish between the act of presentation and 'what is presented.'

Twardowski[4] insisted upon the distinction between two senses of 'what is presented' (*das Vorgestellte*) – object and content – identifying his notion of content with Bolzano's notion of presentation-in-itself, or 'objective presentation.' Bolzano's *subjective* presentation is identified with the 'act.'

One basic feature of Twardowski's theory is the 'double role' attributed to the content of presentations, combining a realist and an 'idealist' interpretation of mental (intentional) contents, a kind of oscillation which returns in recent discussions between philosophers and cognitive scientists on 'mental representations' and their parallels in terms of physical configurations and 'symbols' (marks) in a computer or a brain.[5]

Twardowski makes an effort to dissolve this duplicity in his later theory (see

L. Albertazzi (ed.), The Dawn of Cognitive Science. Early European Contributors, 185–208.

below) by way of a theory of abstraction.[6]

What Twardowski said on mental contents in *Zur Lehre vom Inhalt und Gegenstand der Vorstellungen* (ZL for short) might be summarised as follows:

1. Contents have two roles (functions, tasks, *Aufgaben*):

1.1. To represent an object to the mind (to be its 'mental substitute').
1.2. To constitute the meaning (signification) of a name.

This view is presented rather *en passant,* e.g., in the second footnote on p. 19 of ZL, citing with approval a theory of 'ideas' by G. Noël.[7]

2. Contents are 'aroused'

This seems to apply to the first 'role' of the content. The function is based on the analogy to *names* in language expressed as follows:

[The name] arouses in the hearing person a specific psychic content)

[Der Name] erweckt im Hörenden einen bestimmten psychischen Inhalt (ZL, p. 11).

The question of a 'pre-existence' of contents in relation to the presentations in which they are *first* aroused[8] might be raised. Are we to understand the metaphor of 'arousing' or waking up, as if there were 'sleeping' contents – perhaps in the style of Gestalt?[9] – or is 'creation' taking place? In both cases the content seems to be a singular event 'in' some particular mind.

3. Contents are not private

This seems to apply to the *second* role of the content only. The act is defined as the 'subjective presentation' in Bolzano's sense, the content as 'objective' (ZL, p. 17, footnote). This kind of objectivity and the object itself are of course different, for any given presentation. Meaning is surely normally taken to be *identical* in both the speaker and the listener as presupposed above, that is for several users of a language.

4. Contents are not real

Together with the act, the content constitutes a real, singular whole. The content as such is not a thing, though it always *exists.*[10]

Although it forms together with the act a singular mental reality, while the act of presentation is always something real, the content of the presentation always lacks reality; the object sometimes has reality, sometimes not.[11]

A counterpart to 'real' in traditional philosophical discourse, and in particular in Husserl, is 'ideal.' This discussion is more explicit in the later theory of products and abstraction.

5. Contents are objective

For Twardowski there is no contradiction between being objective and psychical (mental). Interpreting also a text from 1898[12] in the same spirit, *both* 'images' and 'concepts' are objective. This is to say that two persons could have an intuitive (non-conceptual) presentation (*anschauliche Vorstellung*) with the same content. Thus the content might be shared and therefore is not determined as to space and time. This interpretation contradicts a 'picture theory' in a strong sense. Whether it also shakes the privileged (private, subjective) position of 'images' or intuitive presentations, and possibly even of 'inner perception,' for epistemological purposes is another issue – obviously a consequence not desired by Twardowski.

6. The content of *judgements* is the existence of objects

This formula[13] originates in Twardowski's analogy between presentations and judgements, as to a tri-partite structure (act, content, object). The existence of objects, if not interpreted as the *assertion* of existence, is however not naturally classified as 'psychic.' The existence of a material object, as different from the object as such, seems to involve a kind of distinction of types. Twardowski may be taken to have affirmed this. But could the content of a judgement, i.e., existence or non-existence of an object, together with the act of judging, form a psychic reality just as the presentation (act + content) does? The analogy seems to break down. As a matter of fact, the theory does not seem to reappear in later writings of Twardowski.

7. Contents could be independent or dependent of other contents

The dependence or independence of objects in general is formulated by Twardowski in mereological terms: the dependence of an object comes from its being part of a whole, its being 'founded.'[14] Two contents are said to be 'non-detachable,' if they 'could not be presented' without each other. The relation of the content of a colour presentation to the content of a presentation of extension is, for example, one of mutual non-detachability.

8. Contents are directed

Twardowski sees the distinction of content and object as one manner of explicating intentionality or directedness. The content is in Twardowski's terms *related* to an object, or *determines* an object. These terms have different connotations, some of which permit a 'picture theory' involving some kind of similarity, whereas other understandings involve a more abstract relationship of 'representation' involving for example some kind of isomorphy. Twardowski seems to attempt a merger of a theory of intentional directedness with a more general framework of structural analogy of composition. Twardowski's key formula is that the object is presented *through* (or by, *durch*) the content.[15]

9. Contents of presentations stand for (genuine) names

Twardowski solves part of the 'problem of object-less presentations' (notably the interpretation of 'nothing') by understanding 'names' as those 'categorematical' expressions, other than sentences, which have an 'independent' meaning. Some names only *look* like names, they are 'pseudo-names,' as revealed by analysis. And accordingly there is not any corresponding content.

10. All 'material'[16] parts of contents are contents

'Formal parts' – the relations of an object, in Twardowski's terminology – do not constitute parts of the content. Contents are not constituents of the content in the sense of being a content or a partial content.

11. Parts of contents correspond to parts of objects (of the same presentation) of the same *order*

(ZL, p. 71) No part of a part of a content could there-fore correspond to a part of a part of a part of an object.

12. Some parts of the content could be *implicit*

(ZL, p. 76) An example used by Twardowski refers to the points on a continuous line. For *objects* the corresponding assumption is not generally considered controversial (we do not notice everything in an object) but applied to contents as well this thesis might invalidate the *prima facie* uncontroversial thesis of the 'inadequacy'[17] of presentations.

13. Contents could be direct or indirect

(ZL, § 14) This is a thesis rich in consequences for Twardowski's later work.[18] Most of the theory presented in "On Images and Concepts" hinges upon the idea of indirect presentations. On a linguistic level it is related to the idea of an 'inner form' of language, a dominant theme in Anton Marty's work.[19] The distinction between image and concept and the determination of concepts depends on the possibility of *combinations* of presentations (contents). The same applies to Twardowski's theory of generalities (abstraction). Several contents may be combined to yield a complex content, which may however correspond to a (relatively) simple object. An indirect presentation is normally expressed in language as some kind of description – a 'rebus,' to borrow Marty's expression –[20] where the thing asked for is the object.[21]

The notion of indirect presentation might present difficulties for an interpretation of contents of presentations as the *objective* content. Conceived as *images* or pictures it might be reasonable to say that the contents of a presentation of a landscape without mountains one the one hand, and a plain on the other, are the same, but for example in the case of abstract terms, this would seem to be impossible. Twardowski suggests a solution to this problem (ZL, p. 107), which he elaborates in 'On Images and Concepts.' *All* presentations either are, or are built upon (or accompanied by), images. On the other hand, he is also very generous as to what might be considered an image. The possibility of combination explains also why Twardowski finds it so important to reject Kerry's thesis of the equivalence of the 'concept of the concept' and the 'concept'.[22]

The theory of indirect presentations is crucial to the thesis (ZL, p. 102) that a content could never have more than one object. This one object might very well be a second-order object – such as a 'general object' or a set. However, Twardowski (ZL, p. 110) admits the possibility of more perfect psychic 'organizations' able to present general objects directly.[23]

1.2 Twardowski's later theory: Actions and Products

Background

Twardowski's later theory of mental phenomena is presented in an essay from 1912 under the title 'On Actions and Products' (WPF for short).[24] In this essay he abandons the terminology of acts and contents of mental phenomena, and adopts a new conceptual apparatus of actions and products.

Twardowski contends, nevertheless, (WPF, p. 225, foot-note 15) that the new apparatus exactly corresponds to the older apparatus of acts and contents. If this holds, the theses on the content in the earlier theory should be applicable within the new conceptual apparatus.

Essentially, however, the new position rather amounts to a rejection of the earlier theory. Rather the new theory approaches theories of 'traces' of mental acts discussed at about the same time by, among others, Bertrand Russell in terms of theories of 'engrams' into the organic structure,[25] in explicit polemics against Brentano's early theory of intentionality.

Twardowski's ambition in presenting the new proposal is quite clearly to escape from some of Husserl's criticism of mentalism or psychologism. It is also likely that he has been impressed by the later Brentano's 'reism,' thus reformulating on his theory in a less 'mentalistic' language, following trends inspired by the results of experimental psychology and sensory physiology.

Twardowski actually abandoned psychologism[26] in its more pregnant form rather soon after the appearance of Husserl's *Logical Investigations*. • ukasiewicz is ascribed a direct influence on Twardowski in this regard from a logical point of view and Kotarbi• ski, another of Twardowski's prominent disciples, is usually ascribed an influence on his teacher towards 'reism.' Kotarbi• ski elaborates his entire philosophy from theory of action (praxeology) rather than theory of intentionality. Twardowski's theory might also be regarded as a turn from intentionality to praxeology.

The new theory

Twardowski's essay from 1912 carries the sub-title "Some remarks on the frontier between psychology, grammar and logic." Twardowski attaches to Bolzano, Julius Bergmann, Carl Stumpf, Meinong and, most importantly, the psychologist Stephan Witasek, as predecessors. I will summarise the theory in a few short descriptive cuts.

1. The point of departure is *grammatical*, viz. the analysis of pairs of words (verbs and verbal nouns) like to walk–a walk, to run–a run. The second member of these pairs of words – the *fact* of running rather than the action of running – is the subject of Twardowski's investigation.

2. A number of these pairs of words have the property of sometimes using what Twardowski calls 'the etymological figure,' or having an 'inner object' (complement): to dance a dance, to live a life etc. (more common in Polish, Latin and perhaps German than English). This kind of 'object' or complement, expressed by a verbal noun, is in a sense superfluous or pleonastic. One might, alternatively, use a more general term together with the complement and dispose of the verb (to 'perform a dance' or just 'to dance'). It goes without saying that this kind of object has little to do with the concept of object in the early theory. Twardowski, in compliance with later grammatical theory, also prefers the syntactical term 'complement' (*dope• nienie*).

Twardowski observes (WPF, p. 220) that the verbal noun draws the attention to the dance, the life etc. as a phenomenon appearing to someone as a result of a process, rather than an action observed, as it were, along its duration.

In the early theory objects of presentation (such as landscapes) and supposedly also dances, or the dancing, are objects of a *different* act of presentation (judgement etc.) rather than an object of the act of dancing – which as such is difficult to imagine as having an object at all. The presentation of this phenomenon (object) could belong to the person who dances, etc., or to someone observing the dance 'from outside.'

The 'inner complement' exists or occurs *thanks to*, because of, or as a consequence of some *action*. The action is not necessarily the same as the 'act' in the earlier theory. Twardowski calls the kind of object or phenomenon designated by an inner complement, a *product*.[27]

This kind of distinction is related to the grammatical distinction of the category of *aspect* of verbs, the product-designation rather expressing the perfective than the imperfective or durative aspect of the action. Twardowski notes, however, that also verbal nouns might designate both actions (crying, lying, etc. – typically expressed by the nominalised present participle in English, derived from the 'progressive form' of verbs, expressing the imperfective aspect) and end-results (products). Slavonic languages express the difference between perfective and imperfective aspects by way of different (normally rather closely related) verbs, Romance languages by the 'passé simple','imparfait' category in verbal flexion, etc.

3. Actions and products of actions might belong to two different basic kinds – essentially the traditional metaphysical dichotomy between body and mind:

1. Physical (for example a walk)
2. Psychical or mental (a thought, judgement, purpose).

Among the physical actions and products, a particular kind should be singled out, viz. the *psycho-physical actions* and products. These are physical actions/products accompanied by and dependent on mental actions. Examples of such products are: cries, songs, talks, lies, etc. Twardowski does not question the interaction between the mental and the physical spheres.

Twardowski points to (WPF, p. 222) certain ambiguities of terms denoting both actions and products. One most interesting example is 'judgement,' which might denote both the act(ion) of judging and the result of this act(ion).[28]

In very many cases there is no distinction in language between verbal nouns denoting actions and results: in English results of actions might be denoted by verbal nouns derived from participles (perhaps more common in Polish). Slavonic languages like Polish distinguish between process-verbs (mostly imperfective verbs) and result-verbs (perfective) and have verbal nouns derived from both kinds of verbs. Thus, rather elegantly, Twardowski is able to make a distinction between 'theory of knowledge' (*teoria poznania*) and 'theory of knowing' (*teoria poznawania*) – a distinction which might be used for sorting out some ambiguities linked to the notion of 'epistemology' in much contemporary use.

4. Some of the products of actions *last* or endure only during the action itself: the cry only lasts as long as someone cries (with reservations for the time it takes for a sound to cover a distance). Other products are more permanent, or last longer: a drawing lasts longer than the act of drawing, a letter longer than the writing of it, etc. (These kinds of products are perhaps the most natural carriers of the label). Twardowski underscores that these products or 'inner objects' should not be confused with the 'external objects.' Thus to paint a painting is not an action of the same kind as to sell a painting or to think of a painting (or – more complicatedly – the object of the painting). Since some of the 'inner' objects or products last, it is clear that they have an existence equal to that of objects in the 'outer' sense, i.e., ordinary objects that do not exist *thanks to* some action. They somehow remain bound by their constitutive history, or origin. In an interesting foot-note (WPF, p. 228) Twardowski argues that the principal theme of ZL was in fact to distinguish between 'internal' and 'external' objects. Then clearly the notion of 'internal object' replaces, as already noted, the notion of content of a mental act (presentation). This shows that, although Twardowski expresses himself as if the new theory were compatible with his earlier thinking, a rather radical change of approach has taken place.

One of the fundaments of the theory of contents and objects of presentation of ZL is the idea that *all* presentations (and therefore all mental acts) have both contents and objects. Replacing, in the new theory, contents by 'products' or 'internal objects,' one would be obliged – if the fundamental assumption of the general theory of objects should be retained – to find two objects (or one product and one object) for each act. Essentially then the new theory seems to amount to a return to Brentano's theory of two objects for intentional acts – a theory and conceptual apparatus which Twardowski's entire early theory was designed to overcome! In the *grammatical* sense, a theory requiring an external object (together with the internal) for each verb in a sentence would be patently wrong; that would be equal to ignoring the important category of intransitive verbs. Twardowski also (WPF, p. 228) acknowledges that some actions just have 'inner objects.' It is of course possible that Twardowski holds that a large category of actions do not at all have a mental or intentional character. However, the natural conclusion is that the change in Twardowski's outlook is more than a change of terminology. It is difficult to see that some of the actions covered by Twardowski's theory of actions and products would not be 'psychic phenomena' in the sense of ZL. Twardowski appears then to have given the idea of intentionality of all mental acts advocated in ZL quite a new interpretation, or indeed even given up the idea as such. In doing so, he would, not surprisingly, both have followed his teacher Brentano, and joined criticism against Brentano's earlier ideas presented both by behaviourist psychology and philosophers influenced by it, like Bertrand Russell.

5. Psychic products *never last*. They are completely dependent upon their 'host-actions' and do not survive them. This might be regarded as a step towards 'reism': only physical products last. *Prima facie*, 'reism' seems to exclude discussion about 'ideal' existence (which might by definition be interpreted as non-timely) of 'contents' of mental actions. This is, however, not the case, in fact Twardowski offers an explanation. (See below, item 8.)

6. Psycho-physical products may *express* psychical products. Twardowski explains this theory of expression as follows: if a psychic action is the cause (albeit partial) of some psycho-physical product, then this product in its turn might arouse (develop, call for) a similar or the same (type of) psychic product again. Twardowski says that a scream of horror only then expresses horror, if also horror arises in the listener. This rather surprisingly crude account of expression[29] is modified by a reservation about the possibility of a 'substitutive' product, viz. a *presentation* of horror in the listener instead of the actual horror. Psycho-physical products that express some psychical products are *signs* and the psychical products expressed are their *meanings*. The psychic products might be said to exist 'potentially' (WPF, p. 234) in the psycho-physical product which expresses it. The psycho-physical product in its turn 'perpetuates' the non-lasting psychical product.

A corollary of this view, unless modified by some other theory, would be that meanings never persist – although signs last. Twardowski actually does develop a theory to avoid consequences of this kind, referring to Bolzano, though it is doubtful whether he succeeds. (See below, section 9–10.)

The psychic, non-lasting product, which Twardowski intitially identifies with meaning, is a singular object (an individual), something very different from Bolzano's '*Vorstellung-an-sich.*' The situation would be radically different if Twardowski's notion of mental product could in general be understood as an entity common to several persons.[30] *Psycho-physical* products are sometimes lasting and might then be accessible to several persons at the same or different times. If social or cultural products were included (either as a subspecies of psychical or psycho-physical products) the creation or birth of meanings might reasonably be pictured as such. This kind of product must be seen as a lasting product, existing in several persons, and should not be identified with 'its' tokens in various psycho-physical products (signs).

Some of Twardowski's explanations of meanings and signs in terms of products still appear, even if the modifications described below are accepted, highly deficient. The example given, viz. that meaning of a sign (such as a scream of horror) should have its roots in the potential arousal of horror caused by this sign, might be classified as – to paraphrase Twardowski's own words – primitive associationist psychology. At least this judgement seems to be justified if signs should include linguistic signs (and not only what sometimes is called 'natural signs' like smoke from fire); it is difficult to find any empirical support for this thesis. Theories of the different 'functions' of signs, for example that presented by Karl Bühler in 1934,[31] develop some distinctions necessary to avoid these pitfalls.

7. The 'perpetuation' (*utrwalanie*) of psychic products in the form of psycho-physical products, such as texts (recorded, written, printed) is described as follows by Twardowski.

Thoughts, feelings, desires, decisions, etc., are perpetuated by perpetuating non-lasting psycho-physical products which *express* (non-persistent, as always) psychical products. For example, when we think, we perform mental actions, the product of which are our thoughts, which are non-persistent psychical products. 'Due to' actions of speaking non-lasting psycho-physical products (words, sentences pronounced) are produced. These products may be perpetuated, for example in writing them down or recording them.

8. The theory is elaborated by Twardowski – explicitly referring to Husserl's theory of 'ideal meaning' in the *Logical Investigations* – in the following steps.

First: The type-token distinction is introduced by accepting the identity of invariant (Stumpf) *features* of different tokens of psychical products expressed by a particular psycho-physical product. Precisely this set of common features (i.e., universal entities) is the meaning of the expression (on the condition that it corresponds to the *purpose* for which it is used).[32]

Second: This abstraction from the individual psychic products distinguishes meaning from the concrete singular product. The abstracted product might – Twardowski refers to Husserl[33] – be said to be an 'ideal' entity or the 'objective' presentation (*Vorstellung-an-sich*) in as far as we deal with that kind of action.

9. One particularly important kind of product serves to *substitute* for other products. This replaces 'indirect presentations' in the earlier theory. Twardowski cites as an example an artificial foot-print in the earth, which looks like a real foot-print, without being so. This artificial trace might then be

employed for the same purposes (for instance educational or illustrative) as the real trace. A similar idea of substitutivity underlies most theories of representation such as those linked to 'artificial intelligence.'[34]

Twardowski cites as another example the *rôle* played by an actor on stage who represents feelings without really having them (i.e., without having the psychic product of a feeling).

Twardowski underlines the relevance of the kind of products just outlined for various theories, first of all to logic. Judgements or types (classes) of judgements in logical theory are thus not authentic judgements but only examples or, as Twardowski says, *presented* judgements. Now, these presented judgements are substituted by psycho-physical artefacts, serving the same purpose as a real judgement. Twardowski attributes to Bolzano this kind of view on judgements in logic, fundamental to symbolic logic and logical calculus. The '*Satz-an-sich*' is the kind of non-authentic artefact which replaces the real judgement and serves the purpose of treating psycho-physical products in separation from their psychic products, as well as, more importantly, from the psychic action producing the product.

10. Twardowski ends the essay by suggesting that the adoption of the idea of products should become the foundation of a whole new area of scientific research. After the 'liberation' of logic (by Husserl?) from the idea that it concerns psychic actions, it is reasonable, Twardowski suggests, to envisage a definition of the humanities or human sciences as those sciences which treat mental products and their relations to psycho-physical products (direct and indirect, e.g., substitutive products). Twardowski refers to Stumpf's term 'eidology' for this kind of research. The idea also comes close to Husserl's ideas of eidetic science, material (regional) ontologies, etc. as well as structural research (notably in linguistics) inspired from these or similar views (Roman Jakobson and others).

2. MENTAL CONTENT, PRODUCTS AND REPRESENTATIONS IN MACHINES AND OTHER PHYSICAL DEVICES

Does what Twardowski has to say about mental acts, actions, contents and products have a bearing on more recent reflections on the human mind and especially cognition, based on insights into the extended possibilities of data processing, computer-based communication and computation?

I think it has, and – interestingly – the same applies to what might[35] be viewed as the opposite pole of this trend, viz. the hermeneutical and text-theoretical work originating in theology and philology and fortified by phenomenology into research on 'essences' and structures of culture and thought. The latter tendency might be exemplified by Derrida, whose initial philosophical influence came from Husserl, merging it with structuralist linguistics and philosophical anthropology, and of course hermeneutics.

Both these tendencies come together in a common aspiration towards 'metaphysical neutrality,'[36] in opposition to the classical differentiation between body and mind. They represent in certain respects a continuation of the Brentanian programme of a 'Psychology without soul' ('*Psychologie ohne Seele*'). In Derrida's 'deconstructivist' view a text is fundamentally something 'woven together' in a process, loading (and unloading) meaning into matter. The *kind* of matter involved is not at stake; a text might in Derrida be oral, a 'non-persistent' material, to use Twardowski's terms. Texts, as it were, 'live their own lives,' although they by definition are strictly speaking dead matter, or objects. Each new confrontation with a text is viewed as a new event, encompassing elements both of the trace/text 'itself' and its setting or environment, most pregnantly 'use' by those who take up or retrieve it. They are thus not primarily 'mental' in a sense being reserved for experiences (*Erlebnisse*) or subjectivity.

The 'act' of creating the text is, in Biblical contexts often being impossible to identify, less important. After all, the origin of hermeneutics lies in Protestant theology: Luther's slogan was *Sola Scriptura!*[37]

Analogously, the idea of intelligence as something 'neutral,' pertaining to machines as well as living matter and human minds, suggests that mental acts do not necessarily play a vital role in cognition: a most shattering blow to Descartes' belief that intelligent behaviour or thinking in a broad sense proves anything at all about *me* or about anything except itself. In Twardowski's time, researchers like Mach and Avenarius, among the German philosophical positivists, Külpe among the leading psychologists,[38] James and Russell among English-speaking philosophers of a 'pragmatist' leaning, share the ambition to break down the metaphysical border-line between mind and body. It is clear that the same aspiration nourishes the writings of many researchers in cognitive science and philosophers of a 'functionalist' tendency – the project of a unified science has still got a wide circle of advocates.

Russell exploits metaphysical neutrality for epistemological purposes: categorising 'sensations' as being part of both the physical (external) and mental (internal) world, he proposes[39] to have overcome the Cartesian gap between the internal and external. On a metaphysical level Russell's project (just like that of his inspirer Leibniz) comes close to Spinoza's rejection of dualism *sub specie aeternitatis*: the mental and the physical are just two out of the infinitely many 'aspects' of the one and only substance.

I have suggested[40] that this approach to the metaphysical analysis of mind might also be related to Husserl's 'transcendental,' or pure phenomenology.

This theory – aspiring to lay the foundation of all theory as such – rests upon the 'bracketing' of existence as such, studying pure 'objects' of intentional acts. The bracketing of existence, though non-metaphysical in its ambition (its purport is methodological) seems to be dependent upon the idea of an existence-neutral 'object' as suggested by Twardowski and others. The traditional metaphysical categories of mind and body are then transformed into different phenomenal categories within this bracketed horizon of 'neutral' objects. I do not suggest that there is a direct logical link between those different aspects of 'neutralisation,' but it is certainly tempting, from a historical point of view, to examine their mutual connections. 'Unification of science,' or a *mathesis universalis* are, both in the logical monism of Russell, and in Husserl's pure phenomenology, certainly common themes. This movement of unification may also go hand in hand with a rebellion against the uniqueness of the human mind, a tendency which is manifested in our time, both by asserting and investigating the intentionality of animals (and the rights coupled to that intentionality) and by emphasising the phenomenon of intelligence as something pertaining to both biological and inorganic matter. Twardowski does not go this far in his ambitions to break up from a theory of logic and meaning resting upon considerations of mental phenomena and a psychological method. He did, however, share the ambition to create a unified science and regarded metaphysics as a kind of scientific world-view crowning this scientific edifice. These 'monistic' ambitions were perceivable already in his early work but acquired a new form in the later theory. Taking Twardowski's views on contents and products of mental act(ion)s as a point of departure for an analysis of research into cognitive processes – and texts – the following observations might be formulated. As already intimated, the weaknesses of Twardowski's theories to a large degree also seem to be inherent in more recent cognitive research within the 'functionalist' stream. This is no wonder, since, after all, what is a 'product,'[41] if not something determinable as a function of ('dependent on') something else?

1. A product is very often not a product in the 'material' sense of being something completely *new*: at least persisting products depend on some underlying material, out of which the product could be made. You have to have paper and a pencil in order to make a drawing. The product is primarily the new order or arrangement (structure) of the material. What happens in a computer is precisely that after manipulation – input – a new arrangement of the electronic pattern of the memory-store is produced. This notion of product coincides from an ontological aspect by and large with the notion of at least one interpretation of *esse objectivum* of Scholastic ontology – as distinct from the 'unloaded' *esse naturalis.* This is might be taken as a justification for

employing the term 'object' for products also.

2. An act (of presentation, as the 'basic' kind of intentional acts) is a mental process which has or results in a 'product,'[42] for example a thought, a feeling, a desire, a judgement, imagination. This product might also, notably if it is non-persistent, be regarded as the 'content' of the act. The idea of 'inner perception' of Brentano and others is that we do have direct access to 'contents' of our own acts: therefore Brentano is prepared to use the notion of 'secondary' object for this. The *memory* of this act of presentation (or rather the 'content' of it) is not identical with the result, but a distinct kind of result or product, resulting from a new act, viz. an act of remembrance.[43]

3. The product is normally *complex.*[44] It is in this sense a 'text.' The metaphor of the text may for many reasons be preferable to more traditional physicalistic metaphorics of molecule/atom structures.

4. Some texts/products might be entirely undetermined or significantly underdetermined by physical observations and 'laws': in that case they should be considered as 'psychic.' Retrieval of these texts or products is only possible by 'private' or 'inner' methods – notably through memory or some other means of reproduction. (Let it be unsaid whether we reproduce the *same* text; criteria of identity are disputed). Phenomenology is the sphere of investigating these kinds of texts.

5. A text might also be expressed by way of being 'perpetuated' at a material or medium. Or *by* a material: any material might be given any role in this perpetuation. It is not necessary to carve any visual or sensory traces in the material as such in order to give it a meaning: a 'ceremony' suffices to 'endow' meaning to something that was before this sense-less. A piece of cloth becomes a flag, and remains so. Hjelmslev is perhaps the one who has underlined most this 'arbitrariness' of signs. Peirce does the same, essentially, by pointing to the sign-relation as a triadic relation. As a border-line case the 'inner' sphere might be accepted as a medium of expression, but normally the medium has to be physical, for example acoustic waves, electromagnetic waves, and (perhaps secondarily) more down-to-earth matter as paper, canvas, marble, clay, or silicon and copper. A pattern in a computer becomes a programme, or some other 'content' however, by a direct physical manipulation. It is not sufficient to 'imagine' a role for data stored in a computer.

The expressed text or product is an *object in a strong sense*. It has 'objective being,' which means that the 'essence' (substance) or fundamental features of it are distinct from the 'natural' features of it. The 'point' of Nike from Samothrake is not to be a piece of marble, although it is true that she is one. An object is (directly of indirectly) dependent on an intentional (i.e., a meaningful, not necessarily mental in the strict sense) act. Expression 'transsubstantiates' matter into object. Expressions (texts in a narrow meaning) are examined in 'aesthetics' in Croce's sense:[45] the general theory of '*Ausdruck*.' To 'stand for' (or to substitute) is the primitive (triadic) relation of this kind of theory. A token of one expression might substitute for another.

6. In one sense the product *is* (identical with) the act. The identification of an act is done *by way of* describing 'its' product. This should perhaps be modified to say that the identification of an act (e.g. of a presentation) as an *object* in the above meaning is done by way of its content or product. It is crucial that the act *as* an object be distinguished from the object *of* the act – which Twardowski (with Erdmann, Meinong, and perhaps Husserl in some phase of his development) describes as something analogous to Fregean '*Bedeutungen*' ('meanings/references[46]') It could truly be said, too, that an identification of an act is done by way of identifying its object 'under some aspect' or from a perspective. These perspectives or aspects could then be said to be products of an act.

For example: The act of thinking of my bicycle is an act of thinking because there is from now on a product of the thought of my bicycle. This thought is obviously not my bike. I would not dare to ride a thought.

7. The product or the text has a meaning by being established or produced by (dependent on) some intentional act designing it within a structure, determined by substitutivity.

The product itself might also, in phenomenological terms, be said to be directed towards an object – an object different from itself, naturally. Signs refer – at least in an indirect manner - although credit might be given to Peirce's view of signifying as being a triadic relation, which would imply that products do not refer in the same way as an act.

8. The product retains its meaning independently of its substrate, its matter. A sound might have the same meaning as an image or a written text. Meaning could thus be *stored* differently (in a written text, a computer, a recording, a memory).

However, the individual meaning-token is not the same as the meaning. Meaning in the 'ideal' sense is 'abstracted' or retrieved by 'variation' and substitution.

9. On a more fundamental philosophical level, however, the very first question asked, by Husserl and others, on the 'double functions' of contents *and* products seems to remain unanswered, viz.:

Is content/product private, public or both? Is it really possible to aspire to a common entity fulfilling both the function of a 'mental substitute' for the object towards which the act(ion) is directed and the meaning of 'names?' Is it possible to study those two roles or functions in one and the same theory – the theoretical criterion of identity? And, moreover, if we choose to employ the notion of content/product for psychological purposes in an empirical sense, or for computational science purposes, is it still usable for epistemology or for metaphysics, meaning theory, phenomenology of language, culture, etc.?

3. CONCLUDING REMARKS: THE FUTURE OF CONTENT

In my book on Twardowski's philosophy of object and content of presentations I suggested that this pair of terms might have served to the bitter end of their philosophical history. Indeed, looking into the conceptual history of the notion of *object*,[47] the wealth of suggestions offered, the differences of their employments and their entanglements in diverse theories seem to confirm this conclusion, as it were, by walk-over.

As for the history of the notion of *content*, the fact that Twardowski abandons the term might give some support to a similar judgement on this notion, though the material supplied, for example in encyclopaedic works, is much less abundant, indeed even sparse.

Despite severe criticism against some undifferentiated uses of the notion by Husserl, the notion of content and related concepts has continued to play an important role, perhaps less in philosophy than in cognitive science. This circumstance justifies a second look at Twardowski's theories, in particular as the notion of content actually is abandoned, passsing from the first to the second theory.

We have to some degree followed how Twardowski's later theory, constructed on the notion of a product of a psychic(mental) action, was designed in order to transform the earlier theory from a 'psychologistic' theory of acts of presentation, to a theory less tied up to psychological research in the early Brentanian sense of 'inner perception.' The theory thus responds to some of the criticism launched by Husserl against a 'realist psychological' interpretation[48] of the notion of content of a presentation. Instead Husserl suggested a 'functional' (or 'ideal') interpretation of the notion of content.

Actually, much of Husserl's later thinking on constitution, intersubjectivity and 'life-world,' suggesting cultural and social ontologies (material ontologies) as central notions for any theory of intentionality or, more generally, any theory of understanding and communication, is also based on the notion of functional or teleological system.[49]

In analytical philosophy of mind 'functionalism,' or later 'computationalism,' occupy a central place in work inspired by Putnam and discussed among others by Fodor,[50] in the context of general grammar and 'mental' language. In some of these texts the notion of content, or closely related notions, seems to resume much of its initial 'realist' flavour, in other texts the notion of content would hardly make any sense.

'Functionalism' is also sometimes a label given to the later Wittgenstein's view on language and mind, as well as (in quite a different meaning) to radical structuralism in theoretical linguistics, in Hjelmslev for example.

Reflecting whether some notion of content still offers any services in philosophy, it might, finally, be useful to turn away from the traditional linking of content to the sphere of mind in the narrow sense to theories of 'phenomena in general.'

Though the notion of content does not occur very frequently in *philosophical* encyclopaedias, a glance at some general dictionaries presents a slightly different picture.

A general encyclopaedic dictionary, such as the *Oxford English Dictionary*, includes some distinctions of a certain philosophical relevance under the heading 'content.'

As examples the following cuttings might be illustrative (I omit those parts of the articles that do not offer any special philosophical interest):

I. That which is contained in anything

1 (a) A thing contained [...]

1 (d) Psychol. (German translation, *Inhalt*) The totality of the constituents of a person's experience at any particular moment.

[References are taken from James's *Principles of Psychology* (1890), as well as from Stout's *Analytical Psychology* (1896), which is worth quoting:]

"It is an appearance in consciousness. It is what Brentano would call a content (*Inhalt*) of presentation."

Baldwin's *Dictionary of Philosophy and Psychology* (1901) [...]."(2) A constituent of any presented whole, (3) an object meant or intended by the subject [...]"[...]

4. The sum of qualities, notions, ideal elements given in or composing a conception; the substance or matter (of cognition or art, etc.) as opposed to *form*. [References from Huxley and Spencer].[...].

II. Containing capacity, space, area, extent, etc.

One illustrative example of a philosophical notion of content not linked primarily to *mental* phenomena, although certainly having a background in the philosophy of mind, 'rational' or descriptive psychology, is the notion appearing in Husserl's disputes with Schröder in the 1890s,[51] on the logical role of a theory of contents in relation to theories of extensional sets.

This use is primarily related to the formal-ontological theory of wholes and parts (mereology), and secondarily to material ontologies exploiting this kind of theory. Actually, in this kind of use 'content' might be more or less identified with *part* or constituent (all or a selection) – provided a sufficiently wide notion of constituent is allowed. The theory largely builds upon German philosophy at the end of the 19th century, among those Husserl's (early) phenomenology, but also theories by Cornelius and Theodor Lipps (well-known to both Twardowski and Husserl), as well as the general theories of objects (*Gegenstandstheorie*) by Twardowski, Meinong, Husserl, Erdmann and others.[52]

A more recent example of a related employment of the idea of content in formal ontology is provided by the investigations of the original Swedish philosopher Ivar Segelberg. Segelberg exploits the notion in his *The Concept of Property*[53] (BE for short) for a number of analytical purposes, one example being the connection between the notion of content and the very notion of analysis itself – central to most philosophical debates, and a prestigious label for many philosophical schools, notably in this century. Analysis, in the most general sense, displays (articulates) the content of any object.

Segelberg also suggests that theory of content is a better alternative in defining the notion of property (quality), avoiding the pitfalls of nominalistic and 'relational' theories. Segelberg's ontology is constructed on the basis of the notion of 'quality moments,'[54] as the ultimate constituents of being.

Some glimpses of Segelberg's theory might illustrate this general kind of use of the notion of content. He expresses the relationship of 'being the content of' by way of the following, seemingly trivial, thesis (BE, p.75):

S 1: If an object is composed of *a*, *b* and *c*, this object *contains* in some sense *a*, *b* and *c*, and could be *analysed* in those objects.

The central thesis of Segelberg is:

S 2: The relation of containment (content) is a primitive 'material' relation 'I' (*for 'Innehåll', 'Inhalt'*), comparable to the relation 'e', 'being an element in a set' of extensional set theory.

Using this kind of basic relation Segelberg proposes to define a number of other relations usually associated to the notion of content.

Segelberg rejects the idea that there is some kind of 'formal' property (such as transitivity, symmetry, reflexivity or externality-internality as defined by Moore) common to all relations of containment (content).

For another example of a general treatment of the notion of content, Marc-

Wogau, another Swedish philosopher belonging to about the same tradition and generation as Segelberg, might be cited. In a study on the content (*Inhalt*) and extension (*Umfang*) of concepts (*Begriffe*), Marc–Wogau offers a classical definition of the content of a concept as follows:

> The content of the concept – it is stated – consists of characteristics or marks (notae), which are common to the particular objects, falling under the concept. The extension, on the other hand, signifies all that falls under the concept.[55]

This definition still does not treat contents of *mental* phenomena as such.

Actually, Husserl in the *Logical Investigations* (Investigation V, Ch. 6) criticises this kind of definition as dependent on the ambiguity of the notion of content inherent in the notion of presentation (*Vorstellung*) and related concepts among the Brentanists, notably Twardowski.

Whether Segelberg's notion of a 'material' (non–formal) theory of content is tautological is left unexamined here. Ingarden's ontological investigations, for example, rather tend to amalgamate the notions of content and matter as such.[56]

Jens Cavallin
Department of Thematic Studies
Linköping University
Campus Norrköping
SE- 601 74 Norrköping, Sweden
Email: Jens.Cavallin@ituf.liu.se

NOTES

[1] For a general presentation and discussion of Twardowski's theory of contents and objects of presentations see Cavallin 1997.

[2] Whether 'artifacts' (e.g. works of art) – i.e., objects that are not presentations, judgements or any other of the traditional kinds of 'mental phenomena' – might be said to have contents in Twardowski's early sense is a less trivial issue. This kind of of object – the classical scholastic *ens objectivum* – brings us closer to the problems raised by Twardowski's later theory of 'products' of mental acts.

[3] Cf. Höfler, Meinong 1890.

[4] Though strongly criticised for this by Husserl (for example, in Husserl's unpublished review of Twardowski's book in Husserliana XXII, 353, footnote), Twardowski refers to Bolzano's disciple, his own teacher Zimmerman, for forwarding the distinction.

[5] See for example Lindström and Brinck 1993.

[6] A kind of theory which is even stronger criticised by Husserl!

[7] The text referred to is Noël's essay on names and concepts. Cf. Noël 1891.

[8] A related issue is brought up in Plato's *Menon*, presenting a kind of epistemological paradox.

[9] A Gestalt might be conceived as a kind of 'preformed' content in its entirety, waiting only for a stimulus to be triggered. A Gestalt might also be understood as a kind of sensory 'category.'
[10] The term 'real' is explained in ZL, 36. The *object* of an act might be real, irreal, non-existent, perhaps even impossible or contradictory, according to Twardowski's 'liberal' theory of object.
[11] ZL, p. 31: *Wol bildet er mit diesem* [i.e., the act] *zusammen eine einzige psychische Realität, aber während der Vorstellungsact etwas Reales ist, fehlt dem Inhalt der Vorstellung die Realität immer; dem Gegenstande kommt bald Realität zu, bald nicht.*
[12] 'Imageries,' see References. The translation of the title is not too fortunate.
[13] Cf. ZL, 9.
[14] Twardowski refers to Brentano, Höfler and Stumpf for the origin of this notion (ZL, 65, footnote). The translation into English of *ablösbar* follows usage by, e.g., Liliana Albertazzi in Coniglione, Poli, Wolenski 1993. The concept of 'foundation' (*Fundierung*) plays a vital role for Husserl's entire theory of essences (*Wesen*), ideality and 'categorial intuition' – the basis of his phenomenological epistemology.
[15] ZL, p.18.
[16] For the meaning of this notion in Twardowski, cf. Cavallin 1997, 79.
[17] ZL, 83. This thesis plays an important role in Husserl's Sixth Logical Investigation.
[18] Husserl as well pays great attention to a similar problem of indirect and direct acts or intentions, in Sixth Logical Investigation.
[19] Deriving from Humboldt (1837). See p. cvii for a definition, reminding of or perhaps even anticipating Hjelmslev's (1943) theory of two radically different levels of language: content and expression.
[20] ZL, p. 98.
[21] Cf. Frege's category of 'unsaturated' expressions or sentential functions.
[22] Kerry distinguishes, despite Frege's allegations to the contrary, between 'concept in a logical sense' and concepts 'in the head' – although Frege might not accept Kerry's 'concept in a logical sense' as a concept in a non-psychological sense. Cf. Kerry 1885/1891, 458.
[23] Twardowski's remark on these more perfect organisations might lend itself to somewhat frivolous speculations: could this ability, reminding of 'categorial intution' in a Husserlian style, be attributed to '*Übermenschen*,' angels or God?
[24] See References for both the Polish original and the English translation. A lecture in German with by and large the same content was kept in Twardowski's *Nachlaß* (box 21.30) and was edited by Johannes Brandl in *Conceptus* **75**, 1996 together with some comments and editorial notes.
[25] Cf. Russell 1961, 84. It is interesting to note the parallelism, again, between today's ideas of mental functions and the ideas brought forward by early twentieth century physiologists and psychologists. Russell's ideas on the influence of past history on mental events are taken from Richard Semon who formulated two 'mnemic principles,' one of which is the 'Law of Engraphy,' stating that excitements in an organism form 'engram-complexes' – that is, traces. Russell himself takes a reserved attitude to the idea of 'engrams,' preferring what would be termed today a functionalist attitude, involving a recognition of 'mnemic causation' in the explanation of mental events.
[26] Cf. Cavallin 1997, 36. There is a letter from • ukasiewicz to Twardowski (May 17, 1904) where • ukasiewicz says that he has "[...] the theses of Husserl, which I defended," [...] "although I did not, naturally, succeed in convincing 'the psychologicists'" (Pr• gowski, Lilienfeld and Elsner are mentioned among those). My translation. The letter is kept in box K3 in the Library of the Institute of Philosophy of Warsaw University.

[27] An excursus on the choice of expression in any of the three languages involved here might be permitted here. In the German *'Autoreferat'* of his article Twardowski uses the terms *Funktion* and *Gebilde* (configurations) respectively. The term 'product' in English might in some aspects be rather deviant from ordinary use of language. In some examples the notion of *result* would fit better.
The Polish word is *'wytwór,'* derived from *'tworzy•* ,' (create) and the prefix *'wy-'* (out-), marking the end or outcome of an action or process. Twardowski cites a number of sources for his choice of terms: Bolzano talks about a judgement or an assertion (*Behauptung*) as *'etwas, das durch Urteilen* [...] *und Behaupten hervorgebracht ist,'* in *Wissenschaftslehre,* I 1837, 82. (quoted after WPF, 220–1, foot-note). Julius Bergmann in 1879 suggests *'Erzeugnis, Werk, Gebilde,'* Stumpf *in 1907* uses a concept of *'Gebilde psychischer Funktionen.'* The closest predecessor, however, seems to be Stephan Witasek, a disciple of Meinong, professor of psychology in Graz and the author of a number of works, e.g., on the perception of space. Meinong's notion of *'Objektiv'* – to which Russell's concept of 'proposition' should be linked – is also a related notion, but used slightly differently, according to Twardowski. The last term is of course risky – due to the dichotomy of object and content in Twardowski's early theory – but close to Twardowski's own use in the present essay of 'inner object,' and indeed to ancient uses of the term object, and its translations (most interestingly perhaps Eckehart's term *'Gegenwurf.'* Twardowski's sense of the notion of function seems to be different from both Husserl's and that of later thinking on meaning and mind.

[28] Notice here the abandonment of both the term 'act,' together with the term 'content.' The product of the action of judging is the 'judgement in a logical sense.' Twardowski advises to use the term 'judgement' only of the latter. He also recalls two other uses, as a 'disposition' ('Ms. X has got a good judgement,' i.e., she is a reasonable person), and, more importantly, as a psycho-physical 'token' of a judgement, i.e., an enunciation or *'Aussage.'*

[29] The psychologist cited by Russell in *The Analysis of Mind*, Richard Semon, seems to subscribe to some similar view – obviously a relic from associationist psychology.

[30] One way of construing this might be as a 'split' complex individual – like a cloud or a state having several spatially split parts. This way of looking at it, however, might give a somewhat magical impression.

[31] Bühler 1934. Bühler himself was a disciple of Husserl and was certainly inspired by the 'act' conception in psychology when forming his theory of language, which is likely to have played a major role for the later development of the 'speech act theory,' by Wittgenstein and others.

[32] It might be asked whether this purpose is universal or individual, which would only bring the issue one step further, without solving the problem.

[33] Husserl would most likely have rejected this kind of reference, as demonstrated by his harsh criticism of 'modern theories of abstraction' and some notions of representation in the Second Logical Investigation. Twardowski does not directly take up Husserl's criticism of his own work. Only in 1928 (cf. Cavallin 1997, 31) did he read Husserl's unpublished review of ZL, although he, as related here, read and was influenced by the *Logical Investigations* by first explicitly rejecting the idea that the singular psychic products (contents in the first of the 'double functions' above) would be the kind of objects used in grammar and logic. Actually, Twardowski's theory of abstraction as constituting ideality seems to be a good example of what Husserl attacks. Ultimately, the dispute comes close to traditional philosophical disputes on the status of universals. Twardowski, though accepting Husserl's criticism of psychologism, seems, by trying to rescue the idea of contents of mental phenomena under another form, viz. products, to come much closer to a nominalistic position. Cf. footnote 12 above.

[34] 'Physical symbol' is the key term in some texts on AI, for example, cf. Newell and Simon 1976.

[35] Philosophy, perhaps in particular philosophy of mind, often comes into controversies of a more ideological nature. Philosophy itself, at least its Western variety, has been remarkably divided into more or less hostile schools for a rather long time – camps denoted by labels like 'continental,' 'phenomenological,' 'hermeneutic' on one side versus 'Anglo-American,' 'analytic' (positivist), and 'linguistic' on the other hand. The reconciliation of differences in the last decades is to some degree due historical research into the common roots of some of the major doctrines in both schools.

[36] Cf. Cavallin 1996.

[37] It might seem ironical that precisely this 'empire of the text' in Protestant tradition paved the way for a scientific and critical study of Biblical texts – a study which led to a break-down of the authority of the text. Some might consider this development a disproof of the idea of Revelation constructed on the basis of *Sola Scriptura*!

[38] See Kusch 1995, 138 ff. for an overview of some trends of thought in this time.

[39] Unsuccessfully, I have argued elsewhere, in an unpublished paper. Cf. Cavallin 1968.

[40] Cf. Cavallin 1996.

[41] A discussion of Twardowski's theory of products in relation to some theories of thought by Frege and Brouwer is presented by Placek 1996.

[42] As noted, there are much better terms in other languages, '*Gebilde*' or '*Erzeugnis*' in German, or '*alster*' (from a word meaning to 'create' or 'breed') in Swedish, or, of course, the excellent Polish '*wytwór*'. Temporal consecutivity is not necessary: the cry is contemporary with the process of crying, the thought is 'in' the thinking.

[43] 'Memory' is not understood here as the *faculty* of remembering. This kind of discussion brings in the extremely difficult problem of 'inner time,' treated by Brentano, Husserl and others, though with limited success. Questions to be asked are: *when* does the originary act end, what components are required to account for the phenomenology of time as such and what is the relation of memory of more 'distant' events to 'retention,' which is part of the originary act?

[44] Twardowski devotes some reflections on the possibility of simple objects as such (ZL, 68) but satisfies himself with stating that some objects are presented *as* simple. Strictly speaking no object could be simple, if one accepts the concept of 'formal constituent' (for example relations) developed by Twardowski.

[45] A sense radically different from Kant's, which deals in fact with *im*pression rather than with expression.

[46] The translation of the Fregean term into English is always problematic: 'reference', 'referent', 'object', 'meaning' might be proposed, but each brings its own shade or connotation.

[47] Cf. *Historisches Wörterbuch der Philosophie* Vol. VI, 1026–1052. Article on 'Objekt' by Kobusch.

[48] Husserliana XXII, 352 illustrates this view, notably as a criticism of Twardowski.

[49] Husserliana III, 213.

[50] Cf. Fodor 1974, and in particular Block 1978.

[51] Cf. Husserliana XXII.

[52] Cavallin 1997, Section 4.4.

[53] Segelberg 1947. The translation is mine. Segelberg's works have recently been translated into English. See Segelberg 1999.

[54] In other theories the term is 'qualia' – Segelberg's use of 'moment' is not the same as Husserl's but denotes an instantaneous token of a quality.
[55] "*Der Inhalt des Begriffs – so sagt man – besteht aus Kennzeichen oder Merkmalen (notae) die den besonderen unter den Begriff fallenden Gebilden gemein sind. Der Umfang dagegen bezeichnet alles das, was unter den Begriff fällt.*"(Marc-Wogau 1936, 8.) Translation mine.
[56] Ingarden 1962, § 32. Formal-ontological notions of content, where the world is regarded as a complex whole, having a content (parts) of various types, categories, etc. occur in theories of Wilfrid Sellars, Nelson Goodman, Gustav Bergmann and others. Lesniewski viewed this kind of notion as an alternative to Frege's 'extensional' set theory for the avoidance of antinomies. Also Hermann Weyl and Heidegger have used aspects of this theory. A more recent example of ontological theory is represented by Segelberg's Swedish disciple Ingvar Johansson in his work on ontology. See Johansson 1989.

REFERENCES

Andersson, Åke E. and Sahlin Nils-Erik, eds.: 1993, *Huvudinnehåll.* Nora, Nya Doxa.

Albertazzi, L.: 1993, 'Brentano, Twardowski, and Polish Scientific Philosophy,' in Coniglione, Poli and Wolenski, eds. 1993, 12–40.

Bergmann, J.: 1879, *Reine Logik.* Berlin, XXXX.

Block, N.: 1978, 'Troubles with Functionalism', in Savage, Wade, C. ed., 1978, 261–325.

Bolzano, B.: 1837, *Wissenschaftslehre.* Sulzbach, Pustet.

Brandl, J.: 1996, 'Kazimierz Twardowski über Funktionen und Gebilde. Einleitung zu einem Text aus dem Nachlass,' *Conceptus*, **75**, 145-156.

Bühler, K.: 1934, *Sprachtheorie.* Jena, Fischer.

Cavallin, J.: 1968, *Sensibilia in Russell's 'Analysis of Mind'.* Unpublished MA thesis.

Cavallin, J.: 1997, *Content and Object. Husserl, Twardowski and Psychologism.* Phaenomenologica 142, Dordrecht, Kluwer.

Cavallin, J.: 1996, 'The Metaphysics of the Analysis of Mind,' *Axiomathes,* **3**, 337–354.

Coniglione, F., Poli, R. and Wolenski, J. eds.: 1993, *Polish Scientific Philosophy.* Amsterdam, Rodopi.

Croce, B.: 1967, *Aesthetic as Science of Expression and General Linguistic.* En. tr. by Douglas Ainslie, London, Owen.

Fodor, J.: 1974, *The psychology of language.* New York, McGraw Hill.

Goodman, N.: 1966, *The Structure of Appearance.* Indianapolis, Bobbs-Merrill.

Höfler, A. and Meinong, A.: 1890, *Philosophische Propädeutik: Logik.* Wien/Leipzig, Hölder.

Humboldt, W. von: 1836-39, *Über die Verschiedenheit der menschlichen Sprachbaues und ihr Einfluß auf die geistige Entwicklung des Menschengeschlechts.* Berlin, Kgl. Preuss. Akademie der Wissenschaften.

Husserl, E.: 1968, *Logische Untersuchungen.* Halle, Niemeyer.

Husserl, E.: 1950, *Ideen zu einer reinen Phänomenologie und phänomenologischen Philosophie.* Husserliana, Vol. III. Den Haag, Nijhoff.

Husserl, E.: 1979, *Aufsätze und Rezensionen 1890-1910.* Husserliana, Vol. XXII. Den Haag, Nijhoff.

Husserl, E.: 1996, *Logik und allgemeine Wissenschaftstheorie.* Husserliana, Vol. XXX, Den Haag, Nijhoff.

Hjelmslev, L.: 1943, *Omkring sprogteoriens grundlæggelse.* København, Munsksgavard.

Ingarden, R.: 1962, *Spór o istnienie • wiata.* Warszawa, PWN.

James, W.: 1890, *The Principles of Psychology*. Boston, Holt & Co. Repr. 1950, New York, Dover Publications.

Johansson, I.: 1989, *Ontological Investigations*. London, Routledge.

Kerry, B.: 1885/1891, 'Über Anschauung und ihre psychische Verarbeitung,' *Vierteljahrsschrift für wissenschaftliche Philosophie*, **9–11**, 13–15.

Kusch, M.: 1995, *Psychologism*. London, Routledge.

Lindström, S. and Brinck, I.: 1993, *Artificiell intelligens - tankar utan innehåll?* (Artificial intelligence – thoughts without content?), in Andersson and Sahlin, eds., 1993, 121-147.

Marc-Wogau, K.: 1936, *Inhalt und Umfang des Begriffs: Beitrag zur Theorie des Begriffs*. Uppsala, Almqvist & Wiksells. Leipzig, Otto Harrassowitz.

Münch, D.: 1992, *Kognitionswissenschaft*. Frankfurt, Suhrkamp Taschenbuch.

Newell, A. and Simon, H.: 1976, 'Computer Science as Empirical Inquiry. Symbols and Search,' *Communications of the Association for Computing Machinery*, **19**, 113-126.

Noël, G.: 1891, 'Noms et concepts,' *Revue Philosophique*, **31**, 463–483.

Pelc, J. ed.: 1979, *Semiotics in Poland 1894-1969*. Dordrecht, Kluwer.

Placek, T.: 1996, 'Thought as a Product of Thinking,' *Conceptus*, **75**, 191-203.

Ritter, XX. ed.: 1971, *Historisches Wörterbuch der Philosophie*. Basel. Article on 'Objekt' Vol. VI, 1026–1052.

Russell, B.: 1961, *The Analysis of Mind*. London, Allen & Unwin, 6th ed. (1st. ed. London 1921).

Wade, C. Savage ed.: 1978, *Perception and Cognition. Issues in the Foundations of Psychology*. Minnesota Studies in the Philosophy of Science, Vol. 9. Minneapolis, University of Minnesota Press.

Segelberg, I.: 1947, *Begreppet egenskap* (The Concept of Property). Göteborg.

Segelberg, I.: 1999, *Three Essays in Phenomenology and Ontology*. Stockholm, Thales.

J. A. Simpson, J.A. and Weiner, E. S. C. eds.: 1989, *The Oxford English Dictionaryr*. Oxford, Clarendon Press 2nd ed.

Stumpf, C.: 1907, *Erscheinungen und psychische Funktionen*. Berlin, Verlag der Königl. Akademie der Wissenschaften.

Twardowski, K.: 1965, *Wybrane Pisma Filozoficzne* (Selected Philosophical Writings). Warszawa, PWN.

Twardowski, K.: 1995, 'Imageries.' En. tr. of 'Wyobra• enia i poj• cia' (in WPF,114–197) by Agnieszka Lekka-Kowalik, *Axiomathes*, **1**, 79–104.

Twardowski, K.: 1979**,** *On Actions and Products (*En. tr. of *O czynno• ciach i wytw• rach,* WPF, 217–240), in Pelc, ed., 1979. Repr. in Twardowski 1999, 103-132.

Twardowski, K.: 1894, *Zur Lehre vom Inhalt und Gegenstand der Vorstellungen*. Wien, Hölder. En. tr. 1977, by R. Grossmann, The Hague, Nijhoff.

Twardowski, J. 1999: *On Actions, Products and other Topics in Philosophy*, ed. by J. Brandl and J. Wolenski, Rodopi, Amsterdam.

ELIZABETH R. VALENTINE

G.F. STOUT'S PHILOSOPHICAL PSYCHOLOGY

1. INTRODUCTION

My aim in this chapter will be to present some of the chief doctrines of George Frederick Stout's philosophical psychology, particularly those which may have a bearing on current issues in cognitive science. As the main purpose is one of exposition, I shall rely heavily on quotations from his writings. Stout was an armchair psychologist: experimental psychology was for him a source of information rather than a substitute for thinking. It is peculiarly difficult to give an account of his philosophy for several reasons. The various strands are all closely interwoven. For example, embedded in his concept of noetic synthesis are criticisms of association, the doctrine of unity, and the role of thought in sensation. His philosophy is a synthesis. Mace reports that he once gleefully remarked: "I have got them all in my system."[1] Stout favours the via media; thus each statement must be qualified, easily leading to apparent paradox. In my view, these are more serious sources of difficulty than his terminology, though this can be confusing. On the one hand, he expresses new ideas in old-fashioned ways ("He hid his twentieth century light under a nineteenth century bushell").[2] More commonly, he expresses the same ideas in different ways.

2. BIOGRAPHY

George Frederick Stout was born in the north east of England in 1860, the son of a shipbroker. He did well at school and went on to study at Cambridge, where he graduated with first class honours in both Classics in 1882 – achieving special distinction in ancient philosophy – and Moral Sciences the following year, achieving special distinction in metaphysics. Following graduation, he came under the influence of James Ward, who turned his attention to psychology. He was elected Fellow of St John's College Cambridge, appointed University Lecturer in 1887 (where his pupils included G.E. Moore and Bertrand Russell), first Anderson Lecturer in Comparative Psychology at Aberdeen in 1896, first Wilde Reader in Mental Philosophy at Oxford in 1899, and Professor of Logic and Metaphysics at St Andrews in 1903, where

L. Albertazzi (ed.), The Dawn of Cognitive Science. Early European Contributors, 209–223.

he remained until his retirement in 1936. He was editor of *Mind* from 1891-1920. In 1939 he emigrated to Australia, where his son, Alan, had become Professor of Moral and Political Philosophy at Sydney, and died there in 1944.

He had a superb intellect, loving nothing better than to engage in philosophical discussion, at which he was brilliant – despite his deafness, which could be an advantage! He was one of the best read men of a reading generation,[3] with an insatiable interest in books and ideas.[4] He had a freshness and liveliness of mind – a perennial youthfulness. He was a good friend and a shrewd judge of character. He valued simplicity, directness, honesty and integrity, tolerating everything except humbug.[5] Broad (1945) describes him as always "sane, sensible and decent." His life, like Locke's, was "natural, easy and unaffected;" he was a living example of the joy to be found in academic life. His wit, puckish humour, absent-mindedness and engaging eccentricities were known to his friends.

3. INFLUENCE AND ASSESSMENT

Stout was particularly influenced by Herbart's concept of apperception and the possibility of a dynamic theory of knowledge; and by Bradley's and Ward's critiques of the associationist doctrine, for its particularity and atomism. He anticipated modern reactions against the sensationist and associationist traditions, though his arguments were not always original. Perhaps his originality lay in linking dynamic Wundtian voluntarism with the cognitive aspects of Herbartian apperception.[6]

Stout has been described as "the first of the moderns rather than the last of the ancients."[7] He moved with and ahead of his times in every way but one, never conforming to terminological fashion. I believe he became out-moded on account of his old-fashioned terminology and his exclusive reliance on armchair analysis. Kusch (personal communication) has suggested that his old-fashioned terminology may have reduced his influence amongst philosophers while his lack of experimental evidence may have lessened his attraction to psychologists. In addition, he spent most of his life secluded in the remote outpost of St Andrews.

He influenced Samuel Alexander and William McDougall who drew on his doctrine of conation for their own theories; and Spearman who acknowledges his debt to him – his third noegenetic principle of the eduction of correlates was based on Stout's concept of relative suggestion. Many of Stout's ideas anticipated Gestalt psychology. Schaar (1996) has demonstrated that Stout was the mediator between the ideas of Brentano and Twardowski and the realist theories of Moore and Russell, both students of Stout's. Stout pointed to ways of escaping the solipsistic predicament and attempted to work out a theory of perception which would soften the intolerable opposition between the world of phenomenology and the world of physics.

4. PUBLISHED WORKS

Stout wrote four major books and over sixty journal articles, published mainly in *Mind* and the *Proceedings of the Aristotelian Society*. His *Analytic Psychology* (AP), first published in 1896, was an elaborate treatise in two large volumes, a work of reference for psychologists, which went into three editions. By common consent his best book, it established his reputation. Flugel comments that it is:

> well worthy of greater attention than it usually receives from the modern student, for it is at once an acutely penetrating and profoundly satisfying book, and it is astonishing how many anticipations – or, at east, adumbrations – of later twentieth-century developments are to be found in it.[8]

Others complained of the tendency to lapse now and then into an over-refinement of analysis, finding its style rather heavy and slow. It was intended to provide a preliminary analysis of our actual developed consciousness.[9]

> This department of psychology is purely analytical and largely introspective [...] its aim is to discover the ultimate and irreducible constituents of consciousness in general. The only modern writer who appears to have fully realised the importance of this preliminary inquiry is Brentano.[10]

It was to pave the way for the genetic treatment of psychology, "undoubtedly the most important and interesting [...] reserved for a future work."[11]

This exposition of psychology from a genetic point of view was provided by the *Manual of Psychology* (MP),[12] first published in 1898, which became the standard textbook on psychology for students in British universities, going into four further editions (1901, 1913, 1929, 1938). Its unusually wide circulation made it his best known work. It rests on the AP; it is not a textbook in the ordinary sense of the word but an original contribution to psychology.[13] *Mind and Matter* (1931) was the much delayed and incomplete publication of the first series of the Gifford Lectures in Edinburgh from 1919-21, and *God and Nature* (1952) the posthumous publication of the second series.

5. CRITICISM OF ASSOCIATIONISM

Stout was "the most vigorous and influential critic of associationist psychology."[14] He reinforced the powerful attacks developed by Bradley and Ward. "Introspective analysis easily reveals that [the synthesis of the data of different senses in the perception of a single object] cannot be accounted for by association of ideas."[15] "Association of ideas forms no part of the ordinary perception of things."[16] Stout allows that "the associative principle [...] comes into play in so far as the transition from a given sensory experience to a corresponding percept is strengthened by repetition,"[17] but the laws of literal resuscitation, revival, or reinstatement of which Bain speaks are "founded on a false view of the nature both of association and of construction:"[18] "the process as described is a sheer impossibility and absurdity."[19]

He points to three errors:

1. The exclusive emphasis on mere combination. Against this Stout urges that:

> every new synthesis results from the further determination of a psychical whole which in some way already pre-exists. The new synthesis consists in the distinction and definition of the parts and relations within this prior whole [...] It is like the shuffling of a pack of cards; the various combinations which take place presuppose the general and relatively indeterminate combination, in virtue of which the cards constitute a pack at all.[20]

2. Failure on the part of the associationists to recognise the apprehension of a *form* of combination as a distinct psychical element:

> the presentation of a form of synthesis is as distinct from the presentation of the elements combined considered apart from their union, as the presentation of red is distinct from the presentation of green.[21]

3. The disposition to regard mental elements as entering into new combinations, without themselves undergoing transformation in the process. "Elements which enter into a new whole, receive new qualifications from their relations within this whole."[22] Stout provides various colourful examples:

> (a) The whist-player needs not be told that in every fresh deal the several cards become qualified by their relations within the new whole. To fit in to the design and plan of a building the stones must be hewn into shape.[23]

> (b) It would obviously be an absurdity to attempt to account for the organisation of an army merely by the contiguous adhesion of the soldiers *inter se*, apart from the descending scale of subordination to officers. But it is an absurdity of an exactly analogous nature, though much greater in degree, to attempt to account for the systematic unity of the human mind by mere association.[24]

(c) A navigator discovering he is sailing round an island:

> the parts to be put together are not initially apprehended in pure isolation from each other. The navigator starts by considering them under a certain point of view [...] The process of piecing them together is only possible because it is also a process whereby this relatively vague and indeterminate view receives progressive determination in detail.[25]

There are three prongs to his argument: (1) anti-atomism, stressing the general rather than the particular; (2) wholes or unity; and (3) direction.

(1) General versus particular

The ultimate root of Professor Bain's fallacy has been laid bare by Mr. Bradley. It lies in the tacit assumption that association is a link between particulars, as such.[26]

As against the particularism of the associationists, Mr. Bradley lays down the axiom that 'Association marries only Universals' [...] the connection which is operative in the process of revival is not between atomic particulars as such, but between general elements of content which they have in common.[27]

Stout reformulated the Law of Association in two ways, one of which was by introducing the principle of *Relative Suggestion*, which foreshadows stimulus and response generalisation. He introduces the term with the example of Kepler's discovery of the orbit of Mars:

The path of the planet between the observed points had to be mentally supplied by Kepler, in accordance with analogy. A process of this kind is what I propose to call *Relative Suggestion*, a term adapted from Thomas Brown.[28]

He defines it as follows:

The most general formula for it is: *If the presented content b has formed part of a presented whole bc, then the presented content β, when it recurs, will tend to call up a whole βγ formally corresponding to bc*. But the simplicity of this ultimate principle may be complicated in endless ways. The β which corresponds to *b* may be a simple modification of b, as a lighter grey is a modification of a darker grey; or again, it may involve the relation of *b* to a new context, and it may happen that the constituents of this new context have associations of their own, which contribute to determine the productive-reproductive process.[29]

Examples of different types are given, such as the fluctuation of the meaning of words in language with the context; continuing a series; or producing a parallel series, as in melodic transposition.

(2) Wholes or unity

Apprehension of a whole independently of the apprehension of its component details Stout calls *implicit apprehension*. Here his exposition presages that of the Gestalt psychologists:

our cognisance of the form of combination characteristic of a whole is a mode of consciousness distinct from our cognisance of its constituents.[30] I have a prenotion or presentiment of the whole before I resolve it into its components.[31] The nature of any whole is determined, not merely by the nature of its constituent parts, but also by the form of their combination.[32]

Transposition of a tune is again given as an example.

Mace addresses the issue of the relation of Stout to Gestalt psychology in the preface to the fourth edition of the *Manual*:

to claim 'anticipation' would be ungenerous and inaccurate. It would be truer to say that from the time of his earliest writings Professor Stout was in sympathetic accord with the reaction against the sensationalistic tradition, and in part on the grounds which have led to the more developed form of the Gestalt psychology. But to the latter belongs the credit of carrying the campaign in a systematic way into the experimental field, hitherto the stubbornest of strongholds of sensationalist Psychology.

In part, however, the similarity of their doctrine is based on different and independent grounds, and in some respects Professor Stout goes further than the exponents of this school would be prepared to follow. But whilst largely independent, the two developments are on many fundamental points in substantial agreement.[33]

The fifth edition of the *Manual* includes an appendix on Gestalt psychology, written by R.H. Thouless, with a Supplementary Note added by Stout, clarifying his position on the issues raised by the Gestalt psychologists. There are two main points on which he was not satisfied with their psychology of sense-perception. The first is their almost complete denial of the part played by past experience in the perceptual process: "it seems to me that they have thrown out the baby with the bath-water. The bath-water is associationism; the baby is the use of empirical explanation in the psychology of perception."[34] The second is their neglect of the difference between sensation and perception: "the distinction is, I maintain, undeniable and of primary importance both for the theory of knowledge and for psychology."[35]

What constitutes perceptual appearance is that something seems to the percipient to exist objectively, which may or may not really exist, and which may not even be believed to exist. The sensible appearance, on the contrary, is not seeming at all. Neither is it anything which may seem to exist without in fact existing. It is actually experienced and therefore must actually exist.[36]

In Wertheimer's experiments on apparent motion, the sensible appearance of motion is:

unaffected by the knowledge that there are objectively two stationary illuminated points where these seem to be a single moving point. On the other hand the perceptual appearance is not unaffected by knowledge of the objective fact. With such knowledge seeing is no longer believing.[37]

This strikes me as analogous to Pylyshyn's (1984) notion of cognitive penetrability.

On the differences between Stout and Gestalt, Passmore quotes a letter from Koffka to Stout:

As you say in your letter, you derive the supersummative, unitary character of melodies – or any other temporal wholes – from the unity and continuity of interest [...] My system demands that melodies must be whole apart from the interest of the listener.[38]

(3) Direction

The laws of mental grouping [...] are merely modes in which conscious striving seeks satisfaction. To regard them as containing by themselves the explanation of the grouping and sequence of presentations is like the attempt to explain the course of a vessel by reference to the arrangement of the sails, without taking into account the existence and direction of the wind.[39]

The other way in which Stout reformulated the Law of Association was by re-expressing it as the Law of 'Continuity (of Interest)' rather than Contiguity (of Presentation). It thus became a conative, functional, goal-directed process rather than a purely cognitive or mechanical affair. For two items to become associated, mere temporal contiguity is insufficient: they have to be relevant to attaining purpose.

The truth is that the most important condition of association is not mere contiguity in the strict sense of temporal continuity of attention, but also continuity of interest.[40] Throughout it is the reference to the proposed end which at once gives unity to the series and constitutes its *raison d'être*.[41] Trains of ideas, like trains of perceptual activity, have, in general, a certain unity and continuity of interest. They subserve some end, practical or theoretical.[42]

There is a tendency ideally to reinstate those objects which are relevant to the general trend of mental activity at the moment of recall. The sight of rain will suggest an umbrella if we are intending to go out; otherwise it may only suggest the idea of somebody else getting wet. If our minds are occupied with scientific discussion, the word 'proofs' will suggest one group of ideas; if we are engaged in preparing a book for the press, it will suggest something quite different.[43]

Schematic apprehension is the implicit apprehension of a whole combined with successive apprehension of each of its components, so as to control the order of their emergence and exclude intrusion of irrelevant objects. It links the idea of the whole with that of direction.

There is present in the one case a mental synthesis which is absent in the other [...] The vivid, distinct and persistent apprehension of the train of events as a whole, so controls and guides the ideal train as to prevent divergence into cross series which would interrupt and tangle the narrative.[44]

So far then as the implicit idea or perception of a whole determines the successive emergence of its parts in consciousness, we may apply to it the term 'schematic apprehension.' This schematism is not without affinity to that of Kant.[45]

6. NOETIC SYNTHESIS

A central theme in Stout's philosophy is *noetic synthesis*: 'synthesis by the intellect.'

> By *noetic synthesis* I mean that union of presentational elements which is involved in their reference to a single object; or, in other words, in their combination as specifying constituents of the same thought. It is by noetic synthesis that those complex psychical units come into being which we call percepts, ideas, and concepts. All these words imply something which is perceived or conceived, or of which we had an idea; and it is this objective reference which constitutes each of them a unit in mental process.[46]

To the extent that the next step in a train of thought is determined by "the controlling influence of the central idea of the topic with which the whole series is concerned," noetic synthesis is operative; to the extent that it is determined by the special idea which last emerged, the principle of association is operative. The contrast between 'contiguous adhesion' and noetic synthesis is also shown in the passage from automatic processing to 'thought-control:' "In proportion as automatism supervenes association becomes substituted for thought-control."[47]

Stout distinguishes noetic synthesis from Wundtian apperception:

> What I termed noetic synthesis undoubtedly has a rough correspondence with what Wundt called apperception. There is, however, an important distinction. Noetic synthesis owes, in my view, its peculiarity to the introduction of a distinct kind of mental fact, the apprehension of the whole which determines the order and connection of the apprehension of parts [...] Such a conception is widely different from that of the reaction of consciousness upon its own content, which seems to form the essence of Wundt's 'apperception.'[48]

7. RELATION BETWEEN SENSATION AND CONCEPTION

The distinction between thought and presentation, together with the insistence that they must be distinguished only as separable phases of the one cognitive process, was characteristic of Stout's epistemology and set for him his special problems.[49]

8. SENSATION DISTINGUISHED FROM CONCEPTION

In the *Analytic* Stout distinguishes thought-reference from presentation:

> In the process by which we take cognisance of an object two constituents are distinguishable: (1) A thought-reference to something which, as the thinker means or intends it, is not a present modification of his individual consciousness. (2) A more or less specific modification of his individual consciousness, which defines and determines the direction of thought to this or that special object; this special mode of subjective experience we may call a *presentation*.[50]

In the *Manual* Stout distinguishes perceptual appearance (alias activity factor) from sensible appearance:

the development of our knowledge of the material world depends on two main conditions in the most intimate union and interdependence. These may be called the 'sense-factor' and the 'activity-factor.' The activity-factor is essentially involved in our apprehension of the independent reality of physical objects and their qualities [...] By the sense factor is meant the actually experienced sensa which enter as essentially independent into sense-perception.[51]
[T]he sensum and the character of the physical object are not identical [...] nor can we identify the sensum with what we may call the perceptual appearance – with what the physical object *seems* to be to the percipient in the act of perceiving it.[52]

He gives the following examples:

(a) though there is no sensible appearance either of the top of the table where the book covers it or of the under-surface of the book, there is a perceptual appearance of both;[53] (b) Whenever one thing is perceived as being behind another [...] the perceptual appearance is present without any corresponding sensible appearance.[54]

9. INTERDEPENDENCE OF SENSATION AND CONCEPTION

In the *Analytic* Stout allows the possibility of pure 'anoetic sentience' (i.e., simple sensory experience devoid of any form of awareness of anything outside sense experience) but in all subsequent works he denies this. Sensation and thought are inextricably inter-mixed; they are 'inseparably interpenetrate' from the outset. On the one hand, conception implies sensation: thought is "inseparably blended with sense presentation and, so to speak, embedded in it." "Thought is discriminative only in so far as it has *presentation* for its vehicle."[55] On the other hand, sensation implies conception. All sense experience involves thought which "transcends the immanence and immediacy of sense." Stout's final position was that thought is as primary as sensation; every sense experience carried with it reference to something beyond itself.

Our perception of the world is coloured by certain antecedent expectations. We look for things to which appearances and phenomena belong. *Categories* are universal principles of relation holding either for all knowable objects or for all of a certain kind. The awareness of the categories of space, time, thinghood and causality is 'primary' or 'original.' "Our position is that such categories belong even to rudimentary perceptual consciousness as a condition of its further development."[56] They include: spatial unity – "all extended bodies are extended in one and the same space;"[57] temporal unity – "any particular duration or change is, from the outset, apprehended, however vaguely, as having a 'before' and 'after,'"[58] and causal unity – the unity of different attributes as belonging to the same thing.

The view that sensation involves thought is akin to top-down processing, originating in cognitive psychology with the so-called 'New Look' in perception; but the converse position, that thought involves sensation, goes beyond cognitive psychology to Mach and positivism.

The following passage from the *Analytic* will serve to summarise this section and link to the next:

> Whatever is perceived is recognised as such or such; and however vague and rudimentary the recognition may be, it implies a reference to something beyond the given object. The object comes before consciousness as an instance, or example, or particular appearance *of* something which may have *other* instances, or examples, or particular appearances. The word 'other' implies a reference beyond this particular object; a reference to what, for psychological purposes, we may regard as a whole, of which the presented particular is a constituent part. This whole is an object of implicit apprehension, and in all human perception, at least, some such implicit apprehension appears to be involved.[59]

10. EPISTEMOLOGY

Stout never lost his conviction that epistemology was the key both to philosophy and to psychology.[60] The theory of knowledge was always a main theme. At the beginning of the *Analytic* Stout maintains that,

> It is a primary problem of psychology to investigate how such knowledge [of the material world] comes into being [...] Psychology investigates the history of individual consciousness, and this coincides with the history of the process through which the world comes to be presented in consciousness.[61]

11. REALISM

Passmore maintains that Stout tried to steer a via media between absolute idealism and 'new realism,' but in my view he comes much closer to realism than to idealism. Stout challenged the solipsistic premise that our knowledge of the external world is an inference based on sensations and images which alone are directly and certainly known. For him the data of experience are not mere appearances of objects but the objects themselves as they appear. He says: "the physical world [...] has a distinct existence independent of the process of knowing [...] it is immediately known through experience without being actually experienced."[62] The following arguments are adduced:

> (a) [T]he possibility of regarding any object as a mere thought involves a reference to objects which are not regarded as merely identical with the thoughts that think them.[63] (b) If we are to explain illusion, we can do so only by distinguishing between sensations and physical things – a distinction which we should not otherwise have recognised.[64]

Similarly,

In all appearance something real appears; otherwise fiction and error would be impossible [...] To make a mistake is to believe something real to be in some way different from what it really is. But, unless we are cognisant of the real being concerning which the mistake is made, we cannot believe it to be what it is not; we cannot think of it wrongly if we do not think of it at all. If all appearance were mere appearance, everything for us would be equally real or unreal.[65]

12. REPRESENTATION

Passmore states the problem succinctly thus:

Representative perception will not do, for it does not admit that the physical world is immediately known and cannot explain how we could ever know it at all; naive realism will not do, because it cannot account for the difference between the world as it is and the world as we experience it.[66]

The problem is to give an account of the world such that it could include presentations and physical objects as constituents without sacrificing their unity. "A sensum is only a conditioned fragment of a physical object, just as a physical object is only a conditioned fragment of the world."[67]

At the end of the first of two articles on the philosophy of his great disputant, Samuel Alexander, Stout gives the following account of his position on representation:

Am I then committed to a representative theory of sense perception? I admit that I am [...] But though I accept a representative theory of sense-perception I reject any representative theory of knowledge in general. The representative function of sensa, like all knowledge by way of representation, must be founded on an apprehension of some relation between what represents and what is represented; and this must in the long run be apprehended directly and not by way of representation. I have indicated my own positive view on this question in my book on *Mind and Matter*.[68]

In the Unpublished Fragment, Stout provides this wonderful critique of the copy theory:

The preposition 'of' in the phrase 'idea of' is vague and ambiguous. Unless we are careful, we tend to give it a meaning such as it bears when we speak of an effect, or a copy, or a picture, or a reflexion, or a representation of something. An effect is an occurrence distinct from its cause in such a way that we may know it without knowing what produced it. A copy of a picture has an existence distinct from its original; a reflexion has an existence distinct from the thing which casts it. We can see the copy without seeing what is copied; we can see the reflexion without seeing what is reflected.

If we interpret in the same way the relation of an idea to that of which it is an idea, we are plunged into a bottomless abyss of nonsense. We have on the one hand, the experience of an individual self; on the other, the real world beyond it. This real world itself would not enter into individual experience at all but only more or less imperfect copies or representations of it. These copies or representations would be all that the individual is aware of. He would know only his idea, which would intervene, like a painted screen, between him and what really exists. Thus it would for him be exactly the same as if there were nothing else. He could never distinguish what seems to be from what is. He would not be able to apprehend the representations which alone exist for him as being representative of something beyond themselves. Still less would he be able to get outside the inner circle of his ideas so as to determine how far they confirm or fail to conform to the outer circle of real or 'objective' being.[69]

13. FINAL REMARKS AND CONCLUSIONS

My thesis in relation to the question of Stout's anticipation of the cognitive revolution is that he is much too complex to be simply pigeon-holed. In certain respects he did anticipate the cognitive revolution: in his critique of association, his formulation of Gestalt hypotheses, the stress on the role of expectation, top-down processing and directed thinking. But in other ways he goes beyond cognitive science, particularly with regard to the way in which, for him, cognition is imbued with conation. His views on the embodied self adapting to its environment are in many respects reminiscent of the views of Varela, Thompson and Rosch (1991), who espouse an enactive view of cognition. They seek to avoid the scylla of realism – cognition as the recovery of a pregiven outer world, and the charybdis of idealism – cognition as the projection of a pregiven inner world. Both the sensori-motor capacities of the organism and the environmental context (biological, psychological and cultural) need to be taken into account. What constitutes the world of a given organism is enacted by that organism's history of 'structural coupling.' Cognition is inextricably linked to the lived histories which are the result of evolution. Cognitive science has frequently lacked these motivational and historical dimensions.[70]

I conclude with a passage from Passmore's *Memoir*:

Pre-eminently he is a philosopher of the middle way. He would grant you so much, but only so much. Mind is conative – yes, but it is cognitive, too; and cognition and conation are not merely added to one another but mingle in the same process. Characters are particulars, but they belong to kinds; minds are embodied, but mind is not body, nor body mind; things are distinct, but not complete in themselves; we know the world as it is, but it is not only as we know it; to be is not to be known, but whatever is, is known, and whatever is known, is; God is not Nature, but Nature embodies God and God expresses Himself through Nature. The convictions of common sense are preserved, but without disrespect to either science or philosophy. In Stout's philosophy, the idea of Reconciliation finds its most philosophical expression.[71]

Elisabeth R. Valentine
Department of Psychology,
Royal Holloway, University of London,
Egham, Surrey TW20 0EX. UK.
Email: e.valentine@rhbnc.ac.uk

ACKNOWLEDGEMENTS

I should like to thank Dr N.E. Wetherick for his kindly, initial guidance through this thicket. I am grateful to the Bolzano Conference delegates, especially Liliana Albertazzi, Martin Kusch and Robin Rollinger, for comments following presentation of the paper on which this chapter is based.

NOTES

[1] Mace 1945, 313.
[2] Mace 1954, 75.
[3] Wright 1944.
[4] Knight 1946.
[5] Passmore 1952.
[6] I am grateful to Robin Rollinger for this suggestion.
[7] Mace 1954, 75.
[8] Flugel 1933, 152.
[9] AP II, 20–21, footnote.
[10] AP I, 36.
[11] AP I, 37.
[12] All page references are to the fifth edition.
[13] Passmore 1952.
[14] Passmore 1954, xxxviii.
[15] AP II, 24.
[16] AP II, 26.
[17] AP II, 30.
[18] AP II, 44.
[19] AP II, 45.
[20] AP II, 48.
[21] AP II, 48.
[22] AP II, 48.
[23] AP II, 48–49.
[24] AP II, 2.
[25] AP II, 50.
[26] AP II, 45.
[27] AP II, 46.
[28] AP II, 51.
[29] AP II, 52.
[30] AP II, 2–3.
[31] AP II, 25.
[32] MP, 514.
[33] MP, ix–x.
[34] MP, 674.
[35] MP, 676.
[36] MP, 676.
[37] MP, 676.
[38] Passmore 1954, xxxix.
[39] AP II, 82–3.
[40] MP, 506.
[41] AP II, 32.

[42] AP II, 32.
[43] MP, 512.
[44] AP II, 34.
[45] AP I, 96.
[46] AP II, 1.
[47] AP II, 3.
[48] AP II, 401.
[49] Passmore 1954.
[50] AP I, 46–47.
[51] MP, 408.
[52] MP, 409.
[53] MP, 677.
[54] MP, 678.
[55] AP I, 48.
[56] MP, 414.
[57] MP, 414.
[58] MP, 415.
[59] AP II, 6.
[60] Passmore, 1952.
[61] AP I, 7.
[62] Stout 1931, 222.
[63] AP I, 46.
[64] MP, 419.
[65] Unpublished Fragment, cited by Mace 1954, 72. This fragment was a draft of part of chapter 1 for the fifth edition of the *Manual of Psychology*, which in the event was not revised by Stout himself but by Mace in consultation with Stout.
[66] Passmore 1952, xliii.
[67] Passmore 1954, xliii.
[68] Stout 1940, 18.
[69] Unpublished Fragment, cited by Mace 1954, 70–71.
[70] Valentine 1995.
[71] Passmore 1952, l.

REFERENCES

Boyle, D.G.: 1993, *Psychology – the Aberdeen Connection*. Aberdeen, University of Aberdeen.

Broad, C.D.: 1945, 'Professor G.F. Stout (1860–1944),' *Mind*, **54**, 285–88.

Chapman, A., Sheehy, N. and Conroy, W. eds.: 1997, *Biographical Dictionary of Psychology*. London, Routledge.

Flugel, J.C.: 1933, *A Hundred Years of Psychology*. London, Methuen.

Knight, R.: 1946, 'George Frederick Stout: An Appreciation,' *British Journal of Educational Psychology*, **16**, 53–56.

Mace, C.A.: 1945, 'George Frederick Stout 1860-1944,' *Proceedings of the British Academy*, **31**, 306–16.

Mace, C.A.: 1954, 'The permanent contribution to psychology of George Frederick Stout,' *British Journal of Educational Psychology*, **24**, 64–75.

Passmore, J.A.: 1952, 'Memoir: George Frederick Stout,' in Stout 1952, xxv–liv.

Pylkkänen, P. and Pylkkö, P., eds.: 1995, *New Directions in Cognitive Science*. Helsinki, Finnish Artificial Intelligence Society.

Pylyshyn, Z.W.: 1984, *Computation and Cognition*. Cambridge, Mass., MIT Press.

Schaar, M. van der: 1996, 'From Analytic Psychology to Analytic Philosophy: the Reception of Twardowski's Ideas in Cambridge,' *Axiomathes*, **7**, 294–324.

Stout, G.F.: 1896, *Analytic Psychology*. Vols. I and II. London, George Allen & Unwin.

Stout, G.F.: 1898, *Manual of Psychology*. London, University Tutorial Press. 1938, 5th edition.

Stout, G.F.: 1930, *Studies in Philosophy and Psychology*. London, Macmillan.

Stout, G.F.: 1931, *Mind & Matter*. Cambridge, Cambridge University Press.

Stout, G.F.: 1940, 'The Philosophy of Samuel Alexander (I),' *Mind*, **49**, 1–18.

Stout, G.F.: (Unpubl.) 'How Psychology is Concerned with Objects.' Unpublished draft of part of Chapter I for the fifth edition of the *Manual of Psychology*, cited in Mace, 1954.

Stout, G.F.: 1952, *God and Nature*, ed. by A.K. Stout. Cambridge, Cambridge University Press.

Valentine, E.R.: 1995, 'Deconstructing Cognition: Towards a Framework for Exploring Non-conceptualised Experience,' in Pylkkänen and Pylkkö, eds., 1995, 1–9.

Varela, F.J., Thompson, E. and Rosch, E.: 1991, *The Embodied Mind*. Cambridge, Mass., MIT Press.

Wetherick, N.E.: 1997, 'George Frederick Stout,' in Chapman, Sheehy and Conroy, eds., 1997, 547.

Wright, J.N.: 1944, 'Obituary: Prof. G.F. Stout,' *Nature*, No. **3911**, October 14th, 481.

THEO HERRMANN AND STEFFI KATZ

OTTO SELZ AND THE WÜRZBURG SCHOOL

1. Otto Selz was and remains an outsider.[1] His scientific work, although of admirable quality, did not gain the acknowledgement granted to the work of many other psychologists which was, at most, only equally good. Even in the fifty years since the end of the Second World War he has been barely noticed in Germany. This has been somewhat different in other countries. Since 1945 he has found in the USA, the Netherlands, in France and the USSR to some extent, lasting acknowledgement.[2] For instance his work is represented in the history of cognitive science. The field of psychology in Germany has so far hardly considered him. Otto Selz led a wretched life after his professorship was taken from him at the outset of the Nazi regime, with his life in the ghetto and concentration camp and his murder en route to Auschwitz in August 1943.[3]

When Otto Selz is considered at all there are apparently difficulties in classifying him historically. Cognitive scientists sometimes place him with the Gestalt theorists.[4] In Germany Otto Selz is rather classified as a 'late member' of the Würzburg school. He is said to have 'developed further' some of their basic ideas.[5]

We investigate briefly Otto Selz's relationship with the Würzburg school. Our contribution leads to the conclusion that the spiritual relationship of Otto Selz with the Würzburg school is less close than is generally supposed. Rather, Otto Selz was from the start an autonomous and individual scholarly personality who took notice of the Würzburg school for some time. He developed through adoption of some of its ideas and techniques and – more to the point – through friction with their leading exponents and quickly departed – partly in explicit dissociation – from the psychological conceptions of the Würzburg school. We consider it one-sided if not wrong to reconstruct the historical figure of Otto Selz as related to the Würzburg school.

2. In 1909 Otto Selz started a five-year relationship with the Würzburg school. Oswald Külpe was no longer in Würzburg but had been in Bonn for some time. Narziß Ach, probably the member of the Würzburg school who has most strongly influenced the further development of psychology, had been

L. Albertazzi (ed.), The Dawn of Cognitive Science. Early European Contributors, 225–235.

Professor in Königsberg for two years. Otto Selz had been in close contact in particular with Karl Bühler who was at this time *Privatdozent* ('university lecturer') in Bonn. Selz remained close to Bühler throughout his lifetime.

In 1909 Selz was 28 years old. He had finished his studies of law in Munich and had been admitted to the bar. At the same time he had also studied philosophy with Theodor Lipps. Lipps (1851–1914) attempted a psychological-introspective foundation of logic, ethics and aesthetics. In philosophical aesthetics he is accepted as an influential exponent of the theory of empathy (*Einfühlungslehre*). While Otto Selz attended Lipps' lectures on philosophy, Lipps was engaged in theoretical questions of psychology. Lipps gave his opinion on the Gestalt qualities proposed by Ch. v. Ehrenfels and on the theory of objects (*Gegenstandstheorie*), represented by the Graz school at that time and decisively by A. v. Meinong.[6] Lipps characteristically commented at the time on what he called 'mental operations' (*seelische Operationen*) which among others serve to 'cause relationships' between psychical objects.[7] In the scientific surroundings of Selz in Munich operations that are able to affect psychical objects were now the subject of discussion. This may be of biographical interest since Selz later made mental operations an essential part of his psychological theory.

Selz finished his doctorate with Lipps in July 1909. His thesis dealt with English Empiricism and the attempt to solve the so-called 'transcendence problem' (*Transzendenzproblem*). This concerns whether there is an objective world of things outside our consciousness and whether or how we can recognise this 'conscious-transcendental' (*bewußtseinstranszendente*) world.

Oswald Külpe also dealt with the 'transcendence problem' during this time while working on his main work *Die Realisierung. Ein Beitrag zur Grundlegung der Realwissenschaften*.[8] The first volume was published in 1912. The second and third volumes were published by August Messer posthumously and were based on the lecture scripts of Külpe around the turn of the century. Külpe's *Realisierung* is nowadays claimed particularly by the Popper school of critical rationalism. Külpe is called the teacher or predecessor of Popper. They assert that Külpe – like Sir Karl later – already took the view that there is no certain knowledge of an objective outside world, but that this is a metaphysical hypothesis. The acceptance of this hypothesis makes it possible to explain facts discovered by the sciences. The rejection of an objective world of real things, however, as a counter hypothesis, allows only a much poorer explanation of, for instance, peculiarities of our sensory perception.[9]

The historians of the Popper school clearly did not read Otto Selz's dissertation otherwise they would have known that two years before the release of the first volume of Külpe's *Realisierung*, Selz had taken exactly that view commonly attributed to Külpe.[10] For Selz, acceptance of the existence of things independently of consciousness is a 'hypothesis' which allows us 'to explain countless complex state-of-the-case experiences and to predict the phenomena in a countless number of cases' (*zahllose verwickelte Erfahrungstatbestände zu erklären und in einer unbegrenzten Menge von Fällen die Erscheinungen vorherzubestimmen*). No other hypothesis is able to accomplish this.[11] In his thesis Selz did not quote the former work of Külpe treating the problem of the outside world. Missing is any information on Külpe's work *Das Ich und die Außenwelt*, on Külpe's famous *Einleitung in die Philosophie* and on a commemorative publication of Külpe in honour of Kant of the year 1904, where Külpe points out the hypothetical character of philosophical realism.[12] Külpe in any case knew of Selz's dissertation when he was writing the first volume of his *Realisierung*. He quoted Selz in two footnotes, but did not refer to Selz's insight that the assumption of a real outside world would be a good hypothesis.[13]

Otto Selz referred to Oswald Külpe in his *Habilitationsschrift* as an 'admired teacher' but never referred to him at any point again. In later papers Oswald Külpe never turns up as a relevant reference either. By the way, Külpe's assessment of the Selzean *Habilitationsschrift* is of remarkable conciseness, detachment and indifference. We consider it a rather routine assessment. All of this does not indicate a close teacher-student relationship. In the present context further facets of this relationship cannot be dealt with.

The other main exponent of the Würzburg school was Narziß Ach who, as already mentioned, had been Professor in Königsberg since 1907. In 1910, while in Bonn, Selz apparently had nothing more urgent to do than to write a devastating review of Ach's book *Über den Willensakt und das Temperament*.[14] He pointed out that Ach is advancing "assertions of impossibility" (*Unmöglichkeitsbehauptungen*) which can be refuted, is underestimating the value of self-observation in the experiment, did not produce the allegedly investigated act of will in his experiments, confuses the act of will with pure attentional concentration, is at least one-sided with his derivation of temperaments from his doctrine of will and so on.[15]

Ach struck back with the same vehemence and one of the fiercest known scientific controversies between psychologists took place. As Selz's biographer Seebohm writes:

Usually Selz is called a 'Würzburgian.' This classification is only partly right, as can be observed by the fact that he fought his first controversy with a Würzburgian, where both sides stressed the strength of their differences, which can hardly be justified by the factual deviation of opinions.

(*Man ist gewöhnt, Selz als einen 'Würzburger' zu bezeichnen. Wie bedingt richtig diese Einordnung ist, geht schon daraus hervor, daß er seine erste Kontroverse mit einem*

Würzburger durchkämpfte, wobei von beiden Seiten eine Betonung der Stärke der Gegensätzlichkeit zutage kam, die sich aus der sachlichen Abweichung der Auffassungen kaum rechtfertigen läßt.)[16]

In 1913 Külpe accepted a university chair in Munich, where he died in 1915. Selz served in the army in the First World War and returned to Bonn for several years as *Privatdozent*, but a Külpe school no longer existed at that time. In the meantime Karl Bühler was a captain in the Medical Corps on the western front, after which he managed the Institute in Munich after Külpe's death and was offered a chair at the Technical University of Dresden in 1918.

What Selz fundamentally learned from Külpe was experimental work in the laboratory. Immediately after his arrival in Bonn he carried out the self-observation examinations associated with his name, for example tasks of the type 'Co-ordinated concept to farmer?' (*Nebenbegriff zu Landwirt*?), 'Whole to roof' (*Ganzes zu Dach*) and the like. He understood immediately that (1) the task in a narrower sense, for instance finding a co-ordinated concept, and (2) the stimulus word 'farmer' *together* form a 'complete task' (*Gesamt-aufgabe*) that cannot be split (*dekomponiert*). According to Selz the two together form a 'relational fact' (*Sachverhältnis*) which is not decomposable into its elements. Some members of the Würzburg school have worked out similar but less precise conceptions, for instance Karl Bühler.[17] In a congress talk in 1912 Selz introduced his decidedly holistic opinion of 'relational fact' and 'complete task' to the academic public. In this lecture Selz already talked about the actualisation (*Aktualisierung*) of knowledge as one *solution method*; there are certain solution methods 'actualised' (*aktualisiert*) while processing tasks.[18] The latter phrase and the opinion standing behind it are hardly in the traditional line of the Würzburg school.

3. Considering merely the historical facts outlined here, it seems hardly possible to characterise Otto Selz historically as a scientist by his membership to the Würzburg school. This affinity seems even less possible if one looks at his basic scientific convictions and their development and compares them with the ideas held by the Würzburg school. For two different reasons we choose a summary text of Selz from the years 1941–42 for reference.[19] On the one hand it deals with something like a self-promotion of his public image. Here Otto Selz reaps the harvest of the considerations he had made in previous decades. Also, we want to demonstrate the originality and power which Otto Selz was still able to muster in his considerations as he held courses for Jewish teachers in the Jewish ghetto in Amsterdam under the worst personal circumstances. He managed to do all this after living in almost entire scientific and personal isolation in Mannheim from 1933. He had been sent to the concentration camp Dachau in Autumn 1938. He had finally been able to emigrate to Holland.

There he was captured by the German occupying forces and then still survived a few years before his murder on the way to Auschwitz.

The basic idea of Otto Selz, from which he derived many theoretical consequences, can be characterised as follows. Association theory in its different variations, take for instance Hermann Ebbinghaus and Georg Elias Müller, the Wundt school, but also the Würzburg school, really represent one and the same opinion, that is, that the subjects of psychological analysis are 'contents of consciousness' (*Erlebnisinhalte, Bewußtseinstatsachen*), whether dealing with ideas or with thoughts, *Bewußtheiten* or *Bewußtseinslagen* à la Würzburg. Therefore, the common subjects of psychology are for all these schools mental facts, phenomena like mental contents (*Bewußtseinsinhalte*) or experiences (*Erlebnisse*). The Würzburgians stated that mental facts consist not only of imaginal representations but of non-imaginal thoughts as well. The controversy between the different positions concerns in general how the mental contents, like most mental facts, are connected and combined. For someone like Georg Elias Müller the association principle was answer enough to this question, that is the combining of stronger or weaker activating and inhibiting influences between mental contents which is driven primarily through contiguity. Narziß Ach especially, and with him the Würzburg school, backed up the association principle with the 'principle of deliberate manipulation of psychic contents' (*Prinzip willentlicher Manipulation der psychischen Inhalte*), in particular 'determining tendencies' (*determinierende Tendenzen*), and general will. As Wilhelm Wundt had done in another way, the Würzburgians Ach and Henry Jackson Watt stood for a 'superimposition theory' (*Überlagerungstheorie*); the dynamics of the association of psychic contents is superimposed by determining, voluntary events.[20] The expression '*Überlagerungstheorien*' comes, moreover, from Otto Selz.[21]

Selz considered this Würzburgian opinion also to be wrong. He carried out, it can be said today, a veritable change of paradigm through his description of inner life. As an outsider and an author only hesitatingly noticed, he opened the door to a view which is today very familiar to us.

According to Selz, inner life is *not* to be understood as a pattern of psychic experiences (ideas, thoughts and so on), whose connection should be investigated. Inner life is rather, as Selz emphasised repeatedly, a 'self-complementing purposeful structure of activity' (*sich selbst ergänzende zielstrebige Tätigkeitsstruktur*).[22] The subject of psychology is not (imaginal or not-imaginal) mental contents, but what Selz termed at the end of his life 'cognitive modes of behaviour' (*geistige Verhaltensweisen*).[23] For Selz's doctrine the nature of cognitive modes of behaviour is decisive. It does not concern the 'psychic functions' (*psychische Funktionen*) or acts which for instance Karl Stumpf in the tradition of Brentano and Lotze or Karl Bühler compared to psychic phenomena or contents.[24] Selz's cognitive modes of behaviour have nothing to do with the intentionalism of Brentano.[25] Cognitive modes of behaviour can, rather, be compared with 'functional body movements' (*zweckmäßigen Körperbewegungen*); the connection of the cognitive modes of behaviour is 'reflexoidal.'[26]

A person does not, as thought before, acquire a 'system of ideas without dynamics' (*System von richtungslosen Vorstellungen*) but rather 'an organised system of functional cognitive modes of behaviour' (*ein geordnetes System von zweckmäßigen geistigen Verhaltensweisen*).[27] Already at the VIII. International Congress for psychology, Selz summarised his basic idea as follows:

> Intellectual events are no system of diffuse reproductions, as associationism thought would be the case, but are like the system of body movements, especially the reflexes, a system of specific responses in which there is usually a clear assignment of general and specific intellectual operations to definite conditions of elicitation.
>
> (*Das intellektuelle Geschehen ist kein System diffuser Reproduktionen, wie die Assoziationspsychologie es sich dachte, sondern es ist ebenso wie das System der Körperbewegungen, insbesondere der Reflexe, ein System spezifischer Reaktionen, in dem eine für den Regelfall eindeutige Zuordnung allgemeinerer und speziellerer intellektueller Operationen zu ganz bestimmten Auslösungsbedingungen herrscht.*)[28]

Here we may remark that with the fundamental change from mental contents to cognitive modes of behaviour or specific mental actions respectively, an incisive change in the empirical basis of psychology has taken place: mental contents were accessible to self-observation; how is there empirical access to cognitive modes of behaviour? Do the latter rather have the character of theoretical constructs, to use modern terminology? As far as we know, Otto Selz himself hardly recognised and explained this essential methodological implication of his paradigm-change.

According to Selz some cognitive modes of behaviour are available from the moment of birth, based on hereditary provision in an already matured form and on organised connection from the start.[29] In the course of development new operative components become inserted in this ordered cognitive-modes-of-behaviour pattern. Older cognitive modes of behaviour are sometimes replaced by better ones in the course of development. The mental development of a person has something to do with plastic, educable growth. The development of the growing, strictly organised pattern of cognitive modes of behaviour is functional, it serves the mastering of life-tasks which our environment and the culture we are living in demand of us. Behaviour which allows us to be successful in life is effective.

The founding capital of our intellectual life is, therefore, genotype. The influence of tradition is added to this. We adopt skills from other people, through which cultural possession is passed on to single individuals. And finally cognitive modes of behaviour are completed by self-acquired behaviour that represents, as Selz says, our 'personal property' (*persönliches Gut*).[30] Selz distinguishes in this context, probably following Bühler, instinct-dependent, automatic and reasonable learning.[31] Decisive for all learning is the anticipation of success. Learning means, (1) recognising tasks in their structure and (2) finding methods for the solution of the task. More important than finding abstract rules is the analogous application to unfamiliar situations of solution methods from well-known 'model situations' (*Mustersituationen*). A task someone has completely understood is nothing other than anticipation of the goal. Goal anticipation triggers, like an inner cue, specific solution methods. The 'conditions of elicitation' (*Auslösungsbedingungen*) determine which partial solution methods are 'cumulatively' switched on one by one until the goal is attained, and which partial solution methods are chosen as 'substitute operations' (*Ersatzoperationen*) in case of failure. Selz imagines this in the sense of a 'reflex theory' (*Reflextheorie*).[32]

This mental dynamic is considered as a peculiarity of highly complex biological systems. Important for Selz in this are the 'pleasure of function' (*Funktionslust*), the 'delight in work' (*Werkfreude*), the instinctual substantiation of achievements, the transfer of achievements between persons, and the impact of models. In his lectures in the Amsterdam ghetto Selz strictly and precisely marked off his own 'theory of specific reactions' (*Theorie der spezifischen Reaktionen*) from Gestalt theoretical views as well as from Würzburgian conceptions taken by Ach and Watt.[33] His rejection of associationism can be demonstrated by his fierce controversy with Georg Elias Müller, amongst other things.[34] Demarcations of the Berlin Gestalt theory and drafts against it can be found in his published and unpublished papers over decades.[35]

The fathers of cognitive science, Newell, Shaw and Simon have with some justification discussed the theory of Selz as that concept which comes particularly close to their own revolutionary theory of 'Human Problem Solving.'[36] They parallel the Selzean cognitive modes of behaviour with their own algorithmic operations. They identify the fact that in the conceptions of Selz one cognitive mode of behaviour determines the next, with their own concept of strategy. In a similar way they treat the concept of problem, control processes and so on.

As far as we are aware this is the most solid consideration of Selz since the Second World War. As emphasised before, it was not primarily psychologists, and least of all German psychologists, who paid attention to Selz. He is, rather, understood as a predecessor to cognitive science. Newell, Shaw and Simon took note of only half of the Selzean conception, however. They neglect the deeply biological-functionalistic thoughts Selz was developing increasingly plainly until the awful end of his life. The idea of development, the biological functionality, the instinct and motive support of mental activities, but also tradition and social transfer reach far beyond the interest of mainstream cognitive science.

Dutch educational science especially has taken notice of Selz in connection with programs for increasing intelligence up to the present day. In the Amsterdam ghetto and earlier at various university institutions in Amsterdam, Selz lectured on basic ideas of corresponding pedagogical programs.[37] It should not be forgotten that in this context he has presented approaches to a developmental theory of language, which we, at least, consider to be in some parts more plausible than the one presented nowadays on the basis of a linguistic theory of Chomskyan character.[38] Just a short note concerning this; according to Selz language is, on the base of genetic disposition, essentially learned. But this learning is not really rule learning. According to Selz, grammatical rules are rather the retrospective abstract-conceptual fixing of a 'linguistic tradition' (*sprachliche Tradition*) caused by the generation of analogies without knowing rules. Even if we can theoretically reconstruct linguistic behaviour retrospectively by means of grammatical rules, this does not say how language is actually acquired and used. According to Selz, the development of analogies on concrete models is pivotal in language acquisition.[39] We regret that we cannot further explain this point of view in this context.

4. We hope that, in the space available, we have been able to show the relationship of Otto Selz to the Würzburg school in two different ways: (1) by short historical remarks and (2) by a fragmentary sketch of Selz's basic ideas which can be easily compared with the well-known ideas of the Würzburg school. It is apparent that a historical classification of Otto Selz as representative of the Würzburg school gives a one-sided, or even incorrect, view. Because of the necessary limited size of this paper we have not considered his foundation of a psychological theory of space which is up to now completely unknown but highly significant.[40] We consider the essence of Selz's life work to be the early-realised idea that psychology is not about asking how mental contents are combined with each other, but rather about grasping mental life as the determined and organised linking of cognitive modes of behaviour, reactions and solution methods. Around this basic idea he gradually developed his partial theories, here only mentioned briefly.

The impressive independent development of his psychology of mental operations differentiates Selz from every school context of that time. He never established a school of his own and was never appreciated sufficiently. In his lifetime he was alone, an outsider. To investigate historically or by sociology of science the lack of consideration of his work in Germany to date and the reason for this would be an interesting task.

Theo Herrmann, Steffi Katz
Psychologie III
Mannheim Universität
D-68131 Mannheim
Email: fprg@rumms.uni-mannheim.de

NOTES

[1] Métraux 1993.
[2] Humphrey 1963, 132–149; Mandler, Mandler 1964, 225–234. For an overview or a translation of the basic ideas of Otto Selz in English see Humphrey or Mandler, Mandler.
[3] See Métraux 1993.
[4] See Newell, Shaw, Simon 1958, 164.
[5] See also Seebohm 1970.
[6] Lipps 1900, 1905; v. Ehrenfels 1890; v. Meinong 1899; for an overview see among others Herrmann 1976.
[7] Lipps 1902.
[8] Külpe 1912, 1920, and 1923.
[9] Albert 1987, 34 ff.
[10] Selz 1910a.
[11] Selz 1910a, 108 ff.
[12] Külpe 1885, 1892/93; see Albert 1987 for a commemorative publication of Külpe in honour of Kant of the year 1904.
[13] Külpe 1912, 13, 86.
[14] Ach 1910.
[15] Selz 1910b.
[16] Seebohm 1970, 259.
[17] Bühler 1908, 105 ff.
[18] Selz 1912; incidentally, the later Gestalt theorist Kurt Koffka participated in the discussion of this talk.
[19] Selz 1991.
[20] Watt 1904.
[21] Selz 1991, 139.
[22] Note to a letter, Selz 1938.
[23] Selz 1991, 75 ff.
[24] Stumpf 1907; Bühler 1909.
[25] Brentano 1874.
[26] Selz 1991, 75.
[27] Selz 1991, 75 ff.
[28] Cited after Selz 1927, 273.
[29] Selz 1991.
[30] Selz 1991, 85 ff.

[31] Bühler 1927.
[32] Selz 1991, 140 ff.
[33] Selz 1991, 139.
[34] Müller 1919.
[35] See among others Selz 1926; and 1991.
[36] Newell, Shaw, Simon 1958.
[37] Selz 1991.
[38] See among others Szagun 1986.
[39] Selz 1991, 102.
[40] See among others Selz, 1930; and 1941.

REFERENCES

Ach, N.: 1910, *Über den Willensakt und das Temperament*. Leipzig, Quelle & Meyer.

Albert, H.: 1987, *Kritik der reinen Erkenntnislehre*. Tübingen, Mohr.

Balmer, H. ed.: 1976, *Die Psychologie des 20. Jahrhunderts*, vol. 1. Zürich, Kindler.

Brentano, F.: 1874, *Psychologie vom empirischen Standpunkte*. Leipzig, Duncker & Humblot. En. tr. 1973, *Psychology from an Empirical Standpoint*, by A. C. Rancurello, D. B. Terrell and L. McAlister. London, Routledge & Kegan Paul. Repr. 1995.

Bühler, K.: 1908, 'Antwort auf die von W. Wundt erhobenen Einwände gegen die Methode der Selbstbeobachtung an experimentell erzeugten Erlebnissen,' *Archiv für die gesamte Psychologie*, **12**, 93–122.

Bühler, K.: 1909, 'Zur Kritik der Denkexperimente,' *Zeitschrift für Psychologie*, **51**, 108–118.

Bühler, K.: 1927, *Die Krise der Psychologie*. Jena, Gustav Fischer.

Ehrenfels, C. v.: 1890, 'Über Gestaltqualitäten,' *Vierteljahrsschrift für die wissenschaftliche Philosophie*, **14**, 249–292.

Herrmann, Th.: 1976, 'Ganzheitspsychologie und Gestalttheorie,' in Balmer, ed., 1976, 573–658.

Humphrey, G.: 1963, *Thinking. An introduction to Its experimental Psychology*. New York, Wiley.

Külpe, O.: 1885, *Die Einleitung in die Philosophie*. Leipzig, S. Hirzel.

Külpe, O.: 1892/93, 'Das Ich und die Außenwelt,' *Philosophische Studien*, **7**, 394–413; **8**, 311–341.

Külpe, O.: 1912, *Die Realisierung. Ein Beitrag zur Grundlegung der Realwissenschaften*, vol. 1. Leipzig, S. Hirzel.

Külpe, O.: 1920, *Die Realisierung. Ein Beitrag zur Grundlegung der Realwissenschaften*, vol. 2. Ed. by A. Messer, Leipzig, S. Hirzel.

Külpe, O.: 1923, *Die Realisierung. Ein Beitrag zur Grundlegung der Realwissenschaften*, vol. 3. Ed. by A. Messer, Leipzig, S. Hirzel.

Lipps, T.: 1900, 'Zu den Gestaltqualitäten,' *Zeitschrift für Psychologie und Physiologie der Sinnesorgane*, **22**, 383–385.

Lipps, T.: 1902, *Einheiten und Relationen*. Leipzig, Barth.

Lipps, T.: 1905, *Psychologische Untersuchungen I: Bewußtsein und Gegenstände*. Leipzig, Engelmann.

Lück, H.E and Hiller R. eds.: 1993, *Illustrierte Geschichte der Psychologie*. München, Quintessenz.

Mandler, J.M. and Mandler, G.: 1964, *Thinking: From Association to Gestalt*. New York, Wiley.

Meinong, A. v.: 1899, 'Über Gegenstände höherer Ordnung und deren Verhältnis zur inneren Wahrnehmung,' *Zeitschrift für Psychologie und Physiologie der Sinnesorgane*, **21**, 182–272.

Métraux, A.: 1993, 'Otto Selz,' in Lück and Hiller, eds., 1993, 56–59.

Müller, G.E.: 1919, 'Literaturbericht zu: Über die Gesetze des geordneten Denkverlaufs,' *Zeitschrift für Psychologie*, **82**, 102–120.

Newell, A., Shaw, J.C. and Simon, H.A.: 1958, 'Elements of a Theory of Human Problem Solving,' *Psychological Review*, **65**, 131–166.

Seebohm, H. B.: 1970, *Otto Selz. Ein Beitrag zur Geschichte der Psychologie*. Inaugural dissertation, Universität Heidelberg.

Selz, O.: 1910a, 'Die psychologische Erkenntnistheorie und das Transzendenzproblem,' *Archiv für die gesamte Psychologie*, **16**, 1–110.

Selz, O.: 1910b, 'Die experimentelle Untersuchung des Willensaktes,' *Zeitschrift für Psychologie*, **57**, 241–270.

Selz, O.: 1912, 'Experimentelle Untersuchungen über den Verlauf determinierender intellektueller Prozesse,' *Bericht über den V. Kongreß der Gesellschaft für experimentelle Psychologie*, 229–234. Leipzig, Barth.

Selz, O.: 1926, 'Zur Psychologie der Gegenwart,' *Zeitschrift für Psychologie*, **99**, 160–196.

Selz, O.: 1927, 'Die Umgestaltung der Grundanschauung von intellektuellem Geschehen,' *Kantstudien*, **32**, 273–280.

Selz, O.: 1930, 'Von der Systematik der Raumphänomene zur Gestalttheorie,' *Archiv für die gesamte Psychologie*, **77**, 527–551.

Selz, O.: 1941, 'Die Aufbauprinzipien der phänomenalen Welt,' *Acta Psychologica*, **5**, 7–35.

Selz, O.: 1991, *Wahrnehmungsaufbau und Denkprozeß. Ausgewählte Schriften*. Ed. by A. Métraux and Th. Herrmann, Bern, Huber.

Stumpf, C.: 1907, *Erscheinungen und psychische Funktionen*. Aus den Abhandlungen der Königl. Preuss. Akademie der Wissenschaften vom Jahre 1906. Berlin, Reimer.

Szagun, G.: 1986, *Sprachentwicklung beim Kind: eine Einführung*. München, Urban & Schwarzenberg, 3rd ed. (new edition).

Watt, H. J.: 1904, *Experimentelle Beiträge zu einer Theorie des Denkens*. Leipzig, Engelmann.

SERENA CATTARUZZA

THE INSTRUMENTAL MODEL OF LANGUAGE IN KARL BÜHLER

The theme of meaning pertains to various disciplines: logic, linguistics, the philosophy of mathematics, the psychology of perception, gnoseology, the philosophy of language, and so on. The excellence and systematicity of many theories advanced over the last hundred years (roughly from the end of the nineteenth century to the present day: from Frege, Bréal and Husserl to Cassirer, Wittgenstein and Montague), however, is unable to remove the difficulties inherent in the semantics of natural languages. As Putnam observed,

> In the last decades enormous progress seems to have been made in the syntactic theory of natural languages [...] Comparable progress seems *not* to have been made in the semantic theory of natural languages [...] Why is the theory of meaning so *hard?*[1]

Putnam's reply set out in the conclusion to the same work is peremptory in tone:

> The sloppiness, the lack of precise theory and laws, the lack of mathematical rigour, are all characteristics of the social sciences today. A general and precise theory which answers the questions (1) why do words have the different sorts of functions they do? and (2) exactly how does conveying core facts enable one to learn the use of a word? is not to be expected until one has a general and precise model of a language-user; and that is still a long way off.[2]

Definitive solution of these questions may seem utopian, as Putnam suggests. However, more or less exhaustive and noteworthy attempts in this direction were made in the first half of this century.

Here I shall discuss the wide-ranging theory of Bühler, where the semantic relations constitute an obligatory passage for a theory of communication which hinges on the key notions of form, function and sign.

A detailed examination of the historical context of Bühler's thought would require a comprehensive analysis from Brentano's psychology of the act to Dedekind's doctrine of number, from Husserl's *pure* grammar and Marty's

L. Albertazzi (ed.), The Dawn of Cognitive Science. Early European Contributors, 237–250.

general grammar to Meinong's semantics; from Stumpf's theory of musical and verbal sound to Saussure's general linguistics[3] and Wittgenstein's investigations. Here I shall deal only with his concept of the *instrumental model of language* (*Organon Modell der Sprache*), with the emphasis on its semantic features.

A doctor (oculist) and a philosopher (a scholar of Scottish philosophy), K. Bühler worked academically as an experimental psychologist both general and evolutionary in the fields of thought, perception and language. Bühler indeed was one of the brightest thinkers involved in the famous research on thinking carried out by the Institute of Psychology of Würzburg, at the time directed by Külpe.

As for theoretical and experimental psychology research, however, Bühler's program had appeared critical to the wide later spreading Behaviourism and, especially in perception, to Gestalt in its Berlin version, which was establishing its supremacy in the field.[4] Nowadays Bühler is, at most, mentioned in the dictionary of psychology as the discoverer of the phenomenon of immediate comprehension (*Aha Erlebnis*) and as the Author of a not better specified *Organon Modell*.[5]

In fact, he deployed all the components of his discipline to build a general model of language. By the time Bühler dedicated a growing attention to the language-sign event which culminated with the first definition of the *Organon Modell* of 1918.[6]

With this regards, Bühler has always been enigmatic: on the one hand he referred to a classical Platonic paradigm, on the other hand he mentioned convergences with the language theorist Alan Gardiner.[7] As for the actual origin of the communicative structure of this model, he has always referred to that by contrast, underlying his absence in Husserl's *Logische Untersuchungen*, whom he estimated a lot on other matters.[8]

Besides the central role played by the representative module, which applies to linguistic forms a structural criterion similar to the one that operates in the construction of elementary geometric figures, crucial to Bühler's model is the *correlatedness* of the concepts of notificatory information (*Meldung*) and of reception (*Empfang*).

According to Bühler, if we seek the *origin of semantics*, we cannot take the single individual (element, stimulus, speaker) and start from there. In fact, doing so gives rise to:

(i) associative conceptions;
(ii) reproductive schemes;
(iii) mechanical interpretations.

The hypothesis of a system with one single co-ordinate of ego-bound facts as the sole and all-embracing interpretation of all linguistic facts raises major difficulties. As he observes,

> We are thus confronted by the choice between a monadic construction or a universalism of the subject conceptions which are perhaps (but this cannot be taken for granted) justifiable in the ethereal regions of philosophical hypotheses, but which are unacceptable in the lowlands of the empirical phenomena studied by linguistic theory.[9]

Hence, "grammar, as it has developed over two thousand years, presupposes a type of intersubjectivity of language which neither Diogenes in his barrel nor any monad could achieve."[10]

The alternative of a broader scheme, one which does not replace but enriches the former, is indispensable at various levels, from the most abstract level of conceptual linguistics to the lowest but no less fundamental one of animal and human biology. According to Bühler, then, the *perceptive* dimension must be given its proper collocation just as much as the *logico formal* dimension.

The social setting thus emphasised requires a *general model of linguistic exchange*, and vice versa. Working from the bottom up, one reaches the limits of the most simple situation of the non simple mechanism, inherently non simple constitutive semantics, which is grasped as a prototype.

Instructive in this regard is the *wordless psychic exchange* that takes place in everyday contexts, i.e. as in the case of: "drivers [...] who avoid watchful pedestrians: in this case there is reciprocal control,"[11] which is lacking if the driver encounters a pedestrian crossing the street with his eyes blindfolded. Another example is provided by the surgeon and the theatre assistants during an operation. Or in an animal social context, by the courtship behaviour of birds.

The base hypothesis of Bühler's theory can be formulated as follows. *Semantic mechanisms* are not *secondary derived devices*, although they are intrinsic to a minimal communitarian exchange (the transmitter/receiver model). This characterisation is inadequate. The general semantic model requires the *reciprocal control* that comes about in a shared perceptive situation.

From this starts a scale of *intermediate stages*, a labyrinth of modes and possibilities which culminates in a *higher order* contact (to use Meinong's term) where the direction of control is not given by the shared perceptive situation but by semantic mediatory devices (the person who speaks eliciting concepts of the person who listens).

The model therefore does not envisage *mere concordance* in the meaningful behaviour of the community's members. What matters is *regulated concordance*; that is, *dynamic concordance* and *ascertainable regulation*.

In this scheme, and especially in relation to the different role played by *perceptive reference*, a primary function is performed by the *pointing gesture*. This stands "at the threshold of what is specifically human, at the threshold of the representative function of language,"[12] as attested by the comparative history of language and by observations of children. This is because in no other field apart from language does comparison between man and animals find *continuity* and at the same time a *demarcation line* (which does not exclude its modification on the basis of new observational experimental data). Its is therefore *de facto* and not *de jure* that this excludes the *higher order semantic dimension* – that is to say, the *representative linguistic function* – while the notificatory and referential functions are common to animal and human language.

Which brings us to a second axiomatic consideration that derives from the general scheme: if the moods and needs of the members of a community prevail in reciprocal control, then they must be *notified* and *received*.

A third semantic dimension of language establishes *directionality* towards objects and facts, giving a *symbolic valency* to *signs*. In fact, there is no reproductive representation alone, with its relative radial scheme or system of diffuse reproduction. On its basis, the mnestic residues of simultaneous psychic processes are so closely associated that when one of them returns, so too do all the others.

If, for example, one shouts A, this *reawakens* representations 1, 2, 3, etc., of the objects present to the consciousness at the moment of the utterance. Each of these representations in its turn is the *centre* from which associations and so on radiate outwards. The symbolic representation of entirely different nature manifests itself in a variety of forms, ranging from faithful representation very similar to a photogram (as in Meinong)[13] to a more conventional one, like for example the association between the visual signs of the alphabet and the relative sounds. It is from this function that truth criteria are deduced.

The ideal of the pertinent and exact representation largely determines the production of linguistic forms, and it manifests the connection between them and the field of knowledge and science. Hence derives the *connection* between *language* and *logic* already familiar to the Greeks, but also the *diversity* of the *culturally conditioned human languages*. This diversity was first grasped by W. von Humboldt who characterised it in terms of *innere Sprachform* (internal linguistic form).[14] Bühler provides an interpretation of this concept whereby different languages give priority to different media and symbolic fields, in that they do not see the datum 'to represent with the same eyes,' so to speak.

Perhaps the best comparison can be drawn, to a certain extent, with painting technique. However, the model can only be fully defined through comparison with empirical reality. One may, however, provide an outline by stating that certain features of symbolic fields are determined

> on the basis of the fact that the Eskimo languages, for example, can be contrasted, in so far as they are largely *impressionistic*, with the Bantu languages, which are to a large extent *categorial*; or Chinese, with its well-known preference for single entities, with the European languages, all of which treat the *universal* as something to which one can refer.[15] (My translation.)

As Bühler observes, Horace's dictum '*ut pictura poesis*' is at odds with the structural diversity between linguistic and pictorial means of representation. Nevertheless, one perceives an analogy: just as paints require a surface on which to be spread, so *linguistic symbols require a field* in which they can be located.[16]

This notion performs a more general role in Bühler's difficult and unsatisfactory interpretation of the relationship between the *syntactic* and *lexical* components of language. Traditionally, the two levels have been contraposed as *form* and *matter*, an account that does not greatly differ from Aristotle's. However, in the light of research into thought processes, and Gestalt especially, one may attempt to reformulate the problem. One may argue, in fact, that some of the bases of Bühler's theory of language had already been laid in 1907 by his discovery of the 'syntactic patterns' of linguistic thought and, shortly afterwards, by his analysis of the processes involved in comprehension.[17] Still lacking, however, was theoretical experimental study of the *intuitive factor of indication*. No less important was comparison with the theory of aphasia analysed by Head, Gelb and Goldstein and by Isserlin, so that the core of linguistic theory could be tested against studies of the degeneration of the human capacity for language.[18] Moreover, a wide-ranging theory of language could not ignore the results of inquiry into the structure of infant language, with the rich documentation accumulated from recordings of expressions uttered in the first days of life.[19] In this way the *set of axioms* required by a theory of language were completed.

The first axiom of Bühler's *Sprachtheorie* (theory of language) affirms that speaker and the listener are psychophysical systems which act in a exchange of signals. The picture of the situation is the following:

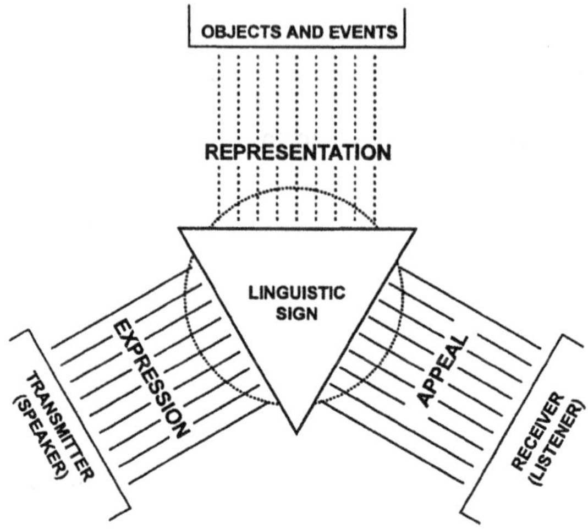

Fig. 1 Bühler's Verbal Communication Model ("Organon-Modell„). The traced lines symbolize three semantic defined relations of the linguistic sign.

The first axiom confirmed the general semantic model (controlled transmission in a bipolar system) and comprised at least three dimensions of meaning: *Ausdruck* (expression), *Auslösung* (triggering) and *Darstellung* (representation), with three types of correlated signs: symptoms, signals and symbols.

In his remarkable *Die Krise der Psychologie* of 1927, he presented a detailed analysis of previously studied topics: his criticisms of both Husserl's overestimation of logic representative functions, and that of Marty's for opposite reasons, i.e., overestimation of the suggestion within the communicative circle.[20]

The formula of the first axiom did not consist of a rigid antithesis, nor of a dialectical resolution of the three dimensions, but of a dynamic distinction which did not exclude a final unitary outcome. This model of language stood in an 'osmotic' relationship with the psychological model of the *Drei Aspektenlehre* (three aspects theory) set out in *Die Krise der Psychologie*. It assumed a minimum base structure which hinged on three correlated features: the aspect *E* of lived experience (*Erlebnis*), the aspect *V* of meaningful behaviour (*sinnvolles Verhalten*), and the aspect *G* which stood for *Gegenstand* (object), *Gebilde* (form), *Geist* (mind).

Peter R. Hofstätter, who studied in Vienna under Bühler and his wife Charlotte, a brilliant mathematician, psychologist and sociologist who afterwards emigrated to the United States, represented Bühler's three-way linguistic semantic and psychological methodological division in a single diagram, as follows:

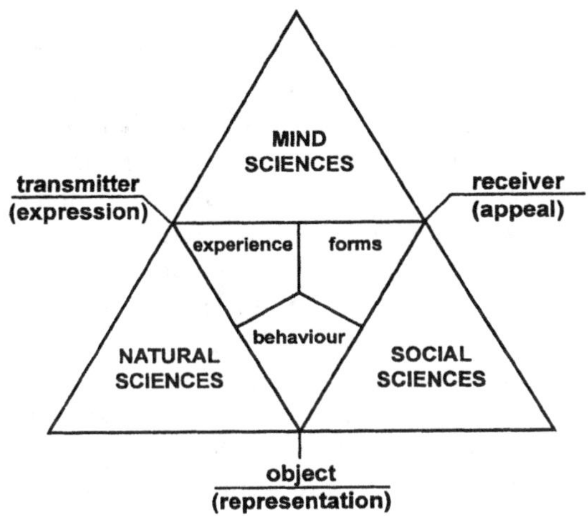

Fig. 2 Bühler's Triangle of Communication according to Hofstätter. (This scheme unifies the "Drei Aspektenlehre" with the "Organon-Modell")

As Hofstätter writes: "Every assertion about experience and behaviour and about the decisive mental functions must therefore in principle comprise the other two aspects as well if it is to achieve its purpose."[21]

Analysis of the correlations and their proof was provided in terms of language, however. This allowed more straightforward axiomatic treatment because the instrumental model exhibited and transposed the connections among the three approaches of psychology.

Conversely, only by considering the *three aspects*, according to Bühler, is it possible to understand the phenomenon of language scientifically. Hence derives the specificity of the instrumental model, which combines logical simplification with unusual theoretical heuristic depth.

Moreover, the general semantic model (*vierpunkte Schema*), implies a set of concepts and fundamental links which warrant close examination. A single aspect, however important, like that of comprehension, for example, or of concordance, does not provide adequate explanation of an elementary exchange situation. The assumption of the dual process theoretically implies the assumption of a more complex one: *contact control comprehension*. Here Bühler imagines a mental experiment which raises the following problem.

On observing two men walking, and if one wishes to establish whether they are in reciprocal psychic contact (in this case, therefore, the *direction* of the control is part of the shared perception), one must employ the concept of the *covariance* of behaviour. There is a specific use of covariance in the exact sciences to establish a presumed relationship between two facts A and B, and it is based on probability. Having gathered a set of data on the two facts, they may be quantified and correlated. If the coefficient of correlation deviates from the compensation value of the probability calculation, then a relation exists.

But there is a *phenomenological* method that is just as efficacious. If A *accelerates* his pace, or *slows down* or *stops*, and B *follows suit*, openly or covertly, the scene is significant; that is to say, a covariance is *perceptively grasped* between the two behaviours. If this is not the case, probabilistic calculation may be used, or at least one may decide to give up. And yet, from the start observers *have assigned meanings* (through expectations, conjectures) to the behaviour of A and B, as they are, for example, passer-by, enemies, relatives. On the basis of these perceptions, more specific projections are made, rather like the plotting of curves.

Here the perceptive capacity is exceptional, and the observer therefore establishes the covariance of the two forms. There is nothing extraordinary about such a sophisticated performance, however. One may cite in this regard the description provided by another experimental phenomenologist, Paolo Bozzi, of a similar, but simplified, scene where a man looks at a tree.

> Describe the small differences in colour among the most minute details that you can see [...] also the expression of the man's face as he inspects the tree (professional curiosity, suspicion, evaluation) [...] The scene that you are looking at can be divided into a multitude of extremely fine details, and you can take into consideration all the differences among these details taken individually or in groups, and categorise them according to a large number of parameters.[22]

The covariance is confirmed in the reciprocal contact of the ongoing dialogue. As he proceeds, each partner is able, to a varying extent, to guide the behaviour and experience of the other. The concept of control therefore has a technical use which implies reciprocal regulation. To summarise, therefore, one starts with a dual *transmitter/receiver system*. The reciprocal contact between the two partners (A and B) involves:

i. a transmitter;
ii. an action system;
iii. a receiver.

If an emotion or aspiration in A's action system reaches receiver B through the communication channel between them, then receiver B will respond through the same channel. But then A responds to this response, so that what matters in this situation is not elementary effects but the *controls of the system*. One thus understands that the feedback from one's actions can in principle be separated from information about the actions of others, although they belong to the same domain of meaning, separated but nevertheless set in new relation.

One thus derives three dimensions of contact and possibilities of comprehension:

(i) A understands his interlocutor B either on the basis of the control effects that he perceives are being exerted on him by the latter; or

(ii) on the basis of the control that he (A) exerts on B and perceives in the latter; or

(iii) there is some objective entity between A and B to which each refers, and which therefore becomes something which enables A to understand B.

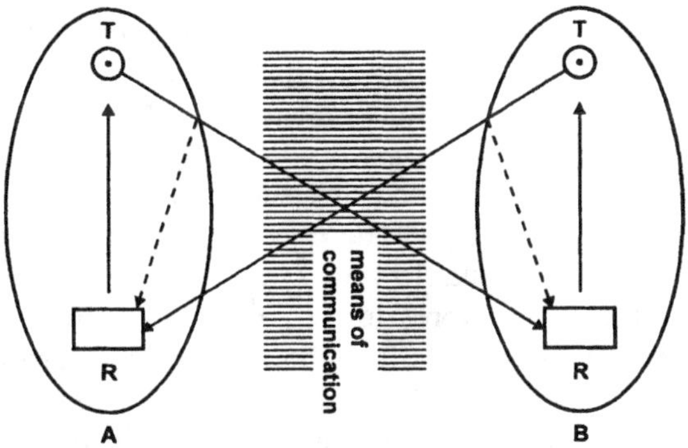

Fig. 3 Bühler's general semantic paradigm (In the reciprocal contact both partners [A and B] have a transmitter [T], an action-system, a receiver [R]

This is the outline from which the instrumental model of language is deduced from the general semantic paradigm.

Bearing these facts in mind, it is not surprising that also the more elaborate form of the instrumental model (that of *Sprachtheorie*) maintains the three largely autonomous and variable sense referents.[23] If the concrete acoustic phenomenon has the character of a sign, in all three cases the latter *expresses a semantic qualification* in the manner specified by the general scheme. In the reverse case, one operates with theoretically uncontrollable heterogeneous concepts, as was for long the case in the critical sector of the theory of sounds.

Despite the numerous merits of a classic work like Stumpf's *Die Sprachlaute*, it was only Trubeckoj's phonological theory that emphasised the functional or sign semantic character of the sound.[24]

The concept of 'sound element' is highlighted by this conceptualisation, which stresses its *diacritical* aspect. This requires evaluation of the authentic meaning of the expression 'linguistic sound' always on the basis of the context or of a specification: whether, that is, one is referring to a phonetic fact, the sound in itself, or a sound mark in the phoneme system of a particular language. Indeed, this entails that an individual phoneme finds different phonetic realisation in two occurrences of the same language, while a single sound unit may have different phonological uses in two different languages.

But once again the functional formulation emphasises the concept of sign. Which brings us to the second axiom of the theory, namely that: *linguistic phenomena are wholly and exclusively signs* (from the phonological level to the representative symbolic one).

But why does the second axiom not state the symbolic nature of linguistic phenomena? Is it correct to lump together symbols, symptoms and signals in the general category 'sign'? Like so many terms in scientific jargon, is not this too an empty husk? This would manifest a 'monstrous misconception' of natural languages. The crucial point is that all three functions of the instrumental model postulate an underlying semantic dimension which is also at the origin of the concept of sign. There is therefore a system of regulated connections which has phenomena flow as if they were perpetual.

Therefore, for example, the expression *it's raining* implies a scenario of the following kind: there are two people in a room – the first person's attention is caught by a shower of rain, s/he looks towards the window and says: *it's raining*; the second looks in the same direction, directly induced to do so by hearing the other's words or by glancing at him/her. There are other implications: semiological analysis relates the objectual meaning to the acoustic image of the word.[25] Moreover, this image comprises a phonemic factor of signalling and, in the case of a context, field signs. By its nature, each sign entity requires beings that it deems such and of which it makes use. In this sense, the psychophysical systems employed act as *detectors* of what is acting by means of signs. When concrete elements satisfying sign functions are predisposed by operating beings and stand with them in the relationship between work and artifice, then these beings can be considered to be the producers of signs. The axiom of the nature of signs here encounters the theoretical model of *homo faber*.

Hence derives the formulation of the third axiom, which in its most elaborate and systematic form displays a cloverleaf pattern (*vierfeld Schema*) consisting of dual polarities.

In fact, the third axiom then distinguishes among:

(i) speaking *activity*/linguistic work;
(ii) *act* of speaking/linguistic form.

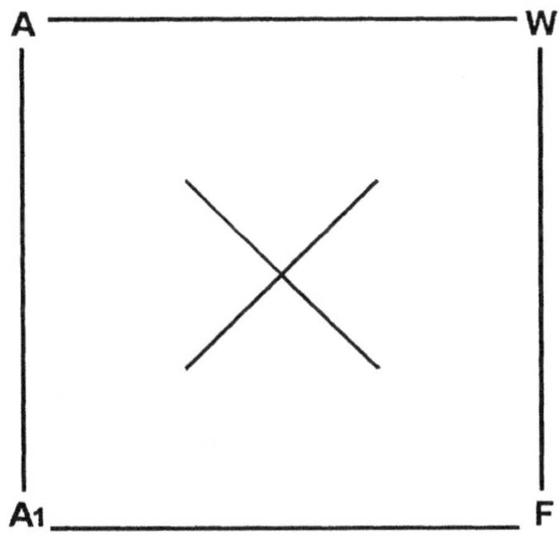

Fig. 4 Bühler's Language-Square
(A= activity; W=work; A1=act; F=form)

This last is governed by two criteria:

(i) *position;*
(ii) *formalisation.*

The first criteria permits to distinguish between phenomena related to a subject (activity, act) and phenomena intersubjectively related (act, form). The second criteria indicates a lower level (activity, work) and an upper level of formalisation (act, form).

In this way an attempt is made to encompass a very broad field of inquiry and to organise it without discarding empirical and formal reasons, in order to show distinction in correlation and vice versa. But here too, in the end the general semantic paradigm prevails as logically preordained, or at least of equal status with theories excessively inclined towards an exclusively objective or subjective treatment. In this axiom, unilaterality is again considered unproductive. And ultimately this predilection is confirmed in the analysis of language from the point of view of the linguistic form. On its basis, the proposition may have just as much preceded the word as the word preceded the proposition, since both are correlated with one and the same level of language.

Thus confirmed is the initial hypothesis: "semantics is not a secondary product but a factor constitutive of all communitarian life."[26]

Serena Cattaruzza
Università di Trieste
Dipartimento di Psicologia
Via dell'Università, 7
I-34100 Trieste
Email: cattarse@univ.trieste.it

NOTES

[1] Putnam 1970, 50.
[2] Putnam 1970, 63.
[3] Eschbach 1984; Albertazzi 1996 and 1998; Cattaruzza 1996; Schnelle 1996; Mulligan 1997.
[4] Boring 1950; Henle 1961; Eschbach 1988; Smith 1988.
[5] Eschbach 1984; Cattaruzza 1990; Conte 1990.
[6] Bühler 1918; Schnelle 1996.
[7] Gardiner 1932; Bühler 1933.
[8] Bühler 1934/1965.
[9] Bühler 1934/1965, 69.
[10] Bühler 1934/1965, xxiv.
[11] Bühler 1927/1965, 41.
[12] Bühler 1927/1965.
[13] Meinong 1899.
[14] Humboldt 1836; Whitney 1875; Marty 1908.
[15] Bühler 1934/1965, 152–3.
[16] Bühler 1927/1965, 1934.
[17] Bühler 1907.
[18] Goldstein 1917; Bühler 1926; Head 1926; Isserlin 1929.
[19] Bühler and Hetzer 1928; Stern and Stern 1928.
[20] Bühler 1927/1965.
[21] Hofstätter 1957, 67.
[22] Bozzi 1989, 12–3.
[23] Bühler 1934.
[24] Stumpf 1926; Trubeckoj 1929.
[25] Bühler 1934/1965, 25.
[26] Bühler 1934/1965, 69.

FIGURES

Figure 1. Bühler's Verbal Communication Model ('Organon–Modell').

Figure 2. Bühler's Triangle of Communication according to Hofstätter.
Figure 3. Bühler's General Semantic paradigm.
Figure 4. Bühler's Language Square.

REFERENCES

Albertazzi, L. ed.: 1996, *The Philosophy of Alexius Meinong*. Special Issue of *Axiomathes*, **7**.

Albertazzi, L. ed.: 1998, 'Perceptual Saliences and Nuclei of Meaning,' in Poli, ed., 1998, 113–138.

Boring, E.G.: 1950, *A History of Experimental Psychology*. New York, Appleton Century Croft.

Bozzi, P.: 1989, *Fenomenologia sperimentale*. Bologna, Il Mulino.

Bühler, C. and Hetzer, H.: 1928, 'Das erste Verständnis für Ausdruck im ersten Lebensjahr,' *Zeitschrift für Psychologie*, **107**, 50–62.

Bühler, K.: 1907, 'Tatsachen und Probleme zu einer Psychologie der Denkvorgänge. I. Über Gedanken,' *Archiv für die gesamte Psychologie*, **9**, 297–365.

Bühler, K.: 1918, 'Kritische Musterung der neueren Theorien des Satzes,' *Indogermanisches Jahrbuch*, **6**, 1–20.

Bühler, K.: 1926, 'Les lois générales d'évolution dans le langage de l'enfant,' *Journal de Psychologie normale et pathologique*, **23**, 597–607.

Bühler, K.: 1927, *Die Krise der Psychologie*. Stuttgart, Fischer. Repr. 1965.

Bühler, K.: 1933, 'Die Axiomatik der Sprachwissenschaften,' *Kant-Studien*, **38**, 19–90. Shorter version 1969, ed. by E. Ströker, Frankfurt a. Main, Klostermann. En. tr. in Innis, 1982.

Bühler, K.: 1934, *Sprachtheorie*. Stuttgart, Fischer. Repr. 1965. En. tr. 1990, *Theory of Language. The representational Function of Language*, by D. Goodwin, Amsterdam, Benjamins Publishing Company.

Cattaruzza, S.: 1996, 'Meinong and Bühler,' in Albertazzi, ed., 1996, 103–111.

Cattaruzza, S.: 1990, 'L'attualità dell'opera bühleriana,' *Lingua e stile*, **25**, 485–95.

Conte, M.E.: 1990, 'La semiotica di Karl Bühler,' *Lingua e stile*, **25**, 471–485.

Eschbach, A. ed.: 1984, *Bühler-Studien*, 2 vols. Frankfurt a. Main, Suhrkamp.

Eschbach, A. ed.: 1988, *K. Bühler's Theory of Language*. Amsterdam, Benjamins Publishing Company.

Gardiner, A.: 1932, *The Theory of Speech and Language*. London, Oxford University Press.

Goldstein, K.: 1917, 'Die transkorticalen Aphasien,' *Ergebnisse der Neurologie und Psychiatrie*, **2**, 352–629.

Graumann, C.F. and Herrmann, T.: 1984, *Karl Bühler's Axiomatik*. Frankfurt a.Main, Klostermann.

Head, H.: 1926, *Aphasia and Kindred Disorders of Speech*, 2 vols. Cambridge, Cambridge University Press.

Henle, M. ed.: 1961, *Documents of Gestalt Psychology*. Berkeley, University of California Press.

Hofstätter, P.R.: 1957, *Psychologie*. Frankfurt a. Main, Fischer.

Humboldt, W.: 1836, *Über die Verschiedenheit des menschlichen Sprachbaues und ihren Einfluß auf die geistige Entwicklung des Menschengeschlechts*. Berlin, Dümmler.

Innis R.E.: 1982, *Karl Bühler: Semiotic Foundations of Language Theory*. New York, Plenum Press.

Isserlin, M.: 1929, 'Die pathologische Physiologie der Sprache,' *Ergebnisse der Physiologie*, **29**, 130–249.

Kiefer, H.E. and Munitz, M.K. eds.: 1970, *Language, Belief and Metaphysics*. New York, State University of New York Press.
Marty, A.: 1908, *Untersuchungen zur Grundlegung der allgemeinen Grammatik und Sprachphilosophie*. Halle s. S., Niemeyer.
Meinong, A.: 1899, 'Über Gegenstände höherer Ordnung und deren Verhältnis zur inneren Wahrnehmung,' *Zeitschrift für Psychologie und Physiologie der Sinnesorgane*, **21**, 182–272.
Mulligan, K.: 1997, 'The Essence of Language: Wittgenstein's Builders and Bühler's Bricks,' *Revue de Métaphysique et de Morale*, **2**, 193–217.
Poli, R. ed.: 1988, *The Brentano Puzzle*. Aldershot, Ashgate.
Putnam, H.: 1970, 'Is Semantics Possible?,' in Kiefer and Munitz, eds., 1970, 50–63.
Schnelle, H.: 1996, *Die Natur der Sprache*. Berlin, De Gruyter.
Smith, B. ed.: 1988, *Foundations of Gestalt Theory*. München, Philosophia Verlag.
Stern, C. and Stern, W.: 1928, *Die Kindersprache*. Leipzig, Barth.
Stumpf, C.: 1926, *Die Sprachlaute*. Berlin, Springer.
Trubeckoj, N.: 1929, 'Zur allgemeinen Theorie der phonologischen Vokalsysteme,' in *Travaux du Cercle Linguistique de Prague*, **1**, 39–67.
Whitney, W.D.: 1875, *The Life and Growth of Language*. New York, D. Appleton.

ALF C. ZIMMER

THE CONCEPT OF PERCEPTUAL 'FIELD' AND THE REVOLUTION IN COGNITION CAUSED BY KÖHLER´S PHYSISCHE GESTALTEN

The development of cognitive science at the beginning of the 20th century is characterised by several seemingly unrelated paradigms:

(i) the formalisation of perceptual and cognitive processes according to the model of Newton's *Principia*;
(ii) the development of theories of reasoning sharing with traditional approaches the distinction between procedures (rules) and content (semantics) but discarding logic as the 'true' form underlying human reasoning instead postulating ratiomorph processes; and finally
(iii) a novel approach to solve the old problem of the relation between sensations and perceptions and the corresponding bottom-up or top-down theories, namely Wolfgang Köhler's field theoretic approach of 1920 and Kurt Koffka's (1935) evolutionary account of the development of cognitive and perceptual processes of 1921 where the concept of 'Prägnanz' (in English: saliency or singularity, see Kanizsa and Luccio, 1986) describes the phenomenological 'pull' of percepts toward Gestalten comparable to the forces in an electromagnetic field.

The first two approaches correspond to the distinction of bottom-up and top-down processes in perception best exemplified by the dispute about colour theory at the beginning of the 19th century: Young postulated three simple filtering processes which transform electromagnetic waves between 380 and 750 nm into colours. In contrast to this position where the physical characteristics uniquely determine the percept, Goethe stressed in addition the importance of the 'entire nature' (*Farbenlehre*, preface) including the world knowledge of the perceiver. Helmholtz (1866) attempted to integrate these opposing approaches by postulating that, for instance, in the perception of objects, 'unconscious inferences' (top-down) are based upon sensations (bottom-up):

L. Albertazzi (ed.), The Dawn of Cognitive Science. Early European Contributors, 251–267.

An astronomer, for example, comes to real conclusions [...], when he computes the positions of the stars in space, their distances, etc., from the perspective images he has had of them at various times and as they are seen from different parts of the orbit of the earth. His conclusions are based on conscious knowledge of the laws of optics [...].

In the ordinary acts of vision [...] knowledge of optics is lacking. Still it may be permissible to speak of the psychic acts of ordinary perception as *unconscious conclusions*, thereby making a distinction of some sort between them and so-called conscious conclusions. And while it is true that there has been, and probably always will be, a measure of doubt as to the similarity between the results of such unconscious conclusions and those of conscious conclusions.[1]

In his influential treatise *The intelligent eye* (1970) Gregory reissued this inferential theory of perception specifying the nature of the 'unconscious inferences' as re-scalation and subsuming the phenomena observed by Gestalt psychologists as special cases. In the most recent edition of this book Gregory depicts the perceptual process as the result of a combination of the above mentioned approaches.

Gregory (1998) classifies the different approaches in the following way:

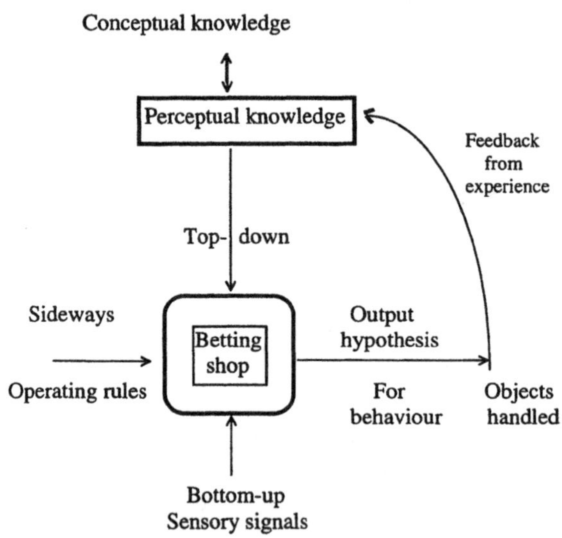

Figure 1: Gregory's (1995) model of the interaction of perceptual processes.[2]

In this classification the misleading impression emerges that Köhler's and Koffka's approach adds only one further level of analysis instead of resolving the traditional dichotomy,[3] furthermore the term 'rules' is misleading because of its Constructivist (top-down) connotation. What catches the Gestalt theoretic intentions better is the term constraints, insofar as perceptual and cognitive processes are the result of the evolution of mechanisms successful in the co-ordination of organisms and their environment; this motivates the following modification of Gregory's model:

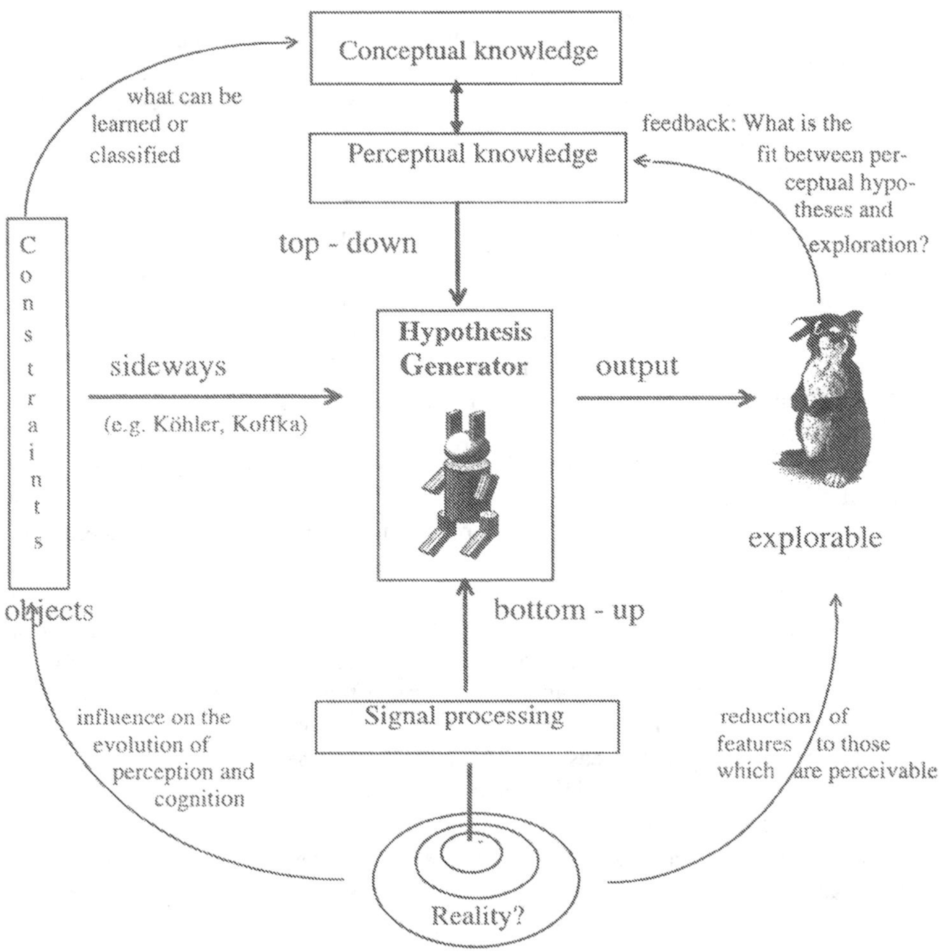

Figure 2: Modification of Gregory's classification of different approaches to perception.

The first formulation of the dichotomy between top-down and bottom-up approaches to perception and the implicit position that this dichotomy cannot be resolved by reducing it to the one or the other alternative can be found in a fragment of Democritus cited by Sextus Empiricus:

The concepts 'coloured', 'sweet', 'bitter' are mere conventions. In reality there are only atoms and the void. 'Poor mind' *thus Democritus lets the senses address the mind* 'from us you get the foundations of your reasoning which you apply to undermine us! Trying to destroy us, you destroy yourself.

By pointing out that percepts (αιγδισιζ) as phenomenologically given have always to be interpreted in relation to a frame of reference which in turn depends on past experience, Democritus refutes any entirely Reductionistic approach to epistemology – from the Miletic School to the foundation of Carnap's 'Einheitswissenschaft' on observational sentences.

In the cited fragment Democritus, furthermore, indicates the principal problem for a unitary theory of perception, that is, reducing all phenomena either to the veridical representation of the world in the sensations (the bottom-up approach) or to the operations of the mind (the top-down approach). The first view is attacked by Democritus when he points out that colours, tastes etc. are mere 'νομοι', that is, classifications depending on cultural traditions or social agreements; the second when he stresses the dependence of cognition upon sensory data.

Since Democritus, analyses of perception and cognition can easily be classified as belonging to one of the two alternatives. Starting with Plato's (*praeter rem*) Idealistic approach according to which the meaning precedes the objects, epistemology found its other extreme in Roscelin's Nominalism according to which meaning is affixed to objects arbitrarily – *post rem*, one can find a similar position in Locke's radical Reductionistic approach to perception and reasoning relying on the mechanistic principle of contiguity which has been refuted by Berkeley's famous example of triangles being perceived as such even if never seen before which in turn led to a radical Idealist position.

The fact that physical measurements and sensory perceptions differ has been a truism in science at least since Galilei who derided their importance as "non sieno altro che puri nomi" (*Il saggiatore*) but the epistemological problem remained how a perception of reality can evolve if the sense data upon which it relies are so unreliable. Following Epicurus, Gassendi stated that all empirical knowledge (in Scholastic terms: intellectus) relies on data from the senses not denying, however, that the mind *ratiocinatio* possesses its own rules and thereby modifies the sensory perceptions (thus Leibniz added to Gassendi´s "nihil in intellectu nisi prius in sensu" the qualification "[...] nisi intellectus ipse").

This sceptical view of the limitations inherent in founding science upon sensations can still be found in Kant's *Prolegomena* where he denies the possibility of psychology as a natural science. Reacting against this view, psychophysics developed in the 19th century with its inherent postulate that sensations depict reality in a way comparable to measuring devices: namely that, in loudness, brightness, or weight, sensations are mere logarithmic transformations of the corresponding physical measurements and that even the apparent arbitrariness of colour and taste can be attributed to the fact that these are not unitary but multidimensional variables.

In general, the development of experimental psychology in the 19th century attempted to answer two problems posed by Kant: (i) Psychophysics, especially Fechner's logarithmic transformation of physical stimuli into well defined sensations, demonstrated successfully that psychology can be treated mathematically,[4] and (ii) the schema underlying the Kantian categories in the *Critique of Pure Reason* was identified as the cognitive process co-ordinating images' as measured in psychophysics with concepts as identified by introspection. Thus defining two methodologically distinct psychologies.

When in 1834 Ernst H. Weber was able to show empirically that for sensory data the general rule seems to hold that the ratio between a just noticeable difference of a physical change and the amount of the physical stimulation is a constant ($k = \Delta R{:}R$) and Fechner in 1860 interpreted this Δ as a differential, the relation between physical data and sensory perception could then be described as $\varphi = c \log \psi$. The consequence of this approach was that Newton's linear model for physics, especially the additivity of forces, could be applied to sensory data, too; this becomes apparent in the vector analytic approach to colour vision by Grassmann and later in Kurt Lewin's vector analysis of the behaviour of people in the environment upon which they act. What this approach could not model in perception, namely the phenomena of constancy, was attributed to the inborn characteristics of the mind (the above mentioned Scholastic ratiocinatio) or – in modern terms – Helmholtz unconscious inferences.

This tradition is still evident in the theories of human information processing where data driven and concept driven processes are assumed to interact in analogy to the computer where programs act upon data (by the way, John v. Neumann, the architect of modern computers, has proposed to use the architecture of computing machines as a model for the human mind, thereby taking up ideas developed for instance by Leibniz and implicit in the theory of thinking of Otto Selz, finally made explicit in the Newell-Simon model of information processing).

However, when Helmholtz (1866) introduced the concept of 'unconscious inferences' in order to explain phenomena like illusory motion (Plateau) and geometric-optical illusions, he took up again the dual view of perception as described by Democritus, namely, that top-down processes (reason, in Aristotelian philosophy) and bottom-up processes (sensations) *together* produce the percept.
In this he was perhaps influenced by Hegel's dialectics:

> The necessity and the wit of man has invented infinitely manifold ways of using and mastering nature.... Whichever forces even nature has produced and released against man, coldness, beasts, water, fire – he knows means against them, these means he takes out of [nature] itself and uses them against it; the cunning of his reason allows that he uses natural objects against natural forces and by pitting them against each other preserves himself.[5]

Hegel's classification of sensory processes which can be seen in analogy to physical measuring devices and therefore open to experimental investigation, and cognitive processes which rely on the active and conscious mind and therefore have to be investigated by introspection, can be interpreted as a first attempt to resolve the classical dichotomy by means of dialectics. A later off-spring of this approach can be found in the Soviet 'Activity psychology' (especially Leontiev, 1973) in which the ascending, that is, the stimulus controlled part of cognitive processes is modelled according to Pavlov's Associationistic theory of the first and the second system of signalling and the descending part of actions is driven by reason which mould and modify nature with the result of an increasingly better fit between reality and cognition. Similar ideas can be found in Roger Shepard's (1981) view of psycho-physical complementarity according to which "(1) The world appears the way it does because we are the way we are, and (2) we are the way we are because we have evolved in a world that is the way it is" (p. 332).

A more direct influence of Hegel´s position can be found in Wundt's seemingly neat solution for the question of top-down or bottom-up processing by locating them as qualitatively different functions on different levels in the hierarchy of mental processes and making them incomparable because of his claim that they have to be investigated by different methods. This claim was challenged by Hermann Ebbinghaus (1885), who was able to show that even 'higher cognitive functions' can be investigated by experimental methods but also that these functions, namely, learning and forgetting processes, can be modelled mathematically. However, in one crucial point Hermann Ebbinghaus deviated from the tradition in psychophysics: He did not borrow his mathematical model from physics but from biology, that is, he interpreted forgetting as decay and learning as growth and chose the Euler function as the appropriate experimental function with a negative exponent, namely time. The computation showed that the entire scope of mental processes can be investigated experimentally and that – a least in principle – they can be modelled formally.

Mach's observation of melodies as invariant forms independent from the physical characteristics of the constituting tones at first seemed to shatter the assumptions underlying psychophysics and Associationism, however Ehrenfels' theory of 'Gestaltqualitäten' and Meinong's (1891) theory of relations seemingly resolved this problem because they identified 'Gestalten' as invariant in relations thereby preserving the linear approach of Associationism by regarding associations between relations as constituent for form perception. Any further inconsistencies, as for instance optical illusions and apparent motion, were discarded as consequences of erroneous top-down-processes, that is, Helmholtzian unconscious inferences.

That the concept of invariants in its strict sense does not resolve the dichotomy between bottom-up and top-down approaches to cognition can be seen from the fact that a melody has to be well formed in order to be transposable, however, what constitutes well-formedness depends not only on physically describable characteristics of the melody but also on the cultural and/or cognitive background of the listener; intercultural comparisons in music show beside a few universals in harmonics a striking variability in rhythm and tonality. Furthermore, the fundamental problem in perception is the separation of signal from noise,[6] that is, why and under which conditions a melody remains recognisable even if it is sung out of tune or against a background of noise.

In order to resolve the impasse created by Wundt's dualistic position towards mental processes and the shortcomings of relationalism it was necessary to give up the mechanistic model underlying association theory and its further developments. As early as 1909, Wolfgang Köhler writes to his gymnasium professor in physics, Hans Geitel, in a comment on Hans Witte's attack against Ernst Haeckel's philosophical monism:

> a unitary system of physics in which all processes in nature are traced back to electrical state transitions, can – in principle – lead as well to 'Materialismus' as the.[traditional system of] mechanics. In our days it would seem only more elegant and therefore more tempting.[7]

Köhler's experiments for his doctoral dissertation '*Akustische Untersuchungen I*', 1909) had forced him to apply modern electro-magnetic technology and to develop some new methods himself (see his contribution to the *Festschrift for Julius Elster and Hans Geitel* of 1915). His background in physics and his experiences as principal investigator at the primate centre of the Prussian Academy of Sciences in Tenerife prepared him to suggest a generalised field concept as the unifying principle for natural philosophy which was published under the title *Die physischen Gestalten in Ruhe und im stationären Zustand. Eine naturphilosophische Untersuchung* in 1920.

Despite the attacks on the implicit equating of psychophysics with perception (Bergson, 1911) the linearity assumptions underlying classical mechanics had remained dominant for theory building in psychology with the exception of Köhler's modelling of psychological and brain processes according to Faraday's and Maxwell's theory of electromagnetic fields. Köhler reports that Wertheimer's experiments with apparent movement, especially the φ–phenomenon, have led him to apply the concept of the electromagnetic field to processes in perception.

Usually, the concept of 'field' is regarded as characteristic of Wolfgang Köhler's attempt to give a consistent theoretical framework for Gestalt theory. However the concept of 'field' pervades the philosophical and psychological treatises on perception in the 19th century. For instance, in Husserl's 'Dingvorlesung' in 1907 a highly detailed systematisation of the interdependency of different perceptual fields was developed (see figure 3).

Köhler's theory of 1920 differs from these earlier field-theoretic approaches to perception and action; he starts from Maxwell's concept of the electromagnetic field, whereas, in contrast, the standard psycho-physical approach of the 19th century had assumed that the basic constituents of a perceptual field are sensations which can be interpreted as forces in a linear Newtonian vector field. As early as 1913 in his *Über unbemerkte Empfindungen und Urteilstäuschungen* Köhler attacked this basic assumption by pointing out that so-called illusions can neither be explained by reference to errors of judgement,[8] nor by Cornelius' unnoticed sensations.[9] Köhler asks:

> How could creatures survive who would be informed about reality by sensations corresponding so little to it? How could these creatures construct a useful physics, since sensations necessarily constitute the primary material of this science? [...] none of us experience possibilities of harm in the Müller-Lyer illusion, the 'apparent' unity of a sound, the octave illusions, whether of judgement or sensation, are biologically completely indifferent, insofar as they are not of tremendous value[...][10]

Following Bergson (1911) he then points out the limited role sense perceptions play in physical laboratories and he finishes:

> from the existence of physics, therefore, no more can be concluded than these cases of perception important for physics permit, namely that these cases correspond to the 'basic assumption'. Whether the whole field of sensations and perceptions conforms to the basic assumption cannot at all be decided by the existence of physics.[11]

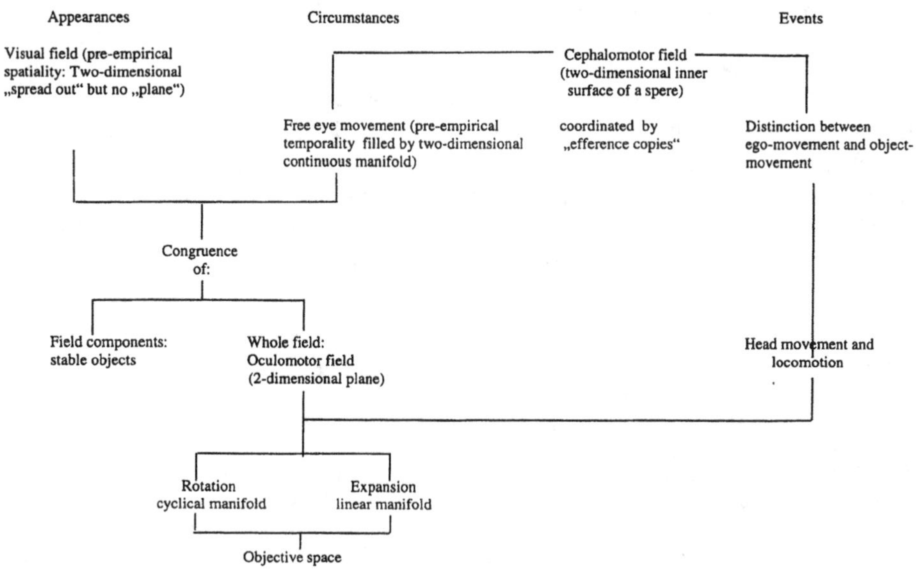

Figure 3: Schema of interrelated perceptual fields as developed in Husserl's Dingvorlesung (Thing and space)

Köhler's solution for the fact that illusions are pervasive and cannot be avoided even by Schumann's 'isolation in consciousness' but at the same time do not prevent a successful interaction with our environment, appears first in a footnote "This fact might easily be explained if we regard as the 'immediately given,' and in any case as the biologically primary 'reality,' not 'sensations,' but (for the most part) 'things.'"[12] Köhler's choice to put this argument into a mere footnote shows that he was aware of the revolutionary character of his explanation. By pointing out that the most simple relations between stimulus and sensation constitute "limiting cases achieved by means of isolation" but do not suffice to explain "the everyday perception of *things*," he finishes his contribution by stating:

> Anyone can see that the theory of perception will at first become less simple in this way. But research so oriented will also find laws and constants in the greater richness of its material and perhaps will finally be able to attain a deeper understanding of the whole field than can be achieved by means of the assumptions which we have opposed.[13]

Köhler's arguments contain the theoretical kernel of Gibson's (1979) later theory of ecological perception which – influenced by Kurt Koffka – James Gibson started to develop just after the end of the critical period for the development of cognitive sciences discussed in his book. By the postulation of *things* as the basic constituents of perception Köhler points into the future of experimental psychology, his reference to Brentano's "unitary nature of judgement experiences" (p. 33) reveals the motivation for Köhler to develop the concept of the perceptual field into a theory which goes beyond the limiting case of pure stimulus-sensation relations as investigated in the field of psychophysics. In Köhler's last scientific paper "Gestalt psychology" published 1967 in *Psychologische Forschung,* he explains which reasons had led him to develop his theory of generalised potential fields of 1920:

> The Gestalt psychologists, we remember, were always disturbed by a thesis which was widely accepted by others. One psychologist, strongly influenced by traditional convictions, had formulated it in the following words: "I do not know whether perceptual fields actually consist of independent local elements, the so-called sensations. But, as scientists, we have to proceed as though this were true. An extraordinary statement – an a priori general conviction about the right procedure in science is assumed to be more important than the nature of the facts which we are investigating.
>
> From its very beginning, Gestalt psychology ignored this thesis and began its work with simple and unbiased observations of facts. Independent local sensations? Consider again what happens in apparent movement. After a first visual object has appeared in one place, a second visual object does not appear in its normal location but nearer the place where the first has just disappeared, and only then moves towards what I just called its normal location. Clearly, therefore, the process corresponding to the second object has been deflected, has been attracted by a remnant of what has just happened in another place, the place of the first object, and has only then approached its 'normal' location. Consequently, under the conditions of such experiments, the second object does not behave as though it were an independent local fact at all. The statement, quoted earlier, that perceptual fields must be assumed to consist of independent local sensations, is therefore at odds with the behaviour of percepts even under

such fairly simple conditions. Or take any of the well-known perceptual illusions, say the Müller-Lyer illusion. Can there be any doubt that in this case two lines of objectively equal length become lines of different length under the influence of the angles added at the ends of the distances to be compared? And so on, in a long list of examples, all of them incompatible with the statement about the nature of perceptual fields.

Ours was an uphill fight. I felt greatly relieved, as mentioned above, to find so fundamentally similar an approach from the side of physics. In his great treatise, *Electricity and Magnetism*, Clerk Maxwell had remarked that we are often told that in science we must, first of all, investigate the properties of very small local places one after another, and only when this has been done can we permit ourselves to consider how more complicated situations result from what we have found in those elements. This procedure, he added, ignores the fact that many phenomena in nature can only be understood when we inspect not so-called elements but fairly large regions.[14]

The field-theoretic concept Köhler (1920) developed for the interaction of forces in biological and psychological systems departed strongly from the linear 'field' concepts which dominated psychology and phenomenology at the turn of the century, because even for the seemingly most simple case of stationary states he postulated non-linear interactions and the minimum principle as the driving force for self organisation. Up to then the fact that light propagates linearly and that its energy dissipates quadratically as gravitation had led to an implicit model of psychological processes which is constituted by independent part interacting like forces in the vector field or – on a qualitative level – which consists of building blocks organised in decomposable hierarchies.[15]

Why the field theoretic concept was so revolutionary can be seen from the fact that even Helmholtz initially reacted strongly against Faraday's description of field processes. The background of his and others' resistance against non-linear processes in physics can be traced back to the fact that these approaches smacked of Cartesian 'vortices,' originally proposed for gravitational processes which were shown by Newton to be linear, but now re-emerging again in the field of electromagnetics. It took the development of Maxwell's formal apparatus to convince physicists like Helmholtz; as late as 1881 he finally wrote:

[...] we see how great a degree of exactness and precision was really hidden behind the words, which to Faraday's contemporaries appeared either vague or obscure [...] I have no intention of blaming his contemporaries, for I confess that many times I have myself sat hopelessly looking upon some paragraph of Faraday's descriptions of lines of force, or of the galvanic current being an axis of power.[16]

Insofar it was not astonishing that Köhler's proposal met with resistance or even in the greater part with indifference because it was generally assumed that even if these non-linear processes existed they were confined to a small area in psychology and therefore not necessitating a general theoretic approach.[17]

One can argue that the ingredients of today's theory of perception as exemplified in the integrative account (in Figure 2) were all present in the first part of the 20th century: (i) the core concept of perceiving invariances as the solution of the problem that the sense impressions vary from time to time and place to place but the perceived world is relatively stable; this solution was implicit in the theoretical framework of the Graz School of Gestalt Theory and can also be found in the writings of Koffka from the Berlin School; (ii) the ability to separate signals from noise, that is, to see patterns (*Gestalten*) behind and out of uninformative local sense impressions could be accounted for by Köhler's field theory for biological and physical phenomena; and finally (iii) the enhancement of contours as shown in Mach bands and Hering's demonstrations of simultaneous contrast could explain why the human perceiver is able to perceive differences where even highly accurate measurement devices fail.

The problem for a development of a full-fledged theory of perception at the beginning of this century was that these different results were not regarded as complementary but as competitive accounts for what perception really is. Modern research in psychophysiology indicates that the physiological processes correlated with these different levels of perception can be regarded as hierarchically ordered: (i) Already in the retina, lateral inhibition occurs which not only accounts for Mach bands and contrast effects but also for the effect that the percept is more fine grained than the receptor structure of the retina, as can be shown in the effect of Vernier acuity. (ii) The innate Gestalt forming mechanisms as postulated by Tembrock,[18] seem to have been identified by Eckhorn and co-workers (1990) and Gray, Engel, König and Singer (1992) as highly correlated excitation patterns in the brain at relatively distant areas of the cortex; these correlations of 40 Hertz oscillations in excitation may constitute the percept of unitary objects which is phenomenological prior to the perception of features. (iii) Due to the feature of self-referentiality [19] these processes can explain why for instance in a movie version of Figure 4 the dog is perceived immediately and separated from the 'noisy' environment without any effort while – in contrast – the inspection of the static form or one where only the parts belonging to the dog are moved homogeneously necessitates 'scrutiny' in order to identify the object.

Figure 4: A hidden figure (Dalmatian dog)

Finally, the tuning of perception to invariants seems to be controlled in the right hemisphere of the brain. Admittedly, the higher the level of form perception is the more disputed are the correlated physiological and brain anatomical problem is that the interactive structure of the brain processes contributing to the stable perceptual representation of the world does not have a simple hierarchical structure with its appealing feature of decomposability but seems to be heterarchical in character; for instance Goodale and Milner (1992) show that the pathways proposed by Trevarthen (1968) and seemingly supported by the experimental results of Mishkin and Ungerleider (1982) do not lead to the proposed anatomically separable modules of form vs. location identification but that even on this level there exist interaction. Also on the phenomenological and behavioural level it can be shown that – for instance – the clear-cut hierarchical model of David Marr (1982) does not hold: Already Schumann's (1904) results that contours are perceivable even if no physical differences exist (which later led to Kanizsa´s 'illusory contours') indicate that the perceivability of patterns (in the sense of *Prägnanz*, that is, being a minimum in a potential field) influences even processes on a lower level of perception.[20]

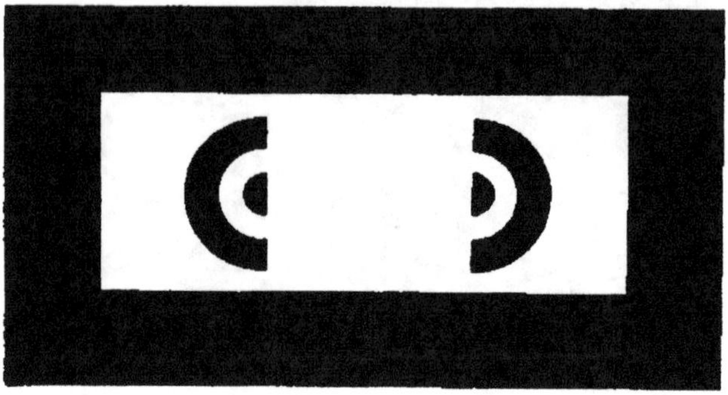

Figure 5: First example of illusory contours analysed in the framework of perception (Schumann, 1904).

Alf C. Zimmer
Psychologie II
Regensburg Universität
Universitätsstraße, 31
D-93053 Regensburg
Email: alf.zimmer@psychologie.uni-regensburg.de

NOTES

[1] Helmholtz 1866, as cited in Johnson-Laird and Wason, 341.

[2] Gregory 1998, 208, comments on this schema "Ins and outs of vision. Not to be taken too literally, this is a way of relating the notions of *bottom-up* signals from the eyes, *top-down* knowledge, and what we are calling *sideways* rules of perception, such as the Gestalt laws of organisation and perspective. This is developed in Chapter 11, 'Speculations,' with an attempt to classify illusions."

[3] See Henle 1987, for the position that the resolution of dichotomies is a central tenet of Gestalt theory.

[4] It should be noted, however, that in his Prolegomena he referred only to the introspective method. In contrast to Wolff's definition of psychology Kant regarded empirical anthropology based upon observable behaviour as a natural science. Insofar, Kant can be regarded as a proponent of the 'methodological behaviourism' of modern cognitive science.

[5] Hegel 1830, 13ff.

[6] Swets,1961.

[7] See Jaeger 1988, 33

[8] See Benussi 1904, or Stumpf 1910.

[9] See Schumann 1912.

[10] Köhler 1913, 76.

[11] Ibidem, 77.
[12] Ivi, footnote 1.
[13] Ibidem, 79–80.
[14] Köhler 1967, 112/113.
[15] How strong this appeal of a strictly linear model for psychology is can be seen from the fact that the dominating theory in cognitive psychology from about 1960 up to the end of the 80's, namely, information processing, is best described by the additive factor model. Cf. Sanders 1998.
[16] Cited after Singer 1959, 433.
[17] David Marr's book of 1980 shows this because in his inherently linear modelling of bottom-up perceptual processes he lists the Gestalt factors of grouping but a uses them as entirely isolated modules, the implicitly non-linear character of them having no bearing on the general linear information processing account.
[18] Tembrock 1973, 123.
[19] See Roth 1992.
[20] v.d. Heydt, Peterhans, Baumgartner, 1984 have been able to identify where in the brain this interaction happens.

FIGURE CAPTIONS

Figure. 1: Gregory's (1995) model of the interaction of perceptual processes.
Figure 2: Modification of Gregory's classification of different approaches to perception.
Figure 3: Schema of interrelated perceptual fields as developed in Husserl's "Dingvorlesung".
Figure 4: A hidden figure (Dalmatian dog).
Figure 5: First example of illusory contours analysed in the framework of perception (Schumann, 1904).

REFERENCES

Basar, E. and and Bullock, T.H. eds.: 1992, *Induced Rhythms in the Brain*. Boston, Birkhäuser.
Benussi, V.: 1904, 'Zur Psychologie des Gestalterfassens,' in Meinong ed., 1904, 302–403.
Bergson, H.: 1911, *Essai sur les données immédiates de la conscience*. Paris, Alcan.
Ebbinghaus, H.: 1885, repr.1971, *Über das Gedächtnis*. Darmstadt, Wissenschaftliche Buchgesellschaft.
Eckhorn, R. and Reitboeck, H.J.: 1990, '*Stimulus-Specific Synchronization in Cat Visual Cortex and its Possible Role in Visual Pattern Recognition*,' in Haken and Stadler, eds., 1990, 99–111.
Ehrenfels, Ch. v.: 1890, 'Über Gestaltqualitäten,' *Vierteljahrsschrift für wissenschaftliche Philosophie*, **14**, 249–292.
Ellis W.D. ed.: 1938, *A Source Book of Gestalt Psychology*. London, Kegan Paul.
Fechner, G.T.: 1860, *Elemente der Psychophysik*. Leipzig, Breitkopf and Härtel.
Gibson, J.J.: 1979, *The Ecological Approach to Visual Perception*. Boston, Houghton Mifflin.
Goethe, J.W. v.: 1808, *Farbenlehre. Vollständige Ausgabe der theoretischen Schriften*. Tübingen, Wissenschaftliche Buchgemeinschaft.
Goodale, M.A. and Milner, A.D.: 1992, 'Separate Visual Pathways for Perception and Action,' *Trends in Neuroscience*, **15**, 20–25.

Gray, C.M., Engel, A.K., König, P. and Singer, W.: 1992, 'Mechanisms Underlying the Generation of Neuronal Oscillations in Cat Visual Cortex,' in Basar and Bullock, eds., 1992, 29–45.

Gregory, R.L.: 1970, *The Intelligent Eye*. New York, McGraw-Hill Books.

Gregory, R.L.: 1995, *The Artful Eye*. Oxford, Oxford University Press.

Gregory, R. L.: 1998, *Eye and Brain. The Psychology of Seeing*. Oxford, Oxford University Press.

Haken, H and Stadler, M. eds.: 1990, *Synergetics of Cognition*. Berlin, Springer.

Hegel, G.W.F.: 1830, *Enzyklopädie der philosophischen Wissenschaften im Grundrisse. Zweiter Teil: Die Naturphilosophie*. In *Theorie Werkausgabe*, vol. 9, Frankfurt a. Main, Suhrkamp 1970.

Heydt, v.d. R., Peterhans, E. and Baumgartner, G.: 1984, 'Illusory Contours and Cortical Neuron Responses,' *Science*, **224**, 1260–1262.

Helmholtz, H. v.: 1867, *Handbuch der physiologischen Optik*. Hamburg, Voss.

Henle, M.: 1987, 'On Breaking out of Dichotomies,' *Gestalt Theory*, **3/4**, 140–149.

Husserl, E.: 1907/1954, *Ding und Raum*. Husserliana, Vol. XVI, Den Haag, Nijhoff (quoted as 'Dingvorlesung.')

Jaeger, S. ed.: 1988, *Briefe von Wolfgang Köhler an Hans Geitel 1907-1920*. Passauer Schriften zur Psychologiegeschichte. Passau, Universitätsverlag.

Johnson-Laird, P.N. and Wason, P.C.: 1977, *Thinking. Readings in Cognitive Science*. Cambridge, Mass., Cambridge University Press.

Kanizsa, G.: 1979, *Organization in Vision, Essays on Gestalt Perception*. New York, Praeger.

Kanizsa, G. and Luccio, R.: 1986, 'Analysis of the Concept of 'Prägnanz,' *Gestalt Theory*, **2**, 99–135.

Köhler, W.: 1913, 'Über unbemerkte Empfindungen und Urteilstäuschungen,' *Zeitschrift für Psychologie*, **66**, 51–80.

Köhler, W.: 1920, *Die physischen Gestalten in Ruhe und im stationären Zustand*. Erlangen, Verlag der Philosophischen Akademie. En. part. tr. *On Physical Forms in Equilibrium and Stationary State* in Ellis, ed., 1938, 17-54.

Köhler, W.: 1967, 'Gestalt psychology', *Psychologische Forschung*, **31**, xviii–xxx.

Koffka, K.:1921, *Die Grundlagen der psychischen Entwicklung*. Osterwieck, Zickfeld.

Koffka, K.: 1935, *Principles of Gestalt Psychology*. New York, Harcourt Brace.

Kubovi, M. and Pomerantz, J.R. eds.: 1981, *Perceptual Organization*. Hillsdale, N.J., Lawrence Erlbaum Associates.

Leontiev, A.N.: 1973, *Probleme der Entwicklung des Psychischen*. Frankfurt a. Main, Athenäum.

Mach, E.: 1886, *Beiträge zur Analyse von Empfindungen*. Jena, Gustav Fischer.

Marr, D.: 1982, *Vision*. New York, Freeman.

Meinong, A. ed.: 1904, *Untersuchungen zur Gegenstandstheorie und Psychologie*. Leipzig, Barth.

Meinong, A. v.: 1891, 'Zur Psychologie der Komplexionen und Relationen,' *Zeitschrift für Psychologie*, **2**, 245–265.

Mishkin, M. and Ungerleider, L.G.: 1982, 'Contribution of Striate Inputs to the Visuospatial Functions of Parieto-preoccipital Cortex in Monkeys,' *Journal of Brain and Behavioural Sciences*, **6**, 57–77.

Roth, G.: 1992, 'Das konstruktive Gehirn: Neurobiologische Grundlagen von Wahrnehmung und Erkenntnis,' in Schmidt, ed., 1992, 277–336.

Sanders, A.F.: 1998, *Elements of Human Performance*. Mahwah, N.J., Lawrence Erlbaum Associates.
Schumann, F.: 1904, 'Beiträge zur Analyse der Gesichtswahrnehmungen IV,' *Zeitschrift für Psychologie*, **36**, 161–185.
Schumann, F.: 1912, 'Über einige Hauptprobleme der Lehre von den Gesichtswahrnehmungen,' *Berichte über den 5. Kongreß für experimentelle Psychologie*. Leipzig, Johann Ambrosius Barth, 179-183.
Schmidt, S.J.: 1992, *Kognition und Gesellaschaft. Der Diskurs des radikalen Konstruktivismus*. Frankfurt a.Main, Suhrkamp.
Shepard, R.N.: 1981, 'Psychophysical Complementarity,' in Kubovy and Pomerantz, eds. 1981, 279–341.
Singer, C.: 1959, *A History of Scientific Ideas*. Oxford, Oxford University Press.
Stumpf, C.: 1910, 'Beobachtungen über Kombinationstöne,' *Zeitschrift für Psychologie*, **55**, 1–142.
Swets, J.A.: 1961, 'Detection Theory and Psychophysics: A Review,' *Psychometrika*, **26**, 49–63.
Tembrock, G.: 1973, *Grundriß der Verhaltenswissenschaften*. Stuttgart, Fischer.
Trevarthen, L.B.: 1968, 'Two Mechanisms of Vision of Primates,' *Psychologische Forschung*, **31**, 299–337.
Weber, E.H.: 1834, *De pulsu, resorptione, auditu et tactu*. Leipzig, Köhler. En. tr. 1978, *De tactu*. New York, Academic Press.
Zimmer, A.: 1991, 'The Complementarity of Singularity and Stability,' *Gestalt Theory*, **4**, 276–282.

DAVID WOODRUFF SMITH

CONSCIOUSNESS AND ACTUALITY IN WHITEHEADIAN ONTOLOGY[1]

1. COGNITION IN THE STRUCTURE OF EXISTENCE

If cognitive science is the science of cognition, what is cognition? Whitehead's answer is drawn directly from his fundamental ontology: cognition is a special form of what he calls 'prehension,' taking more particular forms in conscious perception, thought, imagination and other psychological processes.

From Whitehead's sweeping metaphysics I propose to appropriate key elements and reconstruct from these a fundamental ontology which applies to consciousness in cognitive processes. This Whiteheadian ontology has shaped a novel approach to fundamental ontological categories used in database design or knowledge-based computer systems.[2] And the Whiteheadian approach to conscious processes is instructive for the dynamic systems approach in cognitive science, which has been proposed as an alternative to both the neural-network approach and the language-of-thought approach to modelling cognitive processes.[3]

Whitehead's main concern was not cognitive process, but process in general. Within a fundamental ontology of process, he turned to processes of consciousness, perception, belief, and emotion – almost in passing as he sought to limn the truly ultimate structure of reality. For those raised in empirical science, or in philosophy oriented to science and modern logic, Whitehead's ontology will seem the most extreme of *a priori* speculative metaphysics, written moreover in a philosophical prose that is as complex as Russell's is simple. Still, the mathematical wizardry in *Principia Mathematica* was more Whitehead than Russell, and the purity of Whitehead's ontological vision (shaped by his mathematician's brain) is kindred in soul with the physical vision of an Einstein or a Hawking. It is this pure ontology that I want to frame, in order to construct from it, along lines Whitehead prescribed, an ontology of consciousness in various forms of cognition.

L. Albertazzi (ed.), The Dawn of Cognitive Science. Early European Contributors, 269–297.

The resulting Whiteheadian ontology does not naturalise consciousness by reducing it to physical process. Rather, it places all of nature *cum* consciousness in a world formed by one ontological process of 'becoming' through 'prehension.'

2. WHITEHEAD'S BACKGROUND AND CONTEXT

Alfred North Whitehead (1861-1947) is known as a principal founder of mathematical logic, through the monumental *Principia Mathematica* (1910-13), which he co-authored with the younger Bertrand Russell. Whitehead and Russell exerted an indirect influence through logic on computer science and thus on the computer model of mind that informs cognitive science today. Whitehead's later years, however, saw him engaged in abstract metaphysics, in work that is usually treated quite separately from his early work in mathematics and then mathematical logic. Here I shall address Whitehead's ontology as developed in *Process and Reality* (1929; delivered as a lecture series in 1927–28).

Whitehead posited a basic structure of 'process' applied to all reality. As a mathematician he was keenly attuned to the physics of relativity and quantum mechanics emerging in his day. As a British-born philosopher he had his eye also on the early modern philosophers, from Descartes to Locke and Hume, for whom mental activity was salient. Yet as a philosophical cosmologist he looked to the ancient Greeks, to Plato and the pre-Socratics, especially Heraclitus, who famously proclaimed that everything is flux. Reflecting a further sensibility of the 19th century, Whitehead called his metaphysical-cosmological system 'the philosophy of organism,' and he did indeed think of the cosmos itself as alive in a fundamental sense. Today his system is often called 'process philosophy' and is influential in theology, since Whitehead defined a place for God in the cosmos, as the first being formed by the process of 'becoming.'

With an eye to cognitive science, as opposed to metaphysical system, consider from Whitehead's era three very different approaches to the exploration of the mind. Wundt's experimental psychology used introspection to study sensations. Freud's psycho-analysis used hypotheses about drives and infantile experience to explain adult neuroses. And Husserl's phenomenology used the technique of reflection on one's own experience to analyse the structure of consciousness, including the dynamics of time-consciousness. Here were three starkly different 'sciences' of the mind, differing in both aim and method. Yet each had to address the dynamics of changing and interacting states of mind, and today's cognitive science carries the analysis of mind into mathematical-computational models of the dynamics of cognitive process. Whitehead's ontology of process ought to shed light on the most basic structure of mental process as a flow of cognitive occasions.

Whitehead drew where possible on classical ideas: in Plato, Aristotle, Heraclitus, Descartes, Leibniz, Locke, Hume. Yet the proximate influence on his conception of process was the French philosopher, Henri Bergson. In *Matter and Memory* (1896), Bergson painted a metaphysics featuring 'durations' rather than classical enduring natural objects, what Aristotle called 'substances' (*ousia*). The world is a vortex of passing events in interaction with one another, each bearing within it 'images' of others. Bergson envisioned a monism whereby this metaphysical picture applies equally to physical events and to mental events. Physical events were soon to be described in relativity theory and then in quantum physics, while mental events were soon to be described in phenomenology, in psychoanalysis, and ultimately in what has recently emerged as cognitive science. Bergson would have wished his metaphysics to apply to the dynamics of quantum and relativistic physical phenomena and equally to the dynamics of perception, imagination, thought, and desire. Bergson expressed his vision in vivid impressionistic prose, mirroring in metaphysical language the flow of painting in French impressionism and fauvism. I find Whitehead's metaphysical prose impressionistic yet mathematical. Whitehead's technical terms are often vividly, disconcertingly, mentalistic. Yet his terminology reflects a Bergsonian style of monism rather than the varieties of absolute idealism that ran from Hegel to F. H. Bradley and J. M. E. McTaggart in England. Like absolute idealism, Whitehead's philosophy of 'organism' preached the radical interaction of all things; however, for Whitehead, the ultimately actual entities are neither mental nor physical, but constitute what we know as the mental and the physical alike.

McTaggart taught philosophy at Cambridge when Whitehead taught mathematics at Cambridge. Like McTaggart, as well as Bergson, Whitehead focussed on the process of coming to pass in time, where the truly 'actual' – what deserves the name – is in process, becoming rather than already become. McTaggart (1927) argued that only the present is actual, that past and future are unreal, indeed that time itself is unreal because a time or occurrence cannot be first future, then present, then past, as these temporal attributions are mutually contradictory. Whitehead (1927-28) held, by contrast, that past and future are perfectly real but not 'actual,' or actualising, that is, in process of becoming. Still, Whitehead had his eye on bigger fish when he framed actuality as *becoming*: his concern was not only coming to pass in time, but coming to be an entity at all, of which coming to be an entity in temporal flux is a special case.

Whitehead's fundamental ontology of process, I want to suggest, provides formal tools that can be used to analyse the substantive claims about process, including processes of consciousness, put forth in empirical or experiential studies. By 'formal,' however, I mean pertaining to *formal ontology*, as opposed to formal languages of logic, mathematics, or computer science. Here I draw on Husserl's distinction between formal and material ontology. 'Formal' ontology seeks invariant structures that are realised in various domains of entities, while 'material' ontologies analyse ontological structures within a given domain (such as mind, body, culture, or what have you).[4] Whitehead's account of the process of becoming an entity is, I shall try to show, a formal ontology of 'process,' and consciousness inherits this structure of process.

3. THE OUTLINES OF WHITEHEAD'S ONTOLOGY

Whitehead's ontology in *Process and Reality* is framed by an elaborate system of *categories* (pp. 18–30). I can address only that part of the system which I will appropriate for present purposes.

The most fundamental category, dubbed 'the Category of the Ultimate,' is that of Creativity or Becoming. Becoming is the process by which *many* entities become – come together to form – *one* actual entity. Becoming is also called concrescence (p. 21). Importantly, the category of becoming "replaces Aristotle's category of 'primary substance'" (p. 21). Aristotle's paradigm of a primary substance (*ousia*, or being) is a concrete being such as an individual human organism; such a being is a member of a species, in this case humanity, and species themselves belong to the category Aristotle called secondary substance. For Aristotle, a primary substance is something that endures through time and may change its qualities or 'accidents,' its relations to other things, its location, etc.; and it is a fusion of matter and form, the matter undergoing change in form. Now, Whitehead's leading claim is that such beings are not the most fundamental type of thing in the world. More fundamental are the processes of *becoming* which form all actual entities. Becomings, not 'substances,' are the ultimate constituents of the world, the cosmos. Aristotelian substances are, for Whitehead, the product of many processes of becoming, and this view is an abstract version of modern physical theory, for which everyday objects are products of relativistic, quantum phenomena of a more fundamental and very different kind.

Presupposing the category of Becoming are the Categories of Existence (p. 22). There are eight such categories, of which I shall focus on three: the categories of Actual Entity, Eternal Object, and Prehension. (p. 1) An actual entity, or actual occasion, is a concrete entity rather like a 'point' event. (p. 2) An eternal object, or form, is what Plato called a form (*eidos*), but Whitehead conceives a form, or 'form of definiteness,' as a 'pure potentiality' for being 'determined' in a specific way. For Plato, a form is an actual but eternal entity (residing in a Platonic heaven, as Plato is often parodied, a parody Whitehead rejects). For Whitehead, however, a form is not an actual albeit eternal entity, but rather a pure potential for actual entities: the way an actual entity could be – which is 'eternal' insofar as it is not realised in an actual entity. (p. 3) Now, prehension is the tie that binds everything together; it is the way in which things are, most fundamentally, related. If you will, prehension is simply the fundamental form of relatedness, and particular prehensions are the basic 'concrete facts of relatedness.' Thus, actual entities prehend other (prior) actual entities, and they prehend forms as well. (More on all this shortly.)

Presupposing both the Category of Becoming and the Categories of Existence are two further types of categories. The Categories of Explanation are 27 axiomatic conditions, or types of interaction, among actual entities, eternal forms, prehensions, etc. – basic conditions laid down by axioms in the ontology. And the Categoreal Obligations are nine preconditions laid down, as it were, by further axioms. These categories together weave the story of reality as told in Whitehead's ontology. Importantly, Whitehead *ranks* different types of categories: most basic is the Category of Becoming, next are the Categories of Existence, then the Categories of Condition and then those of Precondition (as I'm calling them).

With this ranking of categories, Whitehead moves well beyond Aristotle, who launched the idea of ontological categories with his famous list:

Primary Substance (Individual), Secondary Substance (Species), Quantity, Quality, Relative, Where, When, Position, Having, Acting, Being-Acted-Upon.

Aristotle's term *kategoriai* originally meant simply 'predicates,' or types of entities. By contrast, what are called categories in biology today – Species, Genus, Family, Order, etc. – are ranks of taxa (groups of living things), defining their place in the classification scheme. Indeed, Whitehead's categorial scheme shows something of the complexity that has since arisen in modern biological systematics, or higher classification. The great biologist Ernst Mayr even refers to Whitehead while distinguishing the evolutionary roles of species as opposed to higher taxa.[5] Where systematics in biology is the science of diversity of living beings, we may conceive *ontological systematics* as the science of diversity of all beings, classifying things of all types and categories. A ranking of categories is then part of the task of ontological systematics, and of 'formal' ontology in the sense that Husserl had in mind.

Assuming Whitehead's scheme of categories, let us turn to his account of actual entities. To be 'actual' is to be active, to be in process, to be part of the flux that constitutes all actuality. As Heraclitus insisted, all is flux, "all things flow" (p. 208). Thus, in Whitehead's vision, the atoms of the universe are not particles but points in the flux, points of flow: processes of becoming, or actual entities or occasions in the process of becoming, coming into actuality. Points along the flow are bound together by prehension: later occasions or actual(ising) entities *prehend* or 'feel' earlier ones. We think of a physical 'feeling' in gravitation or electromagnetic attraction, but prehension is – I should like to say in interpretation – the *formal* connection that is realised as gravitational or electromagnetic connection, or physical causation, in the relevant scenario of physics. Again, we think of an emotional 'feeling' in personal attraction, where prehension is the formal connection that is realised as a bond of friendship, in the relevant scenario of psychology. And we may think of a cognitive 'feeling' in conscious perception of an object, where prehension is the formal connection that is realised as a bond of *intentionality* – to use Husserl's term, which has now gained currency in phenomenology, philosophy of mind, and cognitive science, but was unknown to Whitehead. The root meaning of 'intention' is to aim or reach toward something; this is exactly the metaphor behind Whitehead's term 'prehension,' as he initially defines a prehension as having "a vector character" (p. 19). Whitehead looks to Leibniz's *Monadology* (p. 19) as a model: indeed, his ontology is like Leibniz's but with the monads' windows open. However, while Leibniz looks to mental operations for a model of the physical, producing his unique brand of idealism, Whitehead sees the notion of prehension as striking a 'balance' between mental and physical operations (p. 19). A better way of putting the point, I propose, is to say that prehension is the common formal structure that is realised in operations of both mental and physical 'feeling.'

Now, Whitehead recognises different types of prehension. (p. 1) There are positive prehensions, called 'feelings' (illustrated provisionally above). But there are also negative prehensions, which block or 'eliminate' things from feeling. We shall address only positive prehensions, or feelings. (p. 2) An actual entity E prehends not only other actual entities, but also eternal objects or forms. Prehending an actual entity is called 'physical' prehension, while prehending a form is called 'conceptual' prehension (as discussed below). E's prehending a form F is Whitehead's variation on what we call, since Plato, E's exemplifying F. In Whitehead's idiom, E 'feels' the form F; alternatively, F 'ingresses' into E.

For Plato, forms exist eternally and thus in a different realm than particulars; for Aristotle, contra Plato, forms (species, accidents, locations) exist in the concrete world, but only insofar as instantiated by particular substances. But Whitehead places both concrete actual entities and abstract forms precisely in the flux that is reality. The difference is that a form is a *potential* for definiteness in an actual entity, while an actual entity is a definite realised entity, that is, in process of becoming and thereby being fully 'determined.' Later we shall address, in Whitehead's fine-grained analysis, a hybrid entity that combines the actual and the potential. (I leave the details as a surprise to the reader.)

The prior actual entities prehended by a given actual entity E form the 'datum' of E, or the 'actual world' of E. The constitution of E – its 'real internal constitution' (p. 25) – includes the pattern of prehensions in which E feels the actual entities in its actual world and also feels the forms that are realised in it. Importantly, there is an ontological relativity in Whitehead's scheme: the actual world of E is not the actual world *tout court*, for every actual entity has its own 'actual world.' The point to remember is that Whitehead's ontology is 'tensed,' or rather indexicalised. Recall McTaggarts's distinction between two structures of time, or two types of temporality: times are ordered in what he called the 'A-series' where times are past, present, or future; and times are ordered in the 'B-series' where one time is before another time. It is customary to say that B-series time is time viewed from outside the space-time framework, as it were eternally, whereas A-series time is time viewed from within the space-time framework, and thus indexically. Now, actual entities are ordered by becoming. To add a dose of McTaggart to Whitehead, actual entities are 'present,' or presently actualising. It is this currency that defines 'actuality' for Whitehead. (That is in fact one of the meanings of 'actual,' meaning, in French and German, current events.) Think, then, of flux as a transition from past (formerly) actual entities into present (currently) actual entities and toward future (potentially) actual entities, always informed by the realisation-in-progress of potentials called forms.

There is, however, a second type of flux or 'fluency' that we shall consider shortly: not the *transition* from past actualities to a presently unfolding actuality; but the *concrescence* wherein many actual entities come together to form one actual entity (p. 210).

Whitehead's ontology seems apt for the world of 20th century physics, for points of matter-energy in a field of space-time in a relativistic quantum framework. But how does it apply to the everyday world of objects and properties, the world described by Aristotle's metaphysics, which is a more theoretical version of our common-sense ontology? Whitehead did not simply throw out Aristotelian metaphysics. He said Aristotle's category of substance is not fundamental, but he wanted to preserve everyday objects or 'substances' as emergent from more fundamental 'actual entities.' Moreover, he preserved the Platonic category of form. For Whitehead, actual entities have forms, as did Aristotelian substances, but forms are potentials for actual entities.

4. ONTOLOGICAL VERSUS TEMPORAL 'BECOMING'

All things are in flux, "all things flow," Whitehead proclaims (p. 208), echoing Heraclitus. Yet Whitehead immediately introduces a crucial, and I think profound, distinction:

> [T]here are two kinds of fluency. One kind is the *concrescence* which, in Locke's language, is 'the real internal constitution of a particular existent.' The other kind is the *transition* from particular existent to particular existent. This transition, again in Locke's language, is the 'perpetually perishing' which is one aspect of the notion of time, and in another aspect the transition is the origination of the present in conformity with the 'power' of the past. [p. 210]

Thus, Whitehead distinguishes two fundamentally different types of 'flux' in the process of 'becoming.' We might say that transition from one entity at one time to another entity at a later time defines *temporal becoming*, as the later entity comes into being in time. And we might say that concrescence of one entity out of others defines *ontological becoming*, as the one comes into being on the basis of the others.

Suppose we treat the process of 'flowing' as a *function* or *operation* in the mathematical sense (Whitehead being a mathematician, after all). Then *temporal becoming* is an operation that assigns to one set of occasions $\{O_1, \ldots, O_n\}$ another occasion O^*. The occasions O_i in the original set, which is the argument of the function, are 'prior' in time, 'past' occasions relative to O^*, which is 'present' or 'actual (now).' By contrast, *ontological becoming*, or concrescence, is an operation that assigns to one set of actual entities $\{E_1, \ldots, E_n\}$ another actual entity E^*. The entities E_i compose the 'datum' or 'actual world' of E^*. Notice that the term 'occasion' carries temporal connotations whereas 'entity' does not.

Whitehead assumes 'the ontological principle' (p. 19, p. 24) that every actual entity has its 'reasons' or grounds, the entities out of which it is formed by concrescence: 'no actual entity, then no reason' (p. 19). Clearly, this principle constrains ontological rather than temporal becoming.

Although Whitehead's language of 'process' resonates with the sound of transition through time, his ontology actually treats concrescence, or ontological becoming, as what is fundamental. Indeed, extension in time and space is itself something which must come into being, or become, and its becoming is not temporal. That is, temporality becomes, but its becoming is not temporal. In Whitehead's words:

> [I]n every act of becoming there is the becoming of something with temporal extension; but [...] the act itself is not extensive, in the sense that it is divisible into earlier and later acts of becoming which correspond to the extensive divisibility of what has become. [p. 69]

> This extensive continuum [of time] is one relational complex in which all potential objectifications find their niche. It underlies the whole world, past, present, and future [...] It is not a fact prior to the world; it is the first determination of order – that is, of real potentiality – arising out of the general character of the world. In its full generality beyond the present epoch, it does not involve shapes, dimensions, or measurability; these are additional determinations of real potentiality arising from our cosmic epoch. [p. 66]

> [T]he extensive continuity of the physical universe has usually been construed to mean that there is a continuity of becoming. But if we admit that 'something becomes,' it is easy, by employing Zeno's method, to prove that there can be no continuity of becoming. There is a becoming of continuity, but no continuity of becoming [...] In other words, extensiveness becomes, but 'becoming' is not itself extensive. [p. 35]

Thus, extension in time is a fully 'general' structure of the world. And that structure of temporality must itself *become*. It is in fact 'the first determination of order': first in ontological priority. Time is not, however, the first thing to exist in time, first in temporal priority.

We can clarify Whitehead's distinction of two kinds of flux if we distinguish between formal and material ontology. Temporal becoming is, in these terms, a 'material' process, and ontological becoming is a 'formal' process: concrescence, or ontological becoming, is the formal structure that applies to all actual entities, in particular to the material domain of entities in temporal becoming. Indeed, for Whitehead, concrescence is the most fundamental *formal* ontological structure of the world: the category of 'the ultimate.'

While the process of ontological becoming is the heart of Whitehead's ontology, the most salient application of the ontology is to temporal becoming, or 'process' in the familiar sense, of which mental process is a special case. This dual sense of 'process' plagues the text of *Process and Reality*. May I suggest a resolution of this interpretative quandary? If we analyse the two types of flux in the manner outlined above, then the formal structure of ontological becoming is reproduced at a different level in the material structure of temporal becoming. There is a recursion of basic structure in this Whiteheadian analysis of 'becoming:' temporal becoming is not only the product of ontological becoming (time becomes, takes ontological form), its *form* is the same as that of ontological becoming, or concrescence. However, even if this form does recur at both levels (and I am not sure it really does), it remains misleading to use the same term 'process' for both the becoming of a temporal transition and the transitioning itself.

I share Whitehead's appreciation of the presocratic philosophers of ancient Greece, whose concerns with truly fundamental issues like 'the one and the many' have been eclipsed in more recent philosophy. But where Whitehead nods to Heraclitus on 'flux,' I bow to Anaximander of Miletus on 'origin,' or ontological basis. In Anaximander's metaphysical vision, all types of beings in the world (including the most basic physical-chemical elements as he knew them) have their origin, or *arche*, in something he called the *apeiron*, the 'nonlimited' or 'non-determined.' In this vision I find roots of the notion of ontological becoming, whereby every being or entity (*ousia* in Aristotle) is based in or arises from (what I like to call) primordial *modes-of-being*, which are not entities at all but rather ways in which entities become entities, and so are the bases of beings. I believe Whitehead's 'category of the ultimate' – Becoming – follows this Anaximandrian architecture. Thus, 'determinate' actual entities, extended in time and space-time, are constituted in ways-of-prehending that are not themselves 'determinate' entities.[6]

5. PREHENSION OR 'FEELING'

The ontological structure of an actual entity or occasion, for Whitehead, is that of a complex defined by relations of prehension. Indeed, an actual entity or occasion is a 'cell,' in a sense akin to the biological. As Whitehead puts it:

> [T]he cell is exhibited as appropriating for the foundation of its own existence, the various elements of the universe out of which it arises. Each process of appropriation of a particular element is termed a prehension. The ultimate elements of the universe, thus appropriated, are the already constituted actual entities, and the eternal objects.
>
> The process itself is the constitution of the actual entity; in Locke's phrase, it is the 'real internal constitution' of the actual entity. [p. 219]

An actual entity is in this way 'constituted' by its pattern of prehensions of other actual entities, and also eternal objects or forms. Thereby it *becomes* 'one' actual entity, formed by concrescence out of "many" actual entities and forms.

Prehension – called 'feeling' or 'experience' (if positive) – is not *per se* a psychological phenomenon. Prehending something is not the same as being *conscious of* something, what Husserl dubbed intentionality. As we shall see below, consciousness is a special form of prehension, and not vice versa.

There are two primary species of prehension or feeling: physical and 'conceptual.' Physical prehension is what we think of as causation, while conceptual prehension is what we think of as being informed by a property or form. I say "what we think of" because Whitehead wants to underwrite our familiar, Aristotelian metaphysics by a very different ontology of prehension. In Whitehead's words:

> A simply physical feeling is an act of causation. The actual entity which is the initial datum is the 'cause,' and the simply physical feeling is the 'effect,' and the subject entertaining the simply physical feeling is the actual entity 'conditioned' by the effect. This 'conditioned' actual entity will also be called the 'effect.' [p. 236]

> A conceptual feeling is feeling an eternal object in the primary metaphysical character of being an 'object,' that is to say, feeling its capacity for being a realised determinant of process. [p. 239]

Whitehead thinks of a Platonic form or 'eternal object' – what we call a kind, quality, or relation – as a *potential* or *capacity* for being determined thus-and-so (p. 23), and this is what is felt 'conceptually.' Rather than residing in a Platonic heaven, it exists here in the one Heraclitean flux that is the world. Its prehension Whitehead calls 'conceptual;' 'formal' or 'eidetic' would avoid the suggestion of a mental activity of conceptualisation, but Whitehead's terminology deliberately flouts the distinction between the mental and the nonmental.

Now, every actual entity is formed by concrescence through prehensions of *both* actual entities and forms, as depicted in Figure 1 below. Note the mixing of mentalist and physicalist terminology as Whitehead makes this point:

> In each concrescence there is a twofold aspect of the creative urge. In one aspect there is the origination of simple causal feelings; and in the other aspect there is the origination of conceptual feelings. These contrasted aspects will be called the physical and mental poles of an actual entity. No actual entity is devoid of either pole; though their relative importance differs in different actual entities. Also conceptual feelings do not necessarily involve consciousness; though there can be no conscious feelings which do not involve conceptual feelings as elements in the synthesis. [p. 239]

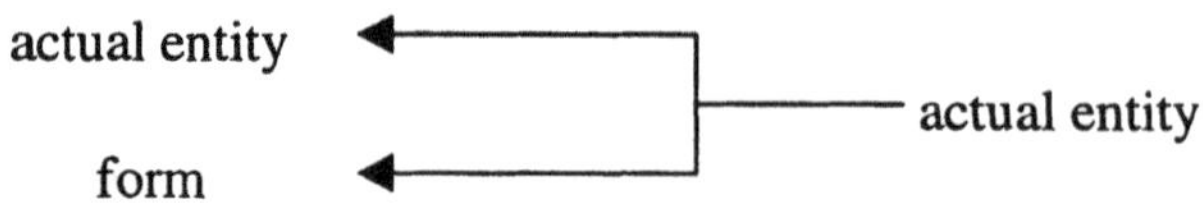

Figure 1: Concresence

So every actual entity 'feels' both physically and conceptually, feeling both (other) actual entities and eternal forms. Feeling a form is called 'mental' but does not involve consciousness, or indeed what we think of as mental activity in thought or perception. Here, in the most elementary structure of the world, Whitehead nails down the monism we noted earlier – under the influence of Bergson rather than Spinoza (who assumed one *substance*). Furthermore, we begin to see the niche Whitehead has in mind for consciousness: whatever is special about consciousness, it will consist in 'conscious feelings' and these always involve feeling forms.

Furthermore, all the world – from actual entities on up – is defined by prehensions of both forms and actual entities:

> Conceptual feelings and simple causal feelings constitute the two main species of 'primary' feelings. All other feelings of whatever complexity arise out of a process of integration which starts with a phase of these primary feelings. [p. 239]

Whitehead holds a kind of atomism (pp. 19, 35, 238), since actual entities are atomic and everything complex is built up from prehensions of actual entities and forms. But the atomism is not that of physics, where elementary particles are bound together by physical-chemical bonds (physical feelings of causation). Nor is the atomism that of logical atomism, as propounded by Wittgenstein in the *Tractatus* (1923) or Russell in "The Philosophy of Logical Atomism" (1918), where objects and properties or relations are bound together by instantiation and the resulting 'states of affairs' or 'facts' are then further bound together by logical operations such as 'and.'

6. PROPOSITIONAL FEELING

The constitution of every actual entity includes prehensions of both actual entities and forms, and the process of concrescence constitutes the entity through these prehensions. Now, concrescence itself takes place in three phases that Whitehead calls reception, supplementation, and satisfaction (pp. 211–215). The concrescence consists in 'reception' of a datum comprising other actual entities and forms, 'supplementation' of the reception by something like interpretation, and completion or 'satisfaction' producing the new actual entity.

(These are not temporal but logical or rather ontological 'phases' of the process.) Concrescence is not a psychological operation, yet Whitehead analyses concrescence into a structure like that of cognition, with a distinctly Kantian twist, and it is in one phase of concrescence that consciousness itself may emerge in some cases. Let us see how.

The supplementation phase of concrescence Whitehead analyses into two sub-phases called 'aesthetic' and 'intellectual' (pp. 213–214). Aesthetic supplementation "has welded the data into a new fact of blind feeling," and then, "This phase requires an influx of conceptual feelings and their integration with the pure physical feelings." This integration of physical and conceptual feelings is 'intellectual 'sight.'' This integrative higher-phase feeling Whitehead calls 'propositional feeling' (p. 214).

Specifically, a form or eternal object is a capacity for 'determination,' that is, being propertied or related in a specific way. When a form is realised in an actual entity, that entity is determinate in respect of that form. And a *propositional* feeling is a higher-phase prehension of the propositional integration of a form in a determinate actual entity:

> An eternal object realised in respect to its pure potentiality as related to *determinate* logical subjects is termed a 'propositional feeling' in the mentality of the actual occasion in question. [p. 214]

In the web Whitehead weaves, we discern the picture in Figure 2.

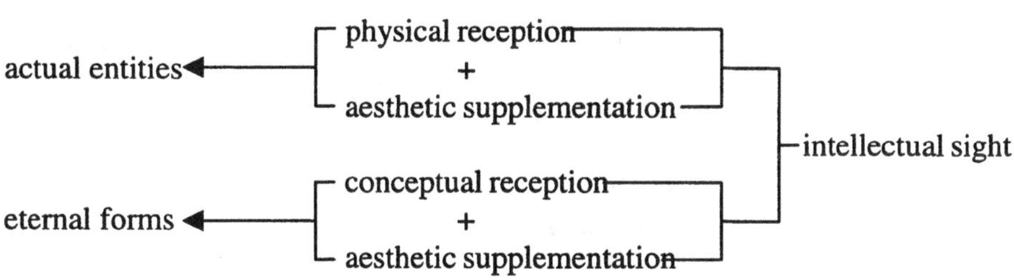

= propositional feeling
= feeling the proposition that those forms are realised in those entities

Figure 2: Propositional Feeling

Thus, propositional feeling is a higher-phase prehension that consists in feeling, with intellectual sight, the integration of the received actual entities and forms. That integration is called a 'proposition' (p. 256) and the prehension of a proposition is a propositional feeling. Figure 3 depicts this form of prehension, whose fine structure is depicted in Figure 2.

(form + actual entity) = proposition ◄——— (propositional feeling)

Figure 3: A Proposition and its Prehension

(We shall return to Whitehead's ontology of propositions.)

In this scheme for propositional feeling we hear the echo of Kant's famous dictum, "Intuition without concepts is blind; concepts without intuition are empty." Kant's concern was the structure of cognition that synthesises sensations and concepts, or sensibility and understanding. This structure applies to objects in nature precisely because, according to Kant's transcendental idealism, natural objects are phenomenal projections of cognition. But Whitehead inverts this order of priority: cognitions carry this structure precisely because all actual occasions have the structure of concrescence, which with intellectual sight begins to constitute an occasion of cognition or consciousness (See the note on p. 215). According to Whitehead, it is in this niche of concrescence, formally defined, that the structure characteristic of consciousness begins to emerge, as we shall see.

If Whitehead's scheme seems to force a mentalistic structure onto concrescence itself (conceptual feeling is 'mental,' propositional feeling is part 'intellectual'), consider a recent program in the philosophy of cognitive science. The Shannon-Weaver notion of 'information' (roughly a conditional probability of this given that) is the basis of Fred Dretske's analysis of knowledge as a 'flow of information.'[7] In Dretske's ontological vision, cognition is a complex transfer of 'information' from the environment to the knower, but this information-processing is conceived as a 'natural' process, like causation, and the 'information' transferred is not what we usually think of as a 'proposition' in a mind. Well, in Whitehead's system, prehension itself is a kind of flow of information, as the datum of actual entities (and forms) that are prehended by an actual entity is the 'reason' or foundation of the entity (the condition of this entity given those). Thus, we could see the *formal* structure of the process of information-flow in either a monistic ontology *à la* Whitehead or a naturalistic ontology *à la* Dretske. (Which is not to reduce Whitehead to Shannon-Weaver-Dretske.)

7. PROPOSITIONS

The 'objective datum' of a propositional feeling – if you will, the object of a propositional feeling – is what Whitehead calls a *proposition* (p. 256). Specifically:

> An 'impure' prehension arises from the integration of a 'pure' conceptual prehension with a physical prehension originating in the physical pole. The datum of a pure conceptual prehension is an eternal object; the datum of an impure prehension is a proposition, otherwise termed a 'theory.' [p. 184]

> A proposition is a new kind of entity. It is a hybrid between pure potentialities [or forms] and actualities. The definite set of actual entities involved are called the 'logical subjects of the proposition;' and the definite set of eternal objects involved are called the 'predicates of the proposition.' The predicates define a potentiality of relatedness for the subjects. The predicates form one complex eternal object: this is 'the complex predicate.' The 'singular' proposition is the potentiality of this complex predicate finding realisation in the nexus of reactions between the logical subjects, with assigned stations in the pattern for the various logical subjects. [pp. 185–186]

Here are echoes of the voice(s) of *Principia Mathematica*.

But Whitehead's later ontology gives 'propositions' an utterly different role than does traditional logic:

> Indeed Bradley does not mention 'propositions' in his *Logic*. He writes only of 'judgements.' Other authors define propositions as a component in judgement. The doctrine here laid down is that, in the realisation of propositions, 'judgement' is a very rare component, and so is 'consciousness.' [p. 184]

Thus, Frege posited a realm of 'thoughts' that serve as 'sense' (*Sinn*) expressed by language but are entities distinct from both the physical and the mental. Again, Husserl posited a realm of 'noemata' that serve as 'sense' (*Sinn*) or content of mental acts including judgements, and are expressible in language, but are distinct in ontological kind from mental acts, physical objects, and cultural institutions (including words). But Whitehead's conception of propositions is strikingly different, and genuinely novel, since a proposition *per se* is neither component nor content nor object of a mental act of judgement or consciousness, nor of a speech act or linguistic expression. Even in causal 'feeling,' a proposition is 'what is proposed' for feeling. Whitehead insists:

> The term 'proposition' suits these hybrid entities, provided that we substitute the broad notion of 'feeling' for the narrower notions of 'judgement' and 'belief.' A proposition is an element in the objective lure *proposed for feeling*, and when admitted into feeling it constitutes *what is felt*. [...] Judgement is the decision admitting a proposition into intellectual belief. [p. 187]

We might call this hybrid entity a *state of affairs*, following a well-established tradition in Austrian philosophy, including Husserl and the early Wittgenstein, for whom a state of affairs (*Sachverhalt*: literally 'things related') is a complex entity formed by a number of objects instantiating a determinate relation (or one object instantiating a determinate property or species). However, Whitehead joins Russell in importing this notion into philosophical English under the term 'proposition.' Unfortunately, we say in English that a proposition is 'true' but a state of affairs 'obtains' or is 'actual.' Nonetheless, Whitehead presses his terminology into the virgin territory of prehension loosed from thought or consciousness:

> The interest in logic, dominating overintellectualised philosophers, has obscured the main function of propositions in the nature of things. They are not primarily for belief, but for feeling at the physical level of unconsciousness. They constitute a source for the origination of feeling which is not tied down to mere datum. A proposition is 'realised' by a member of its locus, when it is admitted into feeling.
>
> There are two types of relationship between a proposition and the actual world [or datum] of a member of its locus. The proposition may be conformal or non-conformal to the actual world, true or false. [p. 186]

Thus, Whitehead finds for 'truth' a distinct role from that of 'actuality' (becoming, concrescence), even while defining a proposition as rather like a state of affairs. Remember, the 'actual world' of an actual entity is the nexus of (other) actual entities prehended by the entity. So a proposition felt by one actual entity is *true* just in case it is conformal to the relevant actual world of the entities that are logical subjects of the proposition felt.

Here Whitehead divines a subtle line between truth and actuality. It is natural to think, in a classical vein, that a proposition is a content of thought or language which *represents* a state of affairs beyond thought or language. Then a proposition is true if and only if it conforms to the actual state of affairs it represents. If there is no actual state of affairs to which a proposition corresponds, then the proposition is false. On a modal variation, a false proposition may yet represent a possible but non-actual state of affairs. However, on Whitehead's approach, a proposition is a hybrid of actual entities and potentialities, i.e., forms as potentialities for determination as thus-and-so in actuality. And this hybrid stands ontologically, modally, between achieved actuality and mere potentiality. A proposition, on this view, is neither a possible state of affairs nor a 'proposed' state of affairs. A possible state of affairs lacks the modal character of potentiality, or *tending toward* actuality; and a proposed state of affairs has still a different modal character, if understood as an object *intentionally contained in* a mental act of judgement or supposition or imagination, or intentionally projected from and so dependent on a propositional act of consciousness. No, for Whitehead, a proposition is 'proposed for feeling' in the sense that it is a not an actual state of affairs but is poised in readiness for being realised, or formed through the process of conscrescence into a nexus of actual entities in determinate relation. And this status is ontologically prior to any conscious act of judgement, imagination, etc.

No stranger to logic, Whitehead recognises quantified as well as singular propositions:

> A 'general' proposition only differs from a 'singular' proposition by the generalisation of 'one definite set of actual entities' into 'any set belonging to a certain sort of sets.' If the sort of sets includes all sets with potentiality for that nexus of reactions, the proposition is called 'universal.' [p. 186]

So Whitehead recognises not only singular propositions (involving definite actual entities as 'subjects') but also universally quantified propositions (involving 'any' definite actual entities) and also existentially quantified propositions (involving 'some set of actual entities' but not a definite set) (see pp. 257-258). Also, Whitehead recognises 'negative' as well as 'positive' prehensions (p. 23) and thence affirmative and negative perceptions and judgements in feeling affirmative and negative propositions (pp. 161, 243, 270–274). So, with truth and the familiar logical forms of propositions, Whitehead articulates the structures of modern logic: all within the realm of propositions defined as ontological structures prior to judgement or thought-content or indeed linguistic expression!

If states of affairs are just 'out there,' determinate matters of fact with no intrinsic connection to consciousness and its logical operations in judgement or language, then it seems gratuitous to posit logically complex states of affairs. Indeed, so it has seemed since Wittgenstein's *Tractatus*. Wittgenstein recognised elementary states of affairs (*Sachverhalten*) of form [aRb] (and arguably situations (*Sachlagen*) of form [aRb and not cQd]). But he did not recognise quantified states of affairs of form [All F's are G's] or [Some F's are G's]. Wittgenstein restricted quantification to propositions (*Sätze*) or thoughts (*Gedanken*), as in propositions of form 'All A's are B' and 'Some A's are B.' Now, Whitehead gives propositions a modal status between that of purely 'objective' actual entities (*faits accompli*) and that of purely 'subjective' prehending entities: as hybrid entities opening toward prehension, or '*proposed for feeling*.' But potentialities may be toward 'some' or 'all' of a set of actual entities, not only particular concrete actual entities, being determined in a certain way. Then it is plausible, surprisingly, that there are in the world 'proposed' states of affairs such as the proposition of form 'Some F's are caused by a.' This hybrid entity integrates the actual entity a with the potentiality of 'some F's' to be related by causation to a. This 'proposition' is just as much a part of the world as is the proposed state of affairs of form 'a causes b,' which integrates two actual entities with the eternal object causation.

Here is a more robust variety of 'logical atomism' than Russell's or Wittgenstein's. For Whitehead's ontology of 'propositions' allows *all* the familiar forms of propositions, not only the singular ones, while placing them all in the world alongside concrete states of affairs and not merely in language or thought or its contents. And thereby Alfred North Whitehead puts logic in its place: in the cosmos!

8. CONSCIOUSNESS

For Whitehead, consciousness emerges, in the order of the cosmos, only with higher phases of prehension: an actual entity is 'conscious' just in case it prehends a *proposition* in a certain way (to be specified below). Fundamentally, then, consciousness is a propositional attitude, to use Russell's idiom. A more systematic phenomenology, I believe, shows that not all acts of consciousness are propositional in form: judging or seeing 'that this is a lizard' is propositional, but seeing 'this lizard' is not.[8] For the moment, though, for the sake of argument, let us grant Whitehead propositionality. What is striking in Whitehead's account of consciousness is the role of his ontology. For consciousness inherits its structure – I would say its phenomenological structure – from the *ontological form* of prehension involved. There lies Whitehead's novel idea.

Consciousness is not Whitehead's central concern, and he would seem to leave the specifics to others: may we say, to phenomenologists in the lineage of Brentano, Husserl, *et al.*, whom Whitehead seems not to have read, and in other respects to neuroscientists in today's 'consciousness studies.' Yet Whitehead explicitly addresses consciousness after he analyses propositional feeling, and he remarks on consciousness all along the way in *Process and Reality*. In his own idiom Whitehead writes:

> [C]onsciousness presupposes experience, and not experience consciousness. It is a special element in the subjective forms of some feelings. Thus an actual entity may, or may not, be conscious of some part of its experience. Its experience is its complete formal constitution, including its consciousness, if any. Thus, in Locke's phraseology, its 'ideas of particular things' are those other things exercising their function as felt components of its constitution. Locke would only term them 'ideas' when these objectifications belong to that region of experience lit up by consciousness. [p. 53]

By 'experience' and 'feeling,' we know, Whitehead means prehension, and the 'subjective form' of a feeling is the way an actual entity receives the 'datum' it prehends. The form of prehension in an actual entity is 'its complete formal constitution,' what defines it ontologically. In some cases an actual entity is "conscious of some part of its [...] constitution." An actual entity has what we call 'ideas of particular things' insofar as it prehends those things and the pattern of prehensions fall in a region of prehension 'lit up by consciousness.'

In short, consciousness is a particular form of prehension. It consists in a pattern of prehensions found in the formal constitution of some but not all actual entities. To be conscious of something, to have an idea of something, is to have this pattern of prehension. And this pattern is the *ontological form* of the act of consciousness: the 'formal constitution' of an actual entity that is conscious.

It is important for Whitehead that consciousness is just another form of prehension, and not something that exists in a world apart from the physical in nature. As he writes:

> Locke inherited [from Descartes] the dualistic separation of mind from body. If he had started with the one fundamental notion of an actual entity, the complex of ideas disclosed in consciousness would have at once turned into the complex constitution of the actual entity disclosed in its own consciousness, so far as it is conscious – fitfully, partially, or not at all. [p. 53]

Here is a dramatic statement of monism in Whitehead's scheme. There is but one fundamental ontological structure in every actual entity (not two, one for the mental and one for the physical): the structure of prehension. And consciousness is a specific variation on this one fundamental structure.

What variety of prehension, then, is consciousness? Whitehead's answer begins after the analysis of propositional feeling:

> The nature of consciousness has not yet been adequately analysed. The initial basic feelings, physical and conceptual, have been mentioned, and so also has the final synthesis into the affirmation-negation contrast. But between the beginning and the end of the integration into consciousness, there lies the origination of a 'propositional feeling.' A propositional feeling is a feeling whose objective datum is a proposition. Such a feeling does not in itself involve consciousness. But all forms of consciousness arise from ways of integration of propositional feelings with other feelings, either physical feelings or conceptual feelings. Consciousness belongs to the subjective forms of such feelings. [p. 256]

The 'subjective' form of a feeling is the form of the prehension as it is absorbed, as it were, by the actual entity (p. 22, p. 24), and consciousness is a prehension of a proposition. But there is more. Consciousness originates, or begins to take form, in awareness of the 'yes' and 'no,' in affirmation and negation. And its final form involves the forms of perception, imagination, judgement, etc.

Whitehead finds the crux of consciousness in something utterly fundamental:

> Consciousness is the feeling of negation: in the perception of 'the stone as grey,' such feeling is in barest germ; in the perception of 'the stone as not grey,' such feeling is in full development. Thus the negative perception is the triumph of consciousness. It finally rises to the peak of free imagination, in which the conceptual novelties search through a universe in which they are not datively exemplified [that is, part of the datum prehended].
>
> Consciousness is the subjective form involved in feeling the contrast between the 'theory' [or proposition] which *may* be erroneous and the fact which is 'given.' Thus consciousness involves the rise into importance of the contrast between the eternal objects designated by the words 'any' and 'just that.' Conscious perception is, therefore, the most primitive form of judgement. [p. 161]

Notice that negative perception is a positive prehension of a negative fact; it is not a negative prehension, or not feeling something.

Why is feeling negation at the foundation of consciousness? Whitehead's motive here lies in his particular conception of eternal objects or forms, and thus the ontology of propositions, and thus the variety of propositional feeling that defines consciousness. An eternal object or form is, for Whitehead, a capacity or potential, and a potential is a tendency that might or might not be realised. Accordingly Whitehead writes:

> In awareness actuality, as a process in fact, is integrated with the potentialities which illustrate *either* what it is and might not be, *or* what it is not and might be. In other words, there is no consciousness without reference to definiteness, affirmation, and negation. Consciousness is how we feel the affirmation-negation contrast. Conceptual feeling is the feeling of an unqualified negation; that is to say, it is the feeling of a definite eternal object with the definite extrusion of any particular realisation. Consciousness requires that the objective datum should involve (as one side of a contrast) a qualified negative determined to some definite situation. [p. 243]

So consciousness, or 'awareness,' always involves some feeling of the *contrast* between actuality and potentiality. This is the fundamental *ontological* structure of consciousness: it is a prehension of this contrast. To embellish Whitehead's example (quoted above from p. 161): when I have a perception of 'the stone as grey,' I have an awareness of the stone's actually being determined as grey and therein potentially being not determined as grey (say, if the stone's temperature changed sufficiently to alter its colour). Husserl would have said this awareness of potentiality is part of the 'horizon' of the awareness of actuality.[9] Presumably, Whitehead thinks this 'horizonal' awareness of potential negative facts is characteristic of consciousness but not other types of prehension.

Specifically, consciousness always involves feeling a proposition. And on Whitehead's analysis, the ontology of propositions is that of a contrast between actuality and potentiality, namely, actual entities and forms:

> [T]he many components of a complex datum have a unity: this unity is a 'contrast' of entities. For example, a proposition is, in a sense, a 'contrast.' The most important of such 'contrasts' is the 'affirmation-negation' contrast in which a proposition and a nexus obtain synthesis in one datum, the members of the nexus being the 'logical subjects' of the proposition. [p. 24]

When 'different' actual entities are formed into a complex, the resulting unity depends on the contrast between the entities combined. A proposition combines actual entities with eternal objects, which contrast with each other precisely because the former are actual and the latter are not. This is an ontologically fundamental contrast: between beings that are realised (affirmation) and beings that are not yet realised (negation). Whitehead claims, then, that consciousness arises most fundamentally in 'feeling' this contrast in the constitution of a proposition. The form of consciousness is depicted in Figure 4 below, where an act of consciousness is a type of actual occasion.

(form + actual entity) ← (propositional feeling) — act of consciousness
proposition
negation-affirmation

Figure 4: The Form of Consciousness

9. FORMAL AND MATERIAL ONTOLOGY OF CONSCIOUSNESS

What else distinguishes consciousness from propositional feeling in general? I have not found an answer in Whitehead, and I suspect there is a reason why I have not.

Whitehead's further analysis of consciousness fans out into analyses of particular types of 'higher phases' of prehension: in perception, imagination, belief, judgement (p. 266 ff). Here, I believe, we should proceed instead into the rich analyses of these phenomena that populate the literature of phenomenology, starting with Husserl's results on such things, which hark back to Hume, Leibniz, and others in ways that Whitehead would approve. In fact, there is a methodological point to be pressed here.

My point assumes the distinction between formal and material ontology, along lines charted by Husserl. I believe Whitehead uses such a distinction in practice, without the terminology. Both were mathematicians turned to a broader canvas, and in this distinction we see the intellectual habits of the pure mathematician. Roughly, formal ontology analyses structure that is invariant, applying to very different domains, ideally to all domains; whereas material ontology analyses structure within a specific domain.

Whitehead approaches the ontology of consciousness through the analysis of prehension, propositional feeling, and feeling the affirmation-negation contrast in a proposition. It is in this latter feeling of contrast, Whitehead holds, that consciousness begins to emerge, that is, as a distinct type of prehension, leading into higher phases of prehension. But here, I believe, the formal ontology ends and the material ontology begins. For the structure of prehension – even of feeling a proposition and feeling the contrast it involves – is purely formal, applying invariantly to all prehension up to that level. But the structure of our own characteristically human forms of consciousness is partly 'material,' in that the forms of feeling in our sensory perceptions, our everyday judgements, our desires and volitions, etc., are partly conditioned by the special structures of our peculiarly human world. Here are forms of prehension that are special to beings like us, in our earthly environment, in this 'epoch' as Whitehead allows. Perception, judgement, imagination, and emotion have their formal constitution, in what Husserl called intentionality and intentional content, but they also have their specific 'material' constitution.

For instance, I see objects around me from the perspective of two eyes looking binocularly toward things I may touch with my two symmetrically placed hands, and some of these things are fellow human beings who share my values and ways of walking and talking and thinking and seeing. There is much to be said about how these 'material' ways of prehending (or intending) realise the formal ways of prehending that Whitehead has analysed. Whitehead the cosmologist has his eyes on the formal structure of prehension, with much less to say about the material structures of perception, imagination, etc.

Reading Whitehead in this way, the formal structure of consciousness is that of feeling a proposition and the contrast it involves, while the material structure of, say, perception or imagination is that of feeling a nexus of actual entities in a region of space-time.

10. REFLEXIVE AWARENESS?

The formal structure of consciousness seems to require one further feature. For an act of consciousness is not only a consciousness of X – a Whiteheadian proposition (we assume for present purposes). It is also *eo ipso* a consciousness *of itself*, of its being a consciousness of X. Without this *reflexive* awareness of itself, the act would not be conscious. Along these lines Brentano and Husserl held that every act includes a primary intention directed toward its object and a secondary intention directed toward itself – but in such a way that no infinite regress ensues (with a third directness toward the second, etc.). Alternatively, I have argued, the reflexive character is a modification of the intention of X, rather than a supplementary intention of the intention-of-X.[10]

Whitehead does not explicitly address this reflexive structure in consciousness, but he comes close to ruling it out on formal ontological grounds. An actual entity is in process of becoming, and the final phase of its concrescence Whitehead calls 'satisfaction.' But the structure of concrescence, Whitehead argues, imposes a certain limitation on consciousness: "No actual entity can be conscious of its own satisfaction; for such knowledge would be a component in the process, and would thereby alter the satisfaction." (p. 85) Whitehead's argument here is elliptic. Becoming cannot be continuous, resolvable into smaller and smaller becomings, Whitehead has argued with reference to Zeno (p. 35). Similarly, the satisfaction of an entity – the final phase of its becoming – cannot be resolved into smaller and smaller component satisfactions. Here the argument may be that satisfaction cannot be resolved into successively nested satisfactions. That is, if an actual entity were to prehend its own satisfaction, then its satisfaction would require the satisfaction (through prehension) of its own satisfaction. With Zeno in mind, this begins to look like a vicious regress.

Whitehead seems to be propounding a formal ontological principle of complete generality: an actual entity cannot prehend its own satisfaction. If we look to the special case of consciousness, however, we find ourselves in more familiar territory: an act of consciousness cannot be conscious of its own becoming. And then Whitehead's principle rules out the reflexive structure that our consciousness seems intuitively to have.

But perhaps we are looking for consciousness in the wrong places. Given Whitehead's 'atomism,' the things we know in everyday life are not simple 'actual entities,' but highly complex groupings or 'nexus' of simple actual entities. Perhaps our everyday conscious experiences are such nexus. Thus, when I see a dog, my conscious perceptual judgement that-this-is-a-dog is a nexus that combines two prehensions: my *perceptual* prehension of the proposition that-this-is-a-dog, and my *reflexive* prehension of that perception. These are distinct prehensions neither of which includes a prehension of itself, of its own becoming. And my conscious perception consists of this nexus. Moreover, the dog and I are ourselves complex nexus of actual entities. Indeed,

Whitehead had his eye on basic physics, and the philosophy of physics must explain the correspondence between basic physical states and everyday affairs.

Husserl's phenomenology of time-consciousness in fact elaborated such a complex form of awareness. On Husserl's analysis, my current phase of perceptual consciousness includes not only a perception of X, but also a spread of 'retentions' of successively prior phases of the ongoing perceptual consciousness, and indeed a spread of 'protentions' of anticipated upcoming phases of perceptual consciousness. My current ongoing experience thus consists of this complex. And my reflexive awareness of the ongoing experience, we might say, consists in the complex of retentions and protentions. But, as Whitehead requires, no 'point' awareness is an awareness of itself.[11]

There remains one further wrinkle by which Whitehead's ontology might capture what we experience as reflexive awareness of experience. Whitehead allows for the 'mediation' of feeling through an actual entity (p. 226): if A feels B (in A's actual world), and B feels C (in B's actual world), then A feels C not directly but mediately. Thus, if B is conscious of a proposition X and A prehends B, then A has a mediate prehension of X. Perhaps then A's prehension of B together with its indirect prehension of X constitutes a 'reflexive' awareness of B's consciousness of X. See Figure 5 below.

proposition X ⟵ act B ⟵ reflexion

Figure 5: Reflexive Awareness

Then the act B does not include a component prehending itself, but is rather accompanied (immediately afterward) by a further 'reflexion' that prehends it.

The remaining problem is how well the phenomenology fits with the ontology.

11. CRITICAL REFLECTIONS

We have sketched on Whiteheadian lines a fundamental ontology and its application to consciousness. I should like to close with a brief appraisal of the chief virtues and problems I see in this Whiteheadian system.

a. The structure of consciousness

As Husserl observed, in the wake of Brentano, conscious experience is paradigmatically intentional: as cognitive science puts it today, mind is representational. Some states of consciousness (like feeling dizzy) are not intentional, but most are. Whitehead captures the structure of intentionality in 'prehension.' So far, so good. He then analyses consciousness as prehension of a proposition. But while it is often assumed that intentional states are essentially propositional in form (like believing that such-and-such is the case), some intentional states (like seeing this dog) are not propositional, and so cannot consist in prehending a proposition. Whitehead touches on specific forms of consciousness including perception, judgement, and imagination, but the details do not really further his basic analysis, and we must look elsewhere for the cognitive or phenomenological structures of perception, thought, desire, etc. What we should learn from Whitehead about cognition is that we need a truly fundamental ontology – in some sense a formal ontology – in order to articulate the structure of cognition and specifically consciousness.

b. The ontology of propositions

On Whitehead's analysis, consciousness begins to emerge in 'feeling' a proposition: a hybrid of actual entities and form (say, that a R b), where the form (the relation R) is a 'potential' for definiteness in the actual entities (a and b). Whitehead may have had in mind a sort of superposition of phenomena as envisioned in quantum mechanics. Alternatively, what is potential is the *possible* state of affairs combining these things (that a R b), which consciousness may prehend indirectly through a propositional meaning entity (the thought that 'a R b'), which itself comes into being with the act of consciousness. Whitehead's intriguing conception of propositions, somewhere in between actuality and potentiality and thereby opening into consciousness, should be parsed, I suspect, into different ontological structures: actual entities, forms, potentiality, and meanings.[12] In any event, all these things will be subject to the 'process' of becoming in Whitehead's overall ontology.

c. The ontology of consciousness

The 'hard' problem of consciousness, it is said in philosophy of cognitive science today, is how to account for subjectivity: the felt quality of experience, what it is like to have a certain form of experience. What is hard is precisely fitting consciousness, in all its phenomenological glory, into a world of particles in the void. Whitehead took modern physics to heart, embracing relativity and quantum phenomena, and he developed a metaphysical vision in which everything consists in interactions of prehension whereby actual entities come into being bearing forms. Consciousness and causality are then two types of prehension, and mind and body find their place as different types of nexus of actual entities, forms, and hybrids thereof. The details of this way of distinguishing consciousness and causality require, shall we say, further work. In particular, are propositions as hybrids a good idea? In any event, the architecture of Whitehead's ontology is admirable: we must look to *fundamental ontology* to solve the mind-body problem, so that the distinctive structures of mind and body emerge respectively from more fundamental ontological structures something like ontological becoming.

d. The principle of becoming

In broadest strokes, Whitehead's ontology is a radical revision of the most basic assumption of classical Western metaphysics: that the world is fundamentally structured into forms and substantial particulars related by exemplification. Plato and Aristotle, and nearly all since, began with this assumption, while differing on the status of the forms. The traditional languages of logic – including quantified predicate logic in *Principia Mathematica* and more recent modal logics – reflect this metaphysics of substances and attributes (including relations). According to Whitehead's ontology, however, all this structure emerges from a more fundamental structure of becoming, or concrescence, through prehension. The principle of ontological becoming, or coming to be an entity, I submit, carries Anaximander's assumption that the order of 'determinate' beings – what Aristotle would later call substances (*ousia*: beings) and their species and attributes – emerges from a different order of the 'non-determinate' (*apeiron*). Whitehead begins with such an assumption in the category of the 'ultimate,' becoming. However, he does not treat substances and attributes in the same way, as 'determinates' formed by concrescence. For while he treats Aristotelian substances as products of concrescence, he leaves Platonic forms as 'eternal' objects (albeit potentials). This disparity between everyday objects and their forms calls for further evaluation.

The overarching lesson from Whitehead, then, is to see how this principle of becoming might ground all ontological structure. The structure of consciousness is then to be a special case.

David Woodruff Smith
University of California, Irvine
Department of Philosophy
CA 92697-4555
Email dwsmith@uci.edu

NOTES

[1] I wish to express my appreciation for the late John D. Goheen of Stanford University. I first studied Whitehead's metaphysics in a course taught by Professor Goheen, and the image of 'prehension' has stayed with me for three decades. In recent years the image has emerged, with a life of its own, in the very practical needs of the PACIS database project of Ontek Corporation. I thank Charles W. Dement of Ontek for many discussions of Whitehead's ontology and its role in prefiguring the ontology of PACIS. My thanks finally to Liliana Albertazzi for inviting me to write this paper, in a sense for seeing its need.

[2] Here I refer to the PACIS system under development by Ontek Corporation in Laguna Hills, California (internet address: www.ontek.com). PACIS is a high-level database system, designed to represent all kinds of entities in a unified way. I think of PACIS, philosophically, as a computational phenomenological ontology, a system of fundamental categories of ontology (what there is) and phenomenology (how we experience or represent what there is) expressed in its own computer language (how we codify our representations of what there is). The ontology has been developed by a team consisting of the present author, Charles W. Dement (Ontek Corporation), and Peter Simons (University of Leeds), with help from Stephen De Witt and John Stanley (Ontek Corporation).

[3] The dynamicist approach to mind is presented in Robert R. Port and Timothy van Gelder. As the editors write (vii), "Dynamicists from diverse areas of cognitive science share more than a mathematical language; they have a common worldview." Their worldview, I suggest, is Whiteheadian. That is to say, the greatest classical metaphysical statement of the dynamicist worldview is found in Whitehead, with precursors tracing to Hume and all the way back to Heraclitus.

[4] The distinction between formal and material ontologies is explicated, in my own terms, in Smith 1995. I do not here restrict myself to the details of Husserl's particular conception of the distinction, as used in Husserl's categorial ontology. Rather, I want to unfold a version of the distinction framed within Whitehead's very different categorial ontology. Whitehead did not draw the distinction himself, nor had he read Husserl (to my knowledge). However, as a mathematician, like Husserl, he would have implicitly understood the distinction. For it is common to think of pure mathematics as developing formal theories that can be applied to very different domains. For instance, the theory of differential equations may be applied to the motion of waves of water in fluid mechanics, or equally well to the flow of waves of radiation in electromagnetism theory. However, formal ontology is not concerned with formal structures of mathematical language that *describe* different domains. Rather, formal ontology is concerned with formal structures of reality that are *realised* in different domains.

[5] Mayr, Ashlock 1991, 256. The PACIS ontology has been influenced by the search for categorical complexity in biological systematics, though the goal of PACIS is that of a 'formal' ontology rather than that of a "material" science like biology.

[6] The current ontology of PACIS develops this ontological architecture, in a particular system of categories, with an eye to both Anaximander and Whitehead. The PACIS scheme, and its Anaximandrian form, are expounded in Smith 1997. There are important differences between the PACIS system and the Whitehead system. I note one problematic feature of Whitehead's treatment of forms: insofar as forms are prehended, they are entities; but insofar as forms are 'pure potentials' for determinateness, they seem to be modes or ways of becoming. The PACIS system clearly distinguishes these phenomena.

[7] See Dretske 1981.

[8] The case for non-propositional intentional acts is detailed in Smith, McIntyre 1982, and Smith 1989. The former book integrates a Husserlian model of intentionality with Jaakko Hintikka's model of propositional attitudes: compare Hintikka 1969. For Hintikka, intentionality is analysed as propositionality, so that an attitude of belief is in effect directed toward a range of possible states of affairs or 'worlds' (compatible with the content of belief). Such doxastically possible states of affairs resemble the 'potentialities' that Whitehead speaks of. Even if seeing a lizard is not a propositional act, though, it may be analysed as embracing a 'horizon' of possible states of affairs in which a lizard is potentially before the perceiver.
[9] See Smith, McIntyre 1982, for a detailed reconstruction of Husserl's notion of horizon.
[10] This reflexive structure of consciousness is studied in Smith 1989, Chapter II.
[11] See Miller 1984, for a precise reconstruction of Husserl's analysis here. On the proposal I am noting, with an eye to Whitehead, what we experience as a reflexive awareness *within* the act of perception is analysed into smaller awareness' in the form of retentions and protentions. The problem for this proposal is that we really are not aware of these distinct phases of retention and protention, and so the reflexive character that we experience in conscious perception is analysed into 'awareness' that we do not actually experience.
[12] The alternative model of intentionality is developed in Smith, McIntyre 1982, integrating a Husserlian model of intention via sense and a Hintikkian model of propositional attitudes as unfolding an array of possible states of affairs or 'worlds.' Compare Hintikka 1969. My concern here is how to understand what Whitehead means by 'potentials,' if not something definable in more recent modal ontology. In the PACIS ontology, sketched in Smith 1997, intentional contents or meanings are treated as entities that are dependent on acts of consciousness and do the work of 'intending' entities within consciousness.

FIGURES

Figure 1. Concrescence
Figure 2. Propositional Feeling
Figure 3. A Proposition and its Prehension
Figure 4. The Form of Consciousness
Figure 5. Reflexive Awareness

REFERENCES

Bergson, H.: 1991, *Matter and Memory*. Translated by Nancy Margaret Paul and W. Scott Palmer from the French 5th ed. of 1908, New York, Zone Books. French original, 1st ed., 1896.
Dretske, F.: 1981, *Knowledge and the Flow of Information*. Cambridge, Mass., MIT Press.
Dretske, F.: 1995, *Naturalizing the Mind*. Cambridge Mass., MIT Press.
Hintikka, J.: 1969, *Models for Modalities*. Dordrecht, D. Reidel Publishing Co.
Mayr, E. and Ashlock, P.D.: 1991, *Principles of Systematic Zoology*. New York, McGraw-Hill, Inc., 2nd ed.
McTaggart, J. and McTaggart, E.: 1927, *The Nature of Existence*, vol. 2, ed. by C.D. Broad. London, Cambridge University Press.
Miller, I.: 1984, *Husserl, Perception and Temporal Awareness*. Cambridge, Mass., MIT Press.
Port, R.F. and van Gelder, T., eds.: 1995, *Mind and Motion: Explorations in the Dynamics of Cognition*. Cambridge, Mass., MIT Press, A Bradford Book.

Russell, B.: 1918, 'The Philosophy of Logical Atomism,' in *Logic and Knowledge: Essays 1901–1950*, ed. by Robert C. Marsh. New York, 1971, Capricorn Books, G.P. Putnam's Sons; prior editions by George Allen and Unwin Ltd. and Macmillan.

Smith, B. and Smith, D.W., 1995, *The Cambridge Companion to Husserl*. Cambridge, Cambridge University Press.

Smith, D.W.: 1989, *The Circle of Acquaintance*. Dordrecht, Kluwer.

Smith, D.W.: 1995, 'Mind and Body' in B. Smith and D. Woodruff Smith, eds., 1995.

Smith, D.W.: 1997, 'Being and Basis.' Unpublished manuscript, extending my presentation under the same title at the Second European Congress for Analytic Philosophy, Leeds England, August 5–7, 1996.

Smith, D.W. and McIntyre, R.: 1982, *Husserl and Intentionality*. Dordrecht, D. Reidel Publishing Company, now Kluwer Academic Publishers.

Whitehead, A. N.: 1978, *Process and Reality*, corrected edition, ed. by David Ray Griffin and Donald W. Sherburne. New York, The Free Press, Macmillan, now Simon and Schuster; 1st ed., 1929.

Wittgenstein, L.: 1994, *Tractatus Logico-Philosophicus,* En. tr. 1961, by D.F. Pears and B.F. McGuinness, London, Routledge. German original,

WOLFGANG WILDGEN

KURT LEWIN AND THE RISE OF 'COGNITIVE SCIENCES' IN GERMANY: CASSIRER, BÜHLER, REICHENBACH

IS THERE A BIRTH DATE FOR 'COGNITIVE SCIENCE?'

If one discusses the rise of cognitive science, one should bear in mind that cognitive science is not a discipline like those separated from the traditional fields of philosophy, the arts or medicine: psychology, linguistics, biology, anthropology, etc., rather it was created by an act of reassembling separated disciplines. The reassembling of separated classes may be done in different ways and have different motives. In fact, since the separation of disciplines (which mainly occurred in the 19th century (psychology was only separated from philosophy at the end of the last century), different initiatives have been taken to form more coherent and larger fields of scientific research.

In the current literature, cognitive science seems to have been created in the early and mid-1950's (almost simultaneously with psycholinguistics) and some authors even propose an exact birth date, birthplace and a couple of parents. Thus George A. Miller is sure that cognitive science saw the light of the scientific world on 11 September 1956 at the M.I.T. (Cambridge) and had the logicians and computer-scientists Newell and Simon, the linguist Chomsky and (implicitly) himself as parents.[1] This birth-story is too naive, too self praising (for the parents and the place) to be seriously considered. But the basic message is that innovative persons from different disciplines with a commitment to scientific rigor came together to open a cooperative enterprise which broke down rigid barriers such as those imposed by behaviourism in psychology, descriptivism in linguistics and abstract mathematics without technological implementation in theorem proving.

Such large-scale innovations and reorganisations of scientific networks occur typically in a period of political or social crisis. In Europe the end of World War I (1919) triggered a political and intellectual reorganisation (mainly in Germany) and World War II was the motivation for the rise of co-operative enterprises not only in nuclear physics and computer technology, but it also created an interdisciplinary network of mathematicians, biologists and linguists

L. Albertazzi (ed.), The Dawn of Cognitive Science. Early European Contributors, 299–332.

which produced information theory, cybernetics (Wiener) and similar programs of integrated scientific endeavours. My claim is that cognitive science (and probably similar enterprises which will follow, e.g., cognitive neuroscience) are not a singular phenomenon, they are the long-range consequence of the inadequate process of disciplinary segregation (which has been a permanent drift in the sciences but was specifically furthered by Cartesian dualism since the 17th century). The 'rise' of cognitive science is thus only a single phase in a series of divergent and convergent moves in the field of scientific disciplines, mainly those concerned with the human mind.

For Germany and parts of Western Europe the 1920's were a period of innovation, rapid change and great hopes. This period was dramatically broken by the economic crisis at the end of the 20's and the rise of fascist regimes (1925 in Italy, 1933 in Germany, 1936 in Spain) and after 1939 their expansion through war over Europe. I will focus here on Kurt Lewin (1890-1947). He attended the philosophical lectures of Ernst Cassirer in Berlin (in 1910), wrote his dissertation in 1914, developed the basic intuitions of his field-theory as a soldier in World War I and started his university career after the war. The context of his academic activity in Berlin was the Berlin school of Gestalt psychology (Köhler, Wertheimer, Koffka). In this period, he was in contact with the logician and physicist Hans Reichenbach. Kurt Lewin, Ernst Cassirer and Hans Reichenbach will be at the centre of my analysis. All of them had to emigrate in the 1930's and continued their work in the USA. As a fourth person I have added Karl Bühler, who was an outstanding psychologist in Vienna in the 1920's.[2] Later, he integrated the linguistic structuralism of Saussure and the Prague School into his *Denkpsychologie*, i.e., cognitive psychology. The four authors and their publications between 1920 and 1933 (Bühler's *Axiomatik der Sprachwissenschaft*, i.e., axiomatic linguistics) will be the reference domain of my treatise. My general claim is that their work represents a major momentum in the constitution of an interdisciplinary and philosophically founded field we may call 'cognitive science' and that the birth-story of Miller, Chomsky et al. is only a further highlight. Hopefully, many further enterprises will follow, because the methodological and theoretical unification of disciplines dealing with aspects of human nature and the human mind is an 'eternal' process, it has neither a birth nor a death (as long as humanity exists).

The major controversy surrounding an integrative program concerns its proper centre. Is it psychology (this was the choice of Gestalt psychology and Bühler), and if it is psychology, what kind of psychology: psychology of perception with a trend to reduce it to neurophysiology, or rather social psychology with an opening to cultural anthropology, linguistics and sociology. If the mathematical rigor of theories is foregrounded, one may ask, if statistical models are favoured or should very simple, algebraic models which allow for a computational implementation be chosen (Chomsky and the so-called birthday party in 1956 would have selected this alternative). In the year 2000 many scholars in the 'cognitive-sciences' would prefer a neuro-cognitive centre of interest, i.e., a theory built on facts we can grasp by techniques of neuro-imaging, neuro-surgery, neuro-computing, etc. In twenty years the cognitive sciences will perhaps be basically neuro-chemical or evolutionary. In this view the cognitive sciences are themselves like a field of bike-racers, where different groups (of disciplines) take the lead, some fall back, others give up, etc.

My analysis of Lewin's contribution and his position in a field which contains his work and the contributions of Cassirer, Reichenbach and Bühler has the aim of highlighting one characteristic period in the recent history of cognitive sciences. Although Lewin was not much concerned with language and linguistic cognition I will concentrate on this aspect. This means that semiotical aspects of human cognition are foregrounded. As a consequence I will discuss the affinity between Gestalt psychology on the one side and another integrating enterprise: Peirce's semiotics and graphical logic at the beginning of the 20th century. The link between Peirce and Lewin is given by the relevance of topology and relational calculus for both.[3]

In the first chapter I shall describe the scientific context for Lewin's work at the beginning of his career, Gestalt psychology in Berlin, and link it to the semiotic thinking of Charles Sanders Peirce at the beginning of the century. Although both directions were rather independent they set the frame for the further development in semiotics and cognitive science. The second chapter will consider two persons immediately relevant for Lewin's philosophy of science: Ernst Cassirer (in his early activities in Berlin) and Hans Reichenbach, who under the influence of Einstein's theory of relativity turned to logical empiricism. In the third chapter I consider Lewin's field theory and compare it to current work on dynamic semiotics. Here Lewin is seen as a forerunner who exploited new conceptual tools (topology, vector analysis) and applied them in the domain of cognitive and social studies. This shows that the germ for such models already existed in the 1920's and 1930's. The last chapter considers developments in the theory of language made by Bühler (1934) and in Cassirer's philosophy of symbolic forms (1923-1929) which continued earlier research lines of these authors but had no tangible effect on Lewin's theoretical thinking.

1. GESTALT THEORY AND LINGUISTIC COGNITION

Modern theories of Gestalt in psychology, philosophy and aesthetics are based on notions developed in the 19th century, e.g., in Goethe's writings on botanics and zoology. His *Urpflanze* (archetypal plant) or basic skeleton can be understood as the search for fundamental morphologies, which control a huge phenomenological domain. The first programmatic statements relevant for Gestalt psychology can be found in the popular lectures by Ernst Mach from 1868 to 1871, but Christian von Ehrenfels was the first person to formulate the general Gestalt principles on a philosophical basis. His favourite example is the melody. Its character is not given by single tones or intervals. The melody is rather an oversummative whole, which can be preserved even if the tones which constitute it are changed in a way which does not destroy the relational structure of the whole (e.g., by transposition or simply by realisation with other instruments). Different schools of Gestalt psychology elaborated this point of view; the best known are the Berlin, the Graz and the Leipzig Schools. In the domain of the Berlin School Wertheimer and his students Köhler and Koffka applied the concept of Gestalt in experimental research on visual perception. Köhler proposed his concept of isomorphy with the goal of linking neurophysiological and psychic states and processes. This was a fundamental step because it bridged the mind-body abyss which had such fatal consequences for modern experimental psychology. More recent research along the lines given by early Gestalt theorists (Köhler, Koffka, Wertheimer) applies models of self-organisation and synergetics to specify the transition between brain dynamics and mental processes (e.g., consciousness).[4]

Wertheimer formulated laws of Gestalt (*Gestaltgesetze*) such as neighbourhood, similarity, closure, symmetry, well-formed continuation, common movement.[5] These 'laws' were applied in the graphic arts, in product design, architecture and urban planning. Even the design of efficient automata requires the application of Gestalt laws.

The Gestalt laws also imply meaning-effects, i.e., the association of meaning with a visual or acoustic Gestalt tends to stabilise and foster this Gestalt in perception and memory. I will come back to this aspect later. Before doing so, I would like to outline the historical development of semiotics since the late 19th century, i.e., in the historical period which witnessed the appearance of Gestalt theory.

1.1 *Gestalts and relational sign-structures*

A specific scientific concern with signs as the major object of research is due to the life-long activity of the American philosopher Charles Sanders Peirce and his lifetime intersects with that of Mach and the first Gestalt theorists. Charles Sanders Peirce (1839-1914) began to publish on semiotics in 1867 but wrote his major treatises after 1903. He is, therefore, the first whose relation to Gestalt theory should be investigated.

Kurt Lewin points out in his lecture "The transition from Aristotelean to Galileian mode of thinking in biology and psychology"[6] that Giordano Bruno, Johannes Kepler and Galileo Galilei abandoned the classificatory procedures of Aristotle, which aimed at an understanding merely by subdivision and abstraction. They favoured a homogenous, unitary view of the world such that the same laws which govern the motion of the planets are operative in the motion of a stone on earth, in bodily locomotion and finally in psychological and semiotic processes.

In modern semiotics Charles Sanders Peirce took up the same thread of ideas. Although his voice was submerged by the neo-Aristotelianism of Carnap's and Morris' semiotics, he is still the main representative of holistic semiotics or, in Lewin's terminology of Galileian semiotics. I can only indicate some features which link Peirce's semiotics and von Ehrenfels' theory of Gestalt.[7]

Von Ehrenfels formulated two criteria for a 'Gestalt:'

a) A complex of *n* impressions is more than the sum of single impressions.
b) If the whole complex is transformed and preserves its relational structure, then the Gestalt remains identical, although the substance of the elements changes.

Köhler expanded these criteria to continuous manifolds and physical Gestalts (e.g., electrical and chemical structures). Since these processes are constitutive for brain dynamics, one can infer from them the continuity between physical and mental processes.

The correlate of this program in Peirce's semiotics is his relational logic. Contrary to an atomistic logic of relations (as proposed by Morgan and later by Russell and Whitehead), a relation is for Peirce more than just the Cartesian product of sets. In the atomistic view, which still dominates logical semantics, two sets $A = \{a, b, c\}$ and $B = \{1, 2\}$ allow the definition of the Cartesian product $A \times B$ which contains the pairs: $\{(a, 1), (a, 2), (b, 1), (b, 2), (c, 1), (c, 2)\}$. A relation is then defined as a subset of $A \times B$. Thus, $R_1 = \{(a, 1),$

(b, 2)} is a strict subset of $A \times B$. If we iterate this procedure we get relations with 3, 4, 5, etc. arguments.

Peirce said that this concept of a relation is too trivial, we need a more restricted and more meaningful concept of relation. As his own proposal is the basis for his classification of signs and thus for his semiotics, I shall elaborate this point.[8]

1. Dyadic relations (with two arguments) cannot be reduced to monadic ones (those with one argument); they may instead be degenerated and, therefore, resemble monadic relations.
2. Triadic relations are reducible neither to dyadic nor monadic relations; they may be degenerated on two levels and thus resemble dyadic or monadic ones.
3. All relations with more arguments may be reduced to one or more of the above mentioned basic relations. As an example Peirce refers directly to linguistic (conceptual) relations.[9]

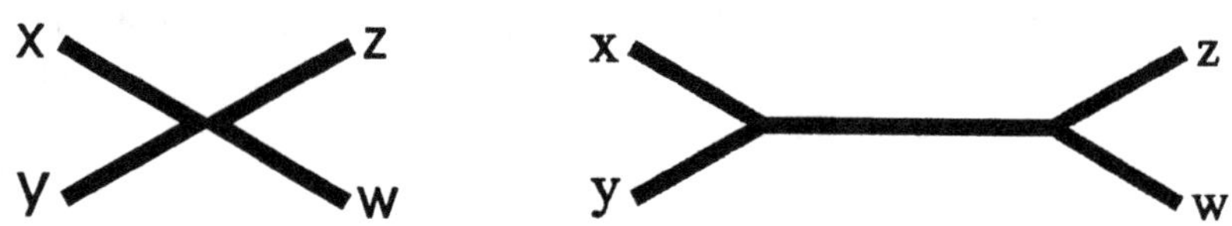

Figure 1. The reduction of a four-fold relation to two triadic relations, which share two arguments; the integrative sum of arguments is: 3 + 3 -2 = 4.

Peirce's relational logic clarifies the central notion of 'relational structure' in von Ehrenfels' definition (see above) and it introduces a limit (cf. point 3).[10] A similar limit of complexity was later proposed by René Thom in his semiophysics and he points to a contemporary of Peirce, the physicist Gibbs (1839-1903) and his rule of phases which formulates a limit of complexity of simple phase transition systems to 3. The ultimate basis of this insight may be found in Leonard Euler's law of polyhedra, which is discussed in Peirce.[11]

Peirce's relational logic has consequences for the classification of signs shown in table 1 and of elementary propositions shown in table 2.[12]

The classification of signs (simplified version)	
the icon:	a monadic relation of similarity
the index:	a dyadic relation of cause and effect

the symbol:	the law which establishes the dyadic relation is called 'interpretant;' it can be a social convention in use, or some specific semiotic medium which controls the dyadic relation between sign-form and sign-content

Table 1. Peirce's classification of signs

The relational aspect of semiotic systems elaborated by Peirce corresponds to the thetic judgements which Anton Marty distinguishes from the categorical ones. As the thetic judgements put forward the relational kernel of a proposition and not the substance of possible arguments (categories) Marty's view is non-Aristotelian. Moreover, it constitutes the starting point of a family of grammars called valence – or dependency-grammars. Tesnière's "Eléments de Syntaxe Générale" (1959) is the classical proponent of this family of grammars and it was used by René Thom to establish his "Topological Semantics" (after 1968).

The dynamical concepts of valence or dependency are very close to basic notions in Gestalt theory and are fundamental for Kurt Lewin's ideas in his topological psychology and field theories. I shall come back to it at the end of our journey through the recent history of semiotics and Gestalt theory.

The classification of elementary propositions	
The monad:	An individual entity has a quality: Jack is tall
The dyad:	An individual entity stands in a binary relation to another individual entity: Jack loves Mary
The triad:	An individual entity stands in a triadic relation to another individual entity on behalf of a third entity: Adam gives an apple to Eve (Adam gives Eve an apple.)

Table 2. Peirce's classification of elementary propositions

1.2 *The meaningfulness of a Gestalt and the semiotic character of Gestalts*

Köhler introduced the term *Prägnanz* (relevance) into Gestalt theory, but he points to Wertheimer as the one who coined the term. Köhler later used the term *prägnante Gestalt* (as a synonym of 'good Gestalt'). Von Ehrenfels' concept of the level of a Gestalt (*Gestalthöhe*) is related to this term and to the concept of complexity of a Gestalt introduced by Rausch (1966), who proposed a list of seven aspects of *Prägnanz*. The final two aspects refer to semiotic features:

6. expressive richness vs. expressive poorness
7. plentifulness of meaning vs. poorness of meaning

Many experiments in visual perception show that the association of meaning with a figure, i.e., the constitution of a semiotic relation, is able to stabilise the perception and even strengthen the memorisation of a pattern. In figure 2 a characteristic examples from bistable perception are given.

Figure 2. Bistable perception as duck/rabbit or cup/human profiles

Kanizsa and Luccio (1986 and 1995) showed that static and dynamic aspects of *Prägnanz* should be distinguished. The association of a figure with a specific meaning tends to stabilise the corresponding perception and it gives a surplus of stability relative to perceptual change. If in a field of transformation between two figures we start from one side, the point of semiotic change/jump is not identical with that adopted in a move from the other side. Thus we regularly observe the hysteresis effect in bistable perception illustrated by figure 3.

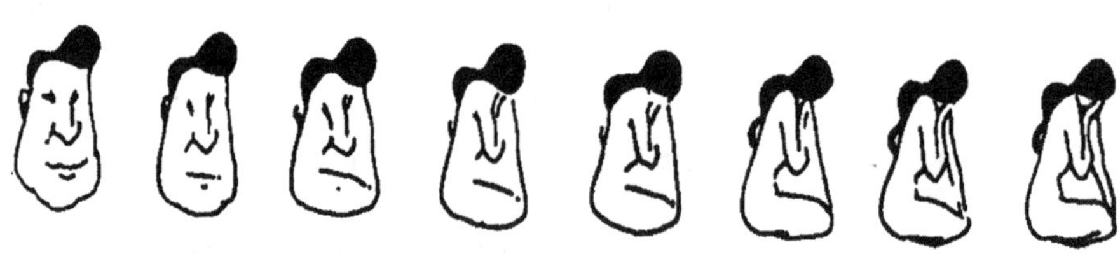

Figure 3. Hysteresis effects in the transition between two meaningful interpretations

The effect of semiotic loading on Gestalt perception was observed throughout the experimental work done by Gestalt psychologists. But since the major orientation was towards techniques and theories in the natural sciences and since semiotics seemed to lead back into the vague and 'unscientific' realm of 'hermeneutics,' the transitions between Gestalt psychology and semiotics were not crowded by excellent researchers in the field. Only later did Karl Bühler and Roman Jakobson consider the consequences of Gestalt thinking for semiotics and linguistics.[13]

I shall come back to Karl Bühler in section 4.1 and shall consider now the specific contribution of Kurt Lewin.

2. KURT LEWIN'S FIELD-PSYCHOLOGY AND DYNAMIC MODELS OF LINGUISTIC COGNITION

Although Kurt Lewin is commonly considered a Gestalt psychologist,[14] mainly of the Berlin group, there are nevertheless specific features which separate Lewin from the triad Wertheimer - Köhler - Koffka. I will just comment on three major features:

a) Lewin's theories are 'psychological' in nature and cannot be reduced to physiological or physical theories. This marks a fundamental difference to Köhler, who aims at the reconstruction of cerebral processes (finally electro-chemical processes) which map into psychological ones and could 'explain' them.
b) Lewin's theories are phenomenological. Thus his intuitions about 'Lebensraum' and other field theoretic concepts were already well developed in his first publication "Kriegslandschaft," where he analyses psychological geography and its change near the line of battle.[15]
c) In order to generalise specific insights Kurt Lewin resorts to new mathematical tools: topology and vector-analysis, and he makes a non-quantitative use of them.

Commonly Lewin's field theory or his topological psychology are related to social psychology and the psychology of learning. How are they related to semiotics? In fact, Kurt Lewin deals with problems of language and communication rather superficially, and we shall see later that Karl Bühler, who may be associated with the Würzburg School of Gestalt psychology, put language into the centre of his scientific undertaking. If, historically, Kurt Lewin's influence in semiotics was not prominent, he, nevertheless, introduced new ideas with a rich potential for current and future semiotics.

2.1 Language as a linking system between the intrapersonal field and the social field

In 1936 Lewin published his "Topological Psychology" (in English). In his chapter on the topology of the person, Lewin says that the motoric domain marks the frontier between the intrapersonal domain and the environment. One of the most fundamental processes in this zone is language. Semiotics is in this perspective an ecological *and* cognitive discipline. Sign-usage has to account for the specific demands of the environment, its mental representation and change *and* it must be able to transmit intrapersonal needs, tensions, feelings, etc.

If we compare this position with de Saussure's semiotics we see that Lewin does not consider the linguistic system as some abstract super-individual and ontologically independent entity, nor does he consider it as some mental device in Chomsky's sense. The nearest position in modern semiotics would be a Gibsonian view.[16]

Between cognitivism and pragmatism, Lewin's ecological semiotics constitute a true alternative. But how can he bring together mental processes, which are more like mechanisms, and social or ecological processes in *one* theoretical format. It is clear that his theory must be qualitative or rough; if not, the different types of description would not fit together. A qualitative but nevertheless systematic and analytic theory is the consequence if language is considered as a phenomenon on the transition line between brain/mind and world/environment.

If we take the sentence as the central linguistic manifestation, we may distinguish two different perspectives. The analyst of a sentence may look at the effect a phenomenological field has on the structure of the sentence. How are action-scenarios, visual scenes and frames mapped into sentence structures or case-roles, into the configuration of nounphrases, syntactic functions (like subject, direct and indirect object) and into types of verbs (intransitive, transitive, bitransitive, etc.)? In a different perspective the analyst can ask how the sentence uttered by a person fits pragmatically into a social field defined by the interlocutors (speaker, addressee, audience) and the situation (actual context, knowledge common to the interlocutors). The field-structure of a sentence has, therefore, two perspectives: the internal perspective of the grammatical organisation and the external perspective of situational fitting. I will give examples for both perspectives, but only the external field has been systematically analysed by Lewin; for the internal field I take an example from catastrophe theoretical semantics. My claim is that this type of model is in the spirit of Lewin's field theory.

2.2 *Valences, forces in an external perspective*

One basic concept of Gestalt psychology is the concept of valence (*Aufforderungscharakter*). The individual in a given situation interprets different objects as having positive valences (they are goals) or negative valences (they are barriers or repellors).[17] Lewin describes valences with the help of vectors. In his classical experiment, shown at the Ninth International Conference of Psychology at Yale,[18] Lewin observed a small child who tried to sit down on a stone, but had to turn its back to the stone in order to achieve the task. The stone (or the place for sitting) has a positive valence and the vector of motion goes towards it. In order to sit down the child must, however, invert the vector of locomotion and turn around. In a phase of perplexity it crawls around the stone, i.e., it moves on a path in equilibrium between approaching and turning back. Figure 4 shows a graphical representation similar to the one used by Lewin.[19]

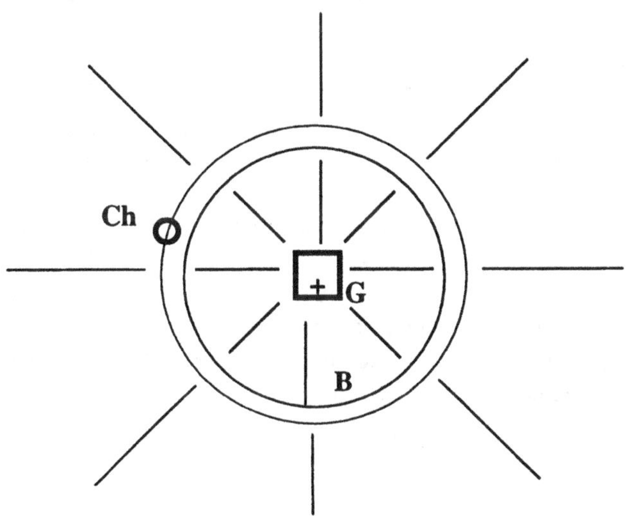

Figure 4. Field of the child who wants to sit down on a stone
(G = goal; Ch = child; B = barrier)

The field shown in Figure 4 shares some objective features with the situation, as it may be observed by an external observer, but it represents basically a cognitive (perceptual and behavioural) field which underlies the observed behaviour of the child. If the child has learned to speak and if communication is relevant for the solution of the behavioural problem, language becomes an integral part of the field. Thus the external field is on the one hand a non-linguistic field which may be described or referred to with linguistic means, but on the other hand language may become a constitutive part of the external field.

2.3 *Attractors and forces in an internal perspective*

Sentences and scenarios associated with verbs (specifically verbs of action) can be understood as coding procedure for mental scenes corresponding to frequently observed and enacted behavioural patterns.

If the agent (e.g., a person) moves, we can define the direction and strength (energy, velocity) of his motion. The direction is given by a goal which is normally a state of rest. In terms of Lewin's vector psychology we have a linear vector-field with a goal (an attractor) which has a positive valence. On its way different barriers may appear which are either objective (physical), as in the case of a labyrinth, or subjective (behavioural) as in the case of the child, who must first turn around and sit-down in a backward motion (in Lewin's experiment the child finally looked at the goal through its own legs and thus re-established a line of motion-control corresponding to the direction of the vector).

If two persons are co-present, they can be either the goal of reciprocal attraction (the meeting of friends, of lovers, etc.), of reciprocal repelling forces (in the case of enemies) or mixed situations may appear. In situations of struggle the opponent becomes an object of aggressive attraction. I will analyse a more complicated case, the exchange of an object M_3 between the agents M_1 and M_2. In this case the topology of the situation is typically changing. Figure 5 shows the basic dynamics in this field.

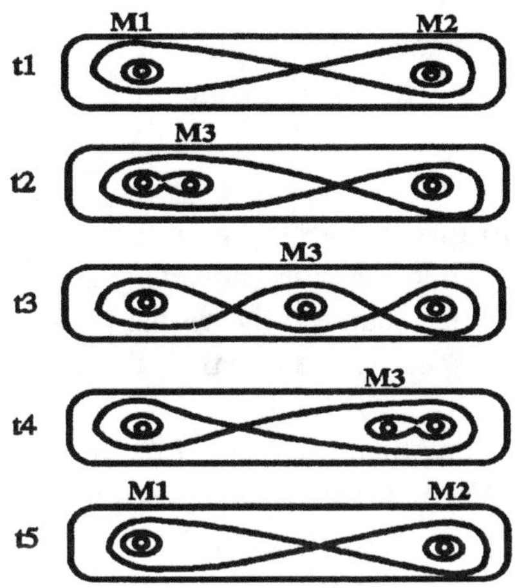

⇒ Starting position: stable state with a neutral vector which links M_1 and M_2.

⇒ Beginning of the move of the object M_3 along the vector between M_1 and M_2.

⇒ Metastable state: the repelling vector centred in M_1 is equally strong as the attracting vector centred in M_2.

⇒ Dominance of the attraction by M_2; M_2 'catches' M_3.

⇒ Final state: the metastable entity M_3 has disappeared in M_2.

Figure 5. A sequence of snapshots in the evolution of the field

This description, which is still phenomenological and uses only the simple field-theoretical concepts introduced by Lewin, can be mathematically

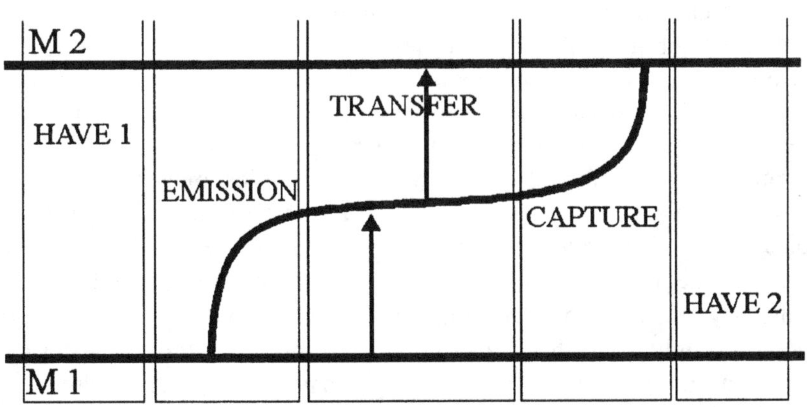

Figure 6. A dynamic-topological description of the action 'TO GIVE'; the lines represent attractors (i.e., minima), the arrows changes of dominance.

sophisticated if we introduce the techniques of catastrophe theory. In figure 6 the discrete situations of figure 5 are integrated into a dynamic-topological description.[20]

This example shows the fundamental affinity between Lewin's vector-psychology and modern dynamical semantics. There remains, however, a bias. If a cognitive scenario underlies a linguistic device found in the verbal lexicon and basic syntax, this does not yet explain the specific grammatical dynamics of the language (English, German, etc.).

The internal perspective may be further differentiated if major functions of language are defined. The prominent function which was presupposed in our example is that of meaning (Meinong calls it *Bedeutung*; Bühler calls it *Darstellung*). The two other major functions[21] are expression (first called *Kundgabe* by Bühler and *Ausdruck* by Meinong; Bühler finally adopted Meinong's term). The third major function is called *Appell* by Bühler and means the suggestive motivating force of language.[22]

2.4 *Beyond vector-psychology: the motor-dynamics of language*

Kurt Lewin characterised human motor-dynamics by 'structural independence' and compared it to the relation between guide and guided group or to that between higher and lower brain centres. Two major examples of learned motor-behaviour are language and action at distance.[23] The grammar of a language can be understood as such a semi-dependent motor-behaviour. It has to be fitted for general cognitive scenarios, such as those described in the previous section, but it also contains independent principles and a domain of freedom, called 'arbitrarity' (vs. motivation) by de Saussure (1916).

Since language is a very complex system to which many generations of speakers have contributed (some basic regularities may have a history of thousands of years), the task of discovering the basic principles, directions of choice and specific organisational pattern is a huge one and defines the core of the discipline called linguistics. Its aims are clearly beyond the field of psychology and the proposals found in Lewin's work or in Gestalt theory in general are only rough hints at a possible solution. In this context I can only propose a subdivision of this task. We may consider the following levels (partial tasks in the sense of Descartes' 'method').

A. Typology. Different language families or languages make a choice as to the prominence of categories or areas on a categorical scale. Thus, a nominative-accusative language will treat subjects as prominent and thus establish a line of communality between the *subject* of a transitive sentence, like: *The master* takes a pencil, and the subject of an intransitive sentence, like: *The master* comes. An ergative language will rather give the prominent role to the causer (the entity which contains the intentional energy) over the non-causer and will, therefore, stress the parallelism between the object (non-causer) in the first sentence (pencil) and the subject in the second sentence, which are both non-ergative. The subject of the transitive sentence is marked as ergative. But already the distinction between verb and nominal roles, which is presupposed in these two cases, and with it the distinction between words and sentences (centred in a verb) may be abolished and the language ductus corresponds rather to a sequence of morphologically complex words (which combine functions distributed over sentences in other languages). This is the case in some American-Indian languages like Cayuga (Iroquois).[24]

B. Word Order. If one distinguishes three major parts of speech (independent of the case-marking system): Subject (S), Object (O) and Verb/predicate (V); six orders are possible: SOV, SVO, OSV, OVS, VSO, VOS. All these orders are documented for languages, but the choice of the first position for the object is rare. Many languages have different word orders in main and dependent clauses (e.g., German), or have a rather free order governed by discourse dynamics (thematic structure, etc.).

C. The specific organisation of a sentence can be dependent on the choice of a lexical item (a verb, a preposition, a conjunction) or a construction. Moreover, the context (previous text and situation of utterance) can replace some elements (in ellipsis) or ask for them (in deixis).

This short sketch of language as culturally transmitted (learned) motor-dynamics shows its degree of independence from cognitive structures on the one hand and the objective (physical) structure of the environment, which we perceive and in which we act, on the other hand.

From Lewin's assertion that language belongs to motor-dynamics it is a long way to a complete (dynamic) theory of language.

3. KURT LEWIN BETWEEN CASSIRER'S NEO-KANTIANISM AND REICHENBACH'S LOGICAL EMPIRICISM

Ernst Cassirer (1874-1945) had been a student of Hermann Cohen and Paul Natorp in Marburg, where he finished his doctoral dissertation in 1899. In 1903 he came to Berlin where he began to work on the history of epistemology and finished his habilitation in 1906. In 1910 he published the book *Substanzbegriff und Funktionsbegriff* which Lewin took as a starting point for his own epistemological reflections.

Kurt Lewin met Cassirer in Berlin in 1910 (Lewin was a student and Cassirer was a 'Privatdozent'). In the same period (1911) Lewin came into contact with Hans Reichenbach (1891-1953), who was a member of the same student union (*Freistudentische Bewegung*). Lewin's relation to Cassirer remained one of distant admiration.[25] His relation to Reichenbach was more personal. They had the same age (Lewin was born in 1890, Reichenbach in 1891) and did identify with the intellectual milieu in Berlin. Lewin did not subscribe the reductionism of Köhler and never was accepted on equal terms by the three leaders of the group: Wertheimer, Koffka and Köhler. Reichenbach followed an independent line influenced by Einstein and was sympathetic to the logical empiricism of the Vienna circle and the logical foundations of mathematics by Whitehead and Russell in England.

3.1 *The influence of Cassirer on Lewin's early scientific work (in Berlin)*

The epistemological position of Cassirer at the beginning of the century is described in his book in 1910 and Lewin systematically refers to it in his early publications. Cassirer tried to reduce the traditional dichotomies like: deduction-induction, object-subject, thought-being, and said that philosophical reflection turns in a circle focusing on one or the other of the terms and thus looses the basic (causal) link between the two parts. Instead of apriori separations, one should consider unity in multiplicity, change and identity. This leads to scales, e.g., of objectivity,[26] to the consideration of continua as underlying observed pieces of reality[27] and to a logical graduation.[28] Lewin elaborated this point of view in his habilitation thesis: "The concept of genesis in physics, biology and developmental history" (printed in 1922).[29] The problem of categorical cuts in physical and biological genetic series is formulated in Lewin's sentence No 8,[30] "The series of existing objects in physics as such do not contain marked cuts."[31] In biology, however, the basic distinction between 'living' and 'dead' establishes a natural cut and physical causality (or genetical series) cannot simply be transferred to biology. Two traits appear in this treatise of Lewin:

a) A basic sceptical reaction as to the continuity between physics and biology, which if generalised to psychology says that psychological series (identities, developments) cannot be reduced to physical laws.

b) Lewin tries to establish a quasi-axiomatic system of hypotheses and to center his arguments around formal definitions and axioms.

The first trait points to the distance Lewin took in relation to Gestalt psychology in Köhler's style (reductionism), the second shows a parallelity to endeavours influenced by the axiomatic work of Russell and Whitehead.[32] Lewin leaves the domain of new-Kantian philosophy and turns to analytical philosophy. In this move he also leaves the area of Cassirer's philosophy of science.

The underlying problem of continuum and the nature of cuts in the continuum is basic for semiotics. Cassirer[33] points out that in a metaphysical view signs relate representations (*Vorstellungen*) to objects, which are considered as underlying them (standing behind them).[34] But if representations (*Vorstellungen*) on one side and things, objects on the other are elements of two independent, ontologically unrelated series, it remains a mystery how signs can establish a stable relation between them, how we can by the use of signs know something about the objects (the world), how truth is possible.[35] Therefore, the problem of genetical series in physics and biology treated by Lewin in his habilitation thesis is also fundamental for semiotics which has to presuppose a substantial unity behind mind and world. Lewin's treatise shows (together with Cassirer's analysis of representation) that a further reflection on disciplinary cuts, object constitution in separate disciplines is a necessary precondition of general semiotics. A rigid separation of the object domains destroys the platform for such a theory.

Cassirer had published an impressive series of books on the history of ideas; the first two volumes appeared in 1906 and 1907 under the title: "The problem of knowledge in the philosophy and science of modern times"[36] which covered the period between Nicholas of Cusa (early Renaissance) and Kant. The period around 1600 with Giordano Bruno, Johannes Kepler and Galileo Galilei was considered the turning point in the development of modern sciences. In his book: *Individuum und Kosmos* (1927) Cassirer describes the attitudes of Leonardo and Galileo, their synthesis between a new Plantonism on one side and empirical (experimental/practical) investigation on the other. The proper form of the new science had to transform the general dynamism present in Bruno's speculative work into new mathematical conceptualisations and new techniques of measurement and experiment (in the work of Kepler and Galilei).

Lewin takes up Cassirer's basic arguments in the treatise *The Transition between the Aristotelian to the Galileian Type of Thinking in Biology and Psychology*.[37] His reference to Cassirer, however, is not very specific as he only refers to Cassirer (1910) in a logical context. Lewin enumerates four features of Galileian physics and psychology summarised in table 3.

Galileian physics	Galileian psychology
The homogeneity of the physical world.	The transition from anthropomorphic distinctions (oppositions) and values to continuous series.
The use of conditional genetic concepts. The distinction between phenotype and genotype.	The generality of psychological laws (as shown by Gestalt psychology).
The prominence of measurement, quantity; in general the preference of concreteness over abstraction.	All 'exceptions' must be properly explained.
The basic laws are more constructive, less 'empirical,' insofar as specific historical constellations are not taken care of.	The reduction of purely statistical results or laws. Chance is not a relevant factor.

Table 3. Four characteristic features of Galileian physics and psychology

As Lewin points out in section B of his paper the proper treatment of the dynamics is characteristic for the Galileian style. In Galilei's experiment with the balls rolling on a gradient, the dynamics is not inherent to the ball, it is a product of the gravity of the object and situation (degree of the slope). Lewin derives his own vectorial psychology from the principles found in Galileian physics and transferred to psychology. The basic ideas of an ecological psychology already sketched in Lewin's earlier papers (as a soldier in the First World War), are now seen as part of a more general scientific development.[38]

At this point it becomes clear that the intellectual lines followed by Cassirer and Lewin only came in contact in an early period of Lewin's career and that the two scientists took different paths later: Cassirer engaged in a philosophy of language and culture, Lewin developed new techniques and fields of analysis in psychology (the withdrawal of his habilitation thesis in 1920 may have been the turning point). In a late commemorative paper published in 1949 Lewin appreciates the immanent methods of Cassirer's historical reconstructions[39] and the importance he gave to conceptual and mathematical innovations. Lewin applies the analysis in physics by Cassirer to psychology (although Cassirer, 1944, is rather reluctant, e.g., to the application of group theory to psychology).[40] In his last section on "Mathematisation and integration in the social sciences,"[41] Lewin says that the social scientist needs analytical tools, which are technically and empirically adapted to the specificity of social processes and that he may use them as a transitory medium towards a full mathematical treatment. The representation of social situations as fields is such a strategy.[42] I will come back to Lewin's concept of 'field' in chapter 4.1.

3.2 *Lewin and Reichenbach's logical empiricism*

In a letter to Reichenbach (written 1940, cited by Wittmann, 1988, 128) Lewin says: "I feel that we are philosophically very close." Reichenbach also referred to Cassirer as his teacher, but he was philosophically rather independent. The work of Hilbert, Whitehead and Russell persuaded him that formal logic was the proper instrument for a critical evaluation of concepts and models in the sciences. After 1920 he called himself a logical empiricist .

The relation between Reichenbach, who worked as a physicist, and Lewin, who was a psychologist, concern mainly theoretical questions such the mathematisation in the sciences. I have already shown that Lewin used a quasi-axiomatic structure in his treatise on the concept of 'genesis' (*Genese*) in physics and biology, published in 1922. In 1921 Hans Reichenbach had analysed the axiomatic in Einstein's theory of time-space and he also wrote a comment on Lewin's treatise (Reichenbach, 1924). It was probably this period that Lewin and Reichenbach were closest to one another. Lewin's concern with dynamics (a basic aspect of the Galileian paradigm) made him prefer geometrical, topological and vectorial tools to logical ones. He acknowledged the advances of mathematical logic (such as in Cassirer's work), but said that they were not very relevant for psychologists and social scientists. The perspective of logicians on science is too abstract, they cannot consider the quick changes in ongoing research and the changing demands for proper conceptualisations in the empirical sciences.[43] This argument characterises Lewin's basic attitude. The ongoing activity of scientists ask for flexible tools which can mediate between practical research (design, experiment, conclusions) and the construction of mathematical models. Logical axiomatisation could be interesting in certain domains of physics which have reached a position of stability; it is not (easily) applicable in the rapidly growing and changing fields of the social sciences.

In the 40's Lewin considered dynamical systems of quasi-stationary equilibrium and models in economics like these proposed by Pareto in 1935 as possible points of synthesis. This means that the interdisciplinary initiatives during and after the Second World War which produced Wiener's cybernetics, Bertalannffy's General Systems theory, Prigogine's theory of dissipative systems (later Thom's catastrophe theory and Haken's synergetics) are in the line of unifying theories Lewin had in mind at the end of his life.[44]

The fact that Reichenbach modelled natural language using mathematical logic[45] (in a treatise on Symbolic Logic) does not contradict Lewin's position, as the dynamic aspects of communication and language use were not the concern of Reichenbach's 'exercise,' which on the one the side showed the interest of logical translations of natural language expressions and on the other (implicitly) marked the limits of such an approach.

4. KURT LEWIN'S FIELD THEORY AND AXIOMATICS COMPARED TO BÜHLER'S SEMIOTICS

4.1 *Field-theoretical concepts in Karl Bühler's 'Sprachtheorie'*

Karl Bühler came from the Gestalt theoretic movement in psychology (School of Würzburg) as his publication of 1913, *Die Gestaltwahrnehmungen* shows. In 1933 he published a first outline of his theory of language, *Die Axiomatik der Sprachwissenschaft* and his major work, *Sprachtheorie* was published in 1934. In chapter IV: "The Constitution of Human Speech. Elements and Compositions" (*Aufbau der menschlichen Rede: Elemente und Kompositionen*), Bühler shows that the oversummativity of a sentence in relation to the words it puts together has been established since Aristotle and was, therefore, never in doubt. Therefore, he is rather critical towards the von Ehrenfels' criterion which cannot be the final stage in a reflection on structures in cognition and language. In this context Bühler introduces the concept of 'field.' The sentence is, on the one hand, a sequence of symbols and, on the other, a field. This sentential field is something observable and cannot be characterised just as 'oversummativity.'

Bühler considers three types of field in language:

1. The field of denomination or deictic field (*Nenn-oder Zeigefeld*).
2. The symbolic field. It is typically an indirect field, since it provides the speaker with instruments which allow a mediation between sound - (shape - in writing) and world-perception/control. The instrumental character of this field is coined by the term 'field-instruments' (*Feldgeräte*).
3. The pictorial field (*Malfeld*). Like de Saussure, Bühler restricts its use to onomatopoeia and rejects it as marginal. This may explain why his semiotics are much closer to traditional philosophy of language than to the psychology of perception and motor-control.

The 'symbolic field,' which is the central concern of Bühler's theory of language, has different aspects, which Bühler calls 'field moments.' They permit the construction of the sentential whole out of its elements. Bühler distinguishes:

2.1 Material field moments (*Stoffhilfen*). They concern the links in a lexical field (as we call it today, following the lexicological tradition).
2.2 Word classes. They allow the construction of a syntactic field, ask for neighbourhood, dependency, or impose restrictions on them.

Now, in comparison with Lewin's theory we can observe two profound differences:

1. In Bühler's two books on language theory Lewin is not mentioned. Of the Berlin group, Bühler mentions only Wertheimer in a footnote. The notion of field remains vague and the proposals for a formal definition of the field by Lewin are ignored.
2. Although language is seen as a medial/instrumental capacity, it is disconnected from motor-behaviour and in general from sensorimotor cognition.

Consequently, Bühler's voluminous theory of language may be considered as a step backward which integrates a lot of philosophical and philological arguments but loses contact with experimental work in Gestalt psychology. The semiotics which could continue Lewin's insights are *not* developed by Bühler.

Cassirer's philosophy of symbolic forms widens the scope of a philosophy of language by considering other types of 'symbolic forms' like myth, religion, art. In the final analysis these forms are perspectively ordered in relation to some ideal of pure knowledge in the Kantian tradition. I will compare some aspects of his semiotic theory in order to illustrate the specific character of Lewin's contributions.

4.2 *Is an axiomatic theory of language possible?*

The *Principia Mathematica* by Whitehead and Russell (first part 1910) established a new target for the reorganisation of the sciences. Although axiomatic structures were only explicitly given for the system of formal logic, Russell presupposed a basic link between laws of nature (as formulated in Newtonian physics) and logical axioms.

In his *Wissenschaftslehre* (theory of science) Lewin distinguished three types of axioms in a hierarchy of relevance:

1. Single facts (rendered by propositions) may be reorganised by finding for every sentence its 'logical position' in the system of sentences. Central laws obtain central (logical) positions and thus disciplines like physics, chemistry, etc., may be reorganised as a hierarchy of propositions. The upper-most propositions may be called axioms. For Lewin, it was not evident that this type of axiomatisation should be complete or unique.
2. Basic principles operative in different disciplines may be stated as axioms. Lewin cites some axioms proposed by Reichenbach[46] such as: "principle of irreversible causality," or "principle of the homogeneity of space." These principles cannot be proved in any discipline, they just represent the paths and goals of scientific inquiry and are a priori.
3. Finally, there are principles which pertain to the nature of the objects in a science. Thus, the term 'nature' in physics or 'language' in linguistics are of this type. As the 'objects' of a science are not naturally given but rather constituted historically as a discipline finds or changes its focus, these principles or axioms are like conventions and cannot be empirically validated.

In the 1920's Bloomfield tried to establish "A set of postulates for the science of language."[47] He used the "method of postulates (that is assumptions or axioms)"[48] to cut off linguistics from psychology and anthropology.[49] Quite the contrary aim was pursued by Karl Bühler (already in his *The crisis of psychology* in 1927). He formulated as basic axioms that language has three necessary aspects: *Ausdruck* (expression), *Appell* (appeal) and *Darstellung* (representation). The last one constitutes the specific nature of human language (separated from animal communication which only shows the first two aspects). As a corollary Bühler deduces that linguistic phenomena fall under the objects of psychology (which has to consider all three aspects). As he considered Darwinian evolution, animal behaviour, and neurophysiology as relevant for psychology, the program contained in his axioms constituted a new psychology with language and cognition in its central scope and with evolutionary and ethological aspects on its fringes (Konrad Lorenz, the founder of animal ethology, attended the lectures given by Bühler in Vienna). We can say that Bühler formulated the program of a cognitive science in 1927 at the same time as he critically discussed behaviourism. Thus, it was not Noam Chomsky in the 1950's, who first defended cognitive psychology against behaviourism; this had already been done 20 years earlier by Karl Bühler. What about Chomsky's axiomatics? In the spirit of the young Russell (at the beginning of the century) Chomsky naively identified the mathematical theory of formal languages with natural language and thus invented a grammar which deduced sentences from the axiom S (= sentence; his generative grammar). Now, we have seen that the axioms which specify the objects of a science[50] are more or less conventions which follow an arbitrary choice of an aspect of the object under consideration. As such, a computer engineer may define 'his' object in language engineering as he likes, and his purpose should make him favour a definition related to program languages, but he should not try to sell this convention as philosophy, psychology or as cognitive linguistics. Thus, the interdisciplinary programs for a broader psychology developed by Bühler, the expansion of experimental methods to social psychology developed by Lewin and Cassirer's philosophy of symbolic forms are more important events for the rise of cognitive science than the mystified M.I.T. conference in 1956.

5. CASSIRER'S PHILOSOPHY OF SYMBOLIC FORMS AS SEMIOTICS

Cassirer refers to Bühler in his *Philosophie der symbolischen Formen* and like Bühler he uses linguistic results from the 19th and early 20th century (W. von Humboldt, Schleicher, Vossler, etc.) as examples to illustrate his theses. Philosophically, Kant and Goethe (and their continuation in Schelling and Hegel) appear as the central points of departure for his philosophy of the mind and of language.

5.1 *The concept of 'field' in Cassirer's philosophy*

In the following I will just outline typical kinds of models, fields, and forces used by Cassirer in order to contrast them with the use Lewin makes of modern mathematical tools like vector analysis. We can distinguish two types of fields and forces in Cassirer's philosophy.[51]

a) In his philosophy of mathematics,[52] Cassirer compares the concept of Gestalt, mainly its central feature 'transposability' to the mathematical concept of 'transformation' and 'group.' The mathematical concept of group permits a grasping of the invariant properties of a geometrical figure. Now, the problem with a correlation between the psychological concept of Gestalt and the mathematical concept of group consists in the multiplicity of geometries and correspondent transformations. With the advent of non-Euclidean geometries different theories of invariance result and it seems impossible to choose the specific (unitary) geometry in correspondence with perception.

> Thus, however, we seem again to be led to a point where the analogy between the invariants of perception and those of geometry disappears. That form of logical *systematisation* which is both possible and necessary in the domain of geometrical thought is once and for all

inaccessible to perception. Here we have to take the phenomenal facts as they present themselves in experience; we cannot go beyond the simple ascertainment of these facts. It is but empirical observations that can tell us in which domains of sense-perception there exist phenomena of constancy and how far their influence extends.[53]

From the point of view of his philosophy of symbolic forms, Cassirer negates a simple mechanistic model of perception based on geometrical transformations. Mathematical group theory is like a manifold of possibilities; which of these possibilities is finally chosen (by a specific species, or in a specific situation) cannot be predicted from the model. There is a genuine space of freedom or arbitrariness (to use a favourite term of de Saussure) and this holds already at the level of perception (a fortiori at the level of linguistic categorisation). Thus, the limits of the mathematical model tell us more than its domain of excellence. Lewin's use of mathematics is more naive and more pragmatic. As far as the models help to clarify methods and procedures they are welcome, but non-formal (hermeneutic) procedures and concepts are used at the fringes of their relevance. Although the mathematical rigor of Chomsky's (early) models of language is at a higher level than Lewin's, he still adopts the purely positive attitude which does not use the limits of a model to understand basic features of the described phenomenon (i.e., language in Chomsky's case). Since Lewin used mathematics (e.g., vector calculus and topology) simply as a heuristic, he is not confronted with the problem of limits in mathematical modelling.

b) As Cassirer has gone or wants to go beyond strict models in his search for the symbolic (which by definition is an activity of creative form-giving) he introduces concepts and schemata developed in the tradition of Herder, Goethe, Kant, Hegel, et al. Although one must acknowledge that the problem of the symbolic was beyond current techniques of experimental psychology or empirical sciences, this move by Cassirer cut him off from current 'science' and threw his philosophy back to a mode of speculative thinking typical for the late 18th and the 19th century (in Germany). Lewin's more pragmatic strategy allows him to expand the methods of Gestalt and field psychology to social psychology and sociology. Later, Cassirer in his book *The myth of state* applied his philosophy to problems of culture and society. Thus, both Lewin and Cassirer pursued the same goal. Lewin approached it by an expansion of experimental methodology, Cassirer on the basis of German idealism. In order to illustrate Cassirer's strategy I will cite some central passages from his programmatic article on "The concept of symbolic form in the constitution of the sciences of the human mind" (*Der Begriff der symbolischen Form im Aufbau der Geisteswissenschaften*, written in 1921, published in Cassirer, 1994).

Cassirer defines the symbolic forms as:"[...] that energy of the mind [...], which maps meaningful content in the mind onto a concrete sign and is internally appropriated by this sign."[54]

The symbolic forms are manifested in the areas of phonetic language, myth, art, religion and pure knowledge ('*reine Erkenntnis*'). This plurality of types of symbolic forms (the list is not complete) functions as a kind of medium ('Medium,' *Vermittlung*) between the objects and ourselves.[55]

> [...] but they designate thereby not only negatively the distance, into which the object moves relative to us, but they create the only possible, adequate mediation and the medium, by which any kind of spiritual being may be grasped and understood.[56]

The symbolic forms establish an intermediate domain, subdivided by different types, in a field between the ego and the outer world.

The typical character of the 'symbolic' form is defined by a vector in this field which transforms passive 'pictures' received by the senses into something which is actively formed by the mind. The force of this formative process transforms the perceptual content into symbolic content. "In it the image has ceased to be just something received from outside; it has become something shaped from within, in which a basic principle of free form-giving governs."[57]

The basic polarity between *self* (consciousness) and *object* is further complicated by the fact that the rather static objects are 'represented' by a steady flow of processes in consciousness. This basic difference precludes any kind of mapping or pictorial representation between the two extremes of the field.[58]

The general feature one may conclude from the positions taken by Cassirer is that his philosophical and semiotic positions are almost independent from new conceptual tools developed in mathematics after Kant and Hegel; these are rather possible objects of philosophical reflection. In the discussion of the different types of symbolic forms it turns out that the forms of pure knowledge as in mathematics constitute a point of culmination in an evolutionary process which shapes the symbolic forms.

5.2 *Religion, myth, art, science etc.: Cassirer's symbolic forms beyond language*

It would be an interesting question in the sociology of science to ask if the kind of conservatism, the strong reference to Kantian traditions, the rather naive integration of current linguistic theories and the rejection of an application of conceptual innovations in mathematics and the natural sciences by the human sciences can be explained by the academic situation in Germany,

Austria (and also in other European countries) at the beginning of the 20th century.

With this background it would be interesting to ask why the Berlin group and specifically Kurt Lewin did not (or not to the same degree) submit to the local 'Zeitgeist.' Ernst Cassirer (1874-1945) had made his academic career in pre-war Germany, he was 45 years old in 1919 and therefore continued along the line of his philosophical thinking shaped by New Kantianism which dominated German philosophy at the end of the 19th century. It is only after his emigration that he changed his style of writing and arguing for an Anglo-Saxon readership.

If Cassirer did not contribute as much to the rise of cognitive science as Lewin, Bühler and Reichenbach did, his philosophy of symbolic forms gave rise to another very forceful interdisciplinary enterprise now called cultural studies (and science of the media). His philosophy can be seen as the rise of a new anthropology which covers ethnology, comparative and typological linguistics, cultural history, religion, history of ideas, etc. Cassirer combined his innovative works in this new field with a thorough knowledge of mathematics and the natural sciences. I think the rising field of 'cultural studies' has to consider Cassirer's work and to cope with the theoretical and philosophical standard he established. In the long run the two enterprises 'cognitive science' and 'cultural studies' will be forced to establish a common platform of empirical methodology and theoretical modelling because the human mind is not subdivided into a cognitive and a cultural hemisphere and any (Cartesian) subdivision of this type must collapse in the long run.

6. CONCLUSIONS

The rise of a cognitive approach to language and other symbolic forms in the 1920's may be characterised by three features:

1 The founder of modern semiotics, Charles Sanders Peirce, developed his holistic, evolutionary and cosmological theory of signs before and independently of Gestalt theory (at the beginning of the 20th century).
2 The theoretically most advanced Berlin School of Gestalt psychology either concentrated on very fundamental problems and looked from psychology towards physiology and physics, or, like Kurt Lewin, favoured a theoretically independent psychology and expanded Gestalt theory towards developmental and social psychology.
3 Those who dealt systematically with language and communication (e.g., K. Bühler and E. Cassirer) left the theoretical level of the Berlin School

> and moved into the more restricted and traditional area of German philosophy of language and philology, dominated by Kantianism and an idealistic theory of language.

As a consequence the effect of Gestalt theory on the international evolution of semiotics remained rather small.[59] The suggestions made by Peirce were too idiosyncratic to fit the intellectual development in American psychology, linguistics or logic.

My main issue is that Kurt Lewin's view of the role of language created something like a program for ecological semiotics. After World War I Lewin turned to fundamental problems of the philosophy of science. His theoretical work was not understood as an important contribution by Stumpf and the other members of the jury for his habilitation.[60] The reason for this incomprehension was probably the very abstract level and the lack of concrete innovations. At the end of his life Lewin said about Cassirer[61] that his work does not show specific concepts or statements which contain a new insight or solve a problem which had not been solved before, thus he applies the criticism he received in 1921 to Cassirer who had been his teacher in 1910. The later experimental and practical work of Lewin was perhaps a response to this criticism. Another reason for Lewin's failure in 1921 was the new almost axiomatic style Lewin used in his treatise. It was a foretaste of later analytic reconstructions of theories and bore witness to an episode in which Lewin was on his way to Analytic Philosophy. In his further research, Lewin did not give up using mathematical concepts and equations in empirical work, although he was eager to embed them in very concrete and intuitively accessible contexts.

Cassirer in his philosophy of symbolic forms abandoned the rigor of his treatise in substance and function in 1910 (which Lewin admired) and did not apply his knowledge of mathematical physics in his foundational work for a science of culture. Thus, the first volume of his philosophy of symbolic forms shows traces of his earlier work in logic, but without any use of modern mathematical logic (as developed by Russell and Whitehead). Some chapters can be read as reports on current research in comparative philology and Cassirer reports their results without any critical comment on their reliability, the presupposed categorical schemata, etc. Basically Cassirer makes clear that symbolic forms belong to a domain of freedom, of culturally shaped and not mechanically determined phenomena. Although the space in which these 'free' forms occur may be mathematically described, the specific forms and their functions ask for a different methodology. If Cassirer was the teacher for the generation of Lewin and Reichenbach, the new generation made other choices: Lewin expanded the experimental methodology and the mathematical modelling in Gestalt psychology to the social sciences, Reichenbach opted for logical positivism in the Vienna style. Karl Bühler (1879-1963) came from a different tradition of Gestalt psychology and developed a program for psychology which integrated the phenomenon of language and for-grounded cognition. Although the four scientists who were born between 1870 (Cassirer) and 1891 (Reichenbach) neither formed a group nor developed a program of cognitive science, their different activities expanded the traditional domains of research in psychology, philosophy, logic, the social sciences, and thus furthered interdisciplinary enterprises of the type called 'cognitive science' and 'cultural studies' today. Their intellectual activity already contained the germs for a cognitive science and in many respects they even went beyond the current scope of the cognitive sciences. Emigration forced Cassirer to adapt again and to leave the complexity of post-Kantian and post-Hegelian philosophy. In the *Myth of the State* Cassirer was able to continue his work on the history of ideas and to apply it to his political experience with the Nazi regime.

Wolfgang Wildgen
Bremen Universität
Fachbereich 10
Sprach-und Literaturwissenschaften
P.O.B. 33 04 40
D-28334 Bremen
e-mail: wildgen@uni-bremen.de

NOTES

[1] Cf. Gardner 1987, 28 ff.

[2] Cf. Bühler 1927.

[3] This whole line of integrative enterprises may be linked to Leibniz's *characteristica universalis* in its geometrical elaboration. It can be opposed to another enterprise referring to Leibniz's program which started with Frege and Peano, was elaborated by Whitehead and Russell, and led to the so-called birth-party of cognitive science in 1956.

[4] Cf. Stadler and Kruse 1995b.

[5] Cf. Metzger 1975.

[6] "Der Übergang von der aristotelischen zur galileischen Denkweise in Biologie und Psychologie," given in Berlin, 1930.

[7] Peirce's thinking contributed to general intellectual development and was not directly dependent on Gestalt theory. In the USA around 1850 the Kantian and Hegelian tradition was very influential; only later (after 1870) did empiricism (e.g., Mach's philosophy) become relevant for US philosophers (Mach gave his lectures between 1868 and 1871). Peirce was a critic of this direction of thought. At the turn of the century two major schools existed in the

USA: the pragmatists (e.g., Peirce) and the logical positivists. Peirce's formal pragmatism comes rather close to the position of the Berlin school of Gestalt psychology.
[8] Cf. Herzberger 1981.
[9] Cf. Ms. 717 in Peirce 1976, vol. 4, 307 ff. and 1999, 295.
[10] This is a fundamental step because it makes use of a basic feature of natural laws to define the domain of a phenomenon (thus the speed of light has a limit and quantum states define limits; these border-lines have a deep effect on all other regularities found in the specific field).
[11] Cf. Peirce 1976, vol. 3, 2, 1046 ff. The mathematician Euler lived from 1707 to 1783.
[12] Cf. Peirce's diagrammatic logic.
[13] It was evident that phonology with its functional abstraction from concrete phonetic realisation was a field of application for the concept of 'Gestalt.' Cf. Holenstein 1975, and Leška 1984.
[14] Cf. Metzger 1975.
[15] Lewin was a soldier in the First World War when he wrote this article. For an application of his concept of 'Lebensraum' in sociolinguistics compare Wildgen 1983.
[16] E.g., the one outlined in Wildgen 1994.
[17] In Lewin 1936/1969, 226 we find the following definition of valence: "Valenz (Aufforderungscharakter): die Valenz entspricht einem Kraftfeld, das nach Art eines Zentralfeldes strukturiert ist. Man kann positive und negative Aufforderungscharaktere unterscheiden."
[18] Cf. Bierbrauer 1983.
[19] Cf. Lewin 1982, 104.
[20] Cf. Wildgen 1982, and 1994.
[21] Cf. Bühler's Organon-model in Bühler 1934/1965, 28.
[22] Cf. also Cattaruzza 1996, 168 ff.
[23] Cf. Lewin 1982, 181.
[24] Thus in Cayuga many words combine a root, which is rather verbal, with a root, which is rather nominal. Cf. Sasse 1993.
[25] Cassirer never got a tenure in Berlin and therefore had no direct influence on Lewin's career (thus he was not in the habilitation's committee in 1919 so disinterested in Lewin's thesis, that he had to withdraw).
[26] Cf. Cassirer "Stufenfolge in den Graden der Objektivität," in Cassirer 1910, 365.
[27] Ibid. 366.
[28] Cf. Cassirer "Logische Abstufung." Ibid. 368.
[29] Original title: "Der Begriff der Genese in Physik, Biologie und Entwicklungsgeschichte." Translation by the author.
[30] Ibid. 95.
[31] Original citation, "Die physikalischen Existenzreihen als solche enthalten keine ausgezeichneten Schnitte." Translation by the author.
[32] In 1926, Bloomfield published his *Set of Postulates for the Science of Language*.
[33] Cf. Cassirer 1910, 373 ff.
[34] Ibid. 373.
[35] Ibid. 374. "Haben wir uns des Daseins zweier verschiedener Grundreihen einmal versichert, so können wir versuchen, kraft eines Analogieschlusses die Verhältnisse, die wir in der einen Reihe finden, auf die andere zu übertragen: unbegreiflich aber bleibt es, wie wir dazu gelangen sollten, die Existenz der einen Reihe aus Daten, die ganz und ausschließlich der anderen angehören, zu fordern."

[36] Original German title: "Das Erkenntnisproblem in der Philosophie und Wissenschaft der neueren Zeit."
[37] Original title: "Der Übergang von der aristotelischen zur galileischen Denkweise in Biologie und Psychologie." A paper read in 1930 and published in 1930/31, cf. Lewin 1971, and Lewin 1981, 233–278.
[38] Lewin's style of historical analysis cannot be properly compared with Cassirer's. The later stands in the tradition of the Warburg school in the history of ideas, where the understanding and reconstruction of historical developments is more important, than consequences for future development. Lewin clearly uses a rather simple analysis of Galileian physics in order to find support for the type of vectorial and dynamic psychology he advocates.
[39] Lewin 1981, 351 ff.
[40] Metzger 1979, 9, says that Lewin's elaboration of Cassirer's philosophy of science (called "Wissenschaftslehre" by Lewin) presented a logically stringent consideration of psychophysical processes.
[41] Ibid. 362, ff.
[42] Ibid. 363.
[43] Ibid. 350.
[44] Some of these approaches are more reductionist than Lewin would have liked. Thus Bertalannffy's 'Fließgleichgewicht' (dynamic equilibrium) applies to physical *and* living system. In a similar vein Prigogine's and Haken's systems expand types of self-organisation in physics to biology and the human sciences. Thom's catastrophe theory is clearly non reductionist, but as Thom (1988) shows, it also returns to a non-Galileian (Aristotelian) type of physics. This chapter of recent history of ideas (after 1945) would merit a systematic analysis in terms of a philosophy of science.
[45] Reichenbach's treatment of natural language in his logic published (1947) was one of the first attempts towards logical semantics, and it may illustrate one way of mathematisation in linguistics (prior to Chomsky).
[46] Lewin 1983, 361.
[47] Cf. Bloomfield 1926/1987, 70–80.
[48] Ibid. 70.
[49] Ibid. 71.
[50] Cf. Lewin, 1981.
[51] Compare my article: "The History and Future of Field Semantics. From Giordano Bruno to Dynamic Semantics," (forthcoming), where Lewin's and Bühler's field concepts are discussed.
[52] Cf. Cassirer's late article "The Concept of Group and the Theory of Perception," 1944.
[53] Ibid. 26.
[54] "[...] jene Energie des Geistes [...], durch welche ein geistiger Bedeutungsgehalt an ein konkretes sinnliches Zeichen geknüpft und diesem Zeichen innerlich zugeeignet ist." Ibid. 175.
[55] Ibid. 176.
[56] "[...] aber sie bezeichnen damit nicht nur negativ die Entfernung, in welche der Gegenstand für uns rückt, sondern sie schaffen die einzig mögliche, adäquate Vermittlung und das Medium, durch welches uns irgendwelches geistige Sein erst faßbar und verständlich wird." Ibid. 197.
[57] "In diesem hat das Bild aufgehört, ein bloß von außen Empfangenes zu sein; es ist zu einem von innen her Gebildeten geworden, in dem ein Grundprinzip freien Bildens waltet." Ibid. 77.
[58] Cassirer introduces another force called 'Zentrierung' or 'Zentrenbildung.' In perception the individual chooses an ideal orientation, in relation to which he constructs his picture of the

world. The identification and contextual integration of the object under inspection depend on such perspectives and more generally on knowledge. Cf. Plümacher 1997, 192 ff.

[59] It was only recently that similar ideas reappeared in Cognitive Semantics. In 1976 (published in Lakoff 1977) Lakoff gave a conference on 'gestalt linguistics,' but the tradition of gestalt psychology and specifically the linguistic proposals which are rather widely distributed over single treatises were ignored. Neither Lewin nor Heider, who had pursued the proposals of Lewin into semantics, are cited by Lakoff.

[60] Cf. Métraux 1992, 32.

[61] Cf. Lewin 1981, 348

FIGURES

Figure 1. The reduction of a four-fold relation to two triadic relations.
Figure 2. Bistable perception as duck/rabbit or cup/human profiles.
Figure 3. Hysteresis effects in the transition between two meaningful interpretations.
Figure 4. Field of the child who wants to sit down on a stone (G = goal; Ch = child; B = barrier).
Figure 5. A sequence of snapshots in the evolution of the field.
Figure 6. A dynamic-topological description of the action 'TO GIVE;' the lines represent attractors (i.e., minima), the arrows changes of dominance.

TABLES

Table 1. Peirce's classification of signs.
Table 2. Peirce's classification of elementary propositions.
Table 3. Four characteristic features of Galileian physics and psychology

REFERENCES

Albertazzi, L. ed.:1996, 'The Philosophy of Alexious Meinong,' *Axiomathes*, 7.

Albertazzi, L. ed.: (forthcoming), *Meaning and Cognition: A Multidisciplinary Approach.* Amsterdam, Benjamins Publishing Company.

Bierbrauer, G.: 1983, *Kurt Lewins Leben, Werk und Wirkungsgeschichte in der Emigration*, Series: Psychologische Forschungsberichte, **32**, Osnabrück.

Bloomfield, L.: 1926/1970, *A Set of Postulates for the Science of Language*, reprinted in Hockett, ed., 1970, 70–80.

Bühler, K.: 1913, *Die Gestaltwahrnehmungen*, vol. 1. Stuttgart, Spemann.

Bühler, K.: 1926/1978, *Die Krise der Psychologie*. Wien, Ullstein. (Partially published 1926 in *Kant-Studien*, **31**(4), 455–526).

Bühler, K.: 1933/1969, *Die Axiomatik der Sprachwissenschaften*. Frankfurt a.Main, Klostermann. (First printed in *Kant-Studien*, **38**, 1933, 24–90).

Bühler, K.: 1934/1965, *Sprachtheorie. Die Darstellungsfunktion der Sprache*. Stuttgart, Fischer, 2nd edition.

Cassirer, E.: 1910, *Substanzbegriff und Funktionsbegriff. Untersuchungen über die Grundfragen der Erkenntniskritik.* Berlin, Verlag Bruno Cassirer.

Cassirer, E.: 1923/1988, *Philosophie der symbolischen Formen. Teil 1, Die Sprache*, 9th ed. Darmstadt, Wissenschaftliche Buchgesellschaft. (1 st ed. 1923.)

Cassirer, E.: 1927/1963, *Individuum und Kosmos in der Philosophie der Renaissance.* Darmstadt, Wissenschaftliche Buchgesellschaft. (Repr. 1963).

Cassirer, E.: 1944, 'The Concept of Group and the Theory of Perception', *Philosophy and Phenomenological Research,* **5** (**1**), 1–36.

Cassirer, E.: 1979, *Symbol, Myth, and Culture. Essays and Lectures of Ernst Cassirer 1935-1945,* ed by D.P. Verene, New Haven, Yale University Press.

Cassirer, E.: 1994, 'Der Begriff der symbolischen Form im Aufbau der Geisteswissenschaften', in Cassirer, E., *Wesen und Wirkung des Symbolbegriffs.* Darmstadt, Wissenschaftliche Buchgesellschaft (special edition), 175–200.

Cassirer, E.: 1995, *Zur Metaphysik der symbolischen Formen.* Hamburg, Meiner. (Series: Nachgelassene Manuskripte und Texte, vol. 1).

Cattaruzza, S.: 1996, 'Meinong and Bühler,' in Albertazzi, ed., 1996, 103–110.

Ehrenfels, C. von: 1890, 'Über Gestaltqualitäten,' *Vierteljahresschrift für wissenschaftliche Philosophie,* **15** (**3**), 249–292.

Eschbach, A. ed.: 1984, *Bühler-Studien.* Frankfurt a.Main, Suhrkamp.

Gardner, H.: 1987, *The Minds New Science. A History of the Cognitive Revolution.* New York, Basic Books.

Goedel, K.: 1986, 'Russells mathematische Logik,' in Whitehead and Russell 1986, v-xxxiv.

Haken H. and Stadler, M. eds., 1990, *Synergetics of Cognition.* Proceedings of the International Symposium at Schloß Elmau, Berlin, Springer.

Heigl-Evers, A. ed.: 1979, *Lewin und die Folgen. Sozialpsychologie. Gruppendynamik. Gruppentherapie* (Series: Psychologie des 20. Jahrhunderts, vol. 20). Zürich, Kindler.

Herzberger, H.G.: 1981, 'Peirce's Remarkable Theorem', in Summer, Slater and Wilson, eds., 1981, 41–58.

Hockett, C.F.: 1987, *A Leonard Bloomfield Anthology.* Chicago, Chicago University Press, Abridged Edition.

Holenstein, E., 1975, *Roman Jakobsons Phänomenologischer Strukturalismus.* Frankfurt, Suhrkamp.

Kanizsa, G. and Luccio, R.: 1986, 'Die Doppeldeutigkeiten der Prägnanz,' *Gestalt Theory,* **8**, 99–135.

Kanizsa, G. and Luccio, R.: 1995, 'Multistability as a Research Tool in Experimental Phenomenology', in Stadler and Kruse, eds., 1995a, 47–68.

Lakoff, G.: 1977, 'Linguistic Gestalts,' *Papers from the Thirteenth Regional Meeting.* Chicago Linguistic Society, Chicago, 236–287.

Lakoff, G.: 1987, *Women, Fire, and Dangerous Things. What Categories Reveal About the Mind.* Chicago, Chicago University Press.

Lang, A.: 1979, 'Die Feldtheorie von Kurt Lewin', in Heigl-Evers, ed., 1979, 51–57.

Leška, O.: 1984, 'Karl Bühler und die Prager Schule,' in Eschbach, ed., 1984, 263–272.

Lewin, K.: 1936/1969, *Grundzüge der topologischen Psychologie.* Bern, Huber. First published in 1936 under the title *Principles of Topological Psychology,* New York, McGraw-Hill.

Lewin, K.: 1951/1963, *Feldtheorie in den Sozialwissenschaften. Ausgewählte theoretische Schriften,* ed. by D. Cartwright, Bern, Huber (first published in English, New York, Harper and Brothers).

Lewin, K.: 1971. *Der Übergang von den aristotelischen zur galileischen Denkweise in Biologie und Psychologie.* Darmstadt, Wissenschaftliche Buchgesellschaft (repr. in Lewin, 1981, 233–278).

Lewin, K.: 1981, *Wissenschaftstheorie I*, Werkausgabe, edited by F. Graumann, vol. 1, Bern, Huber.

Lewin, K.: 1982, *Feldtheorie*, in *Werkausgabe*, ed. by F. Graumann, vol. 4, Bern, Huber.

Lewin, K.: 1983, *Wissenschaftstheorie II*, in *Werkausgabe*, ed. by F. Graumann, vol. 2, Bern, Huber.

Marr, D.: 1982, *Vision: A Computational Investigation Into the Human Representation and Processing of Visual Information*. New York, Freeman.

Marty, A.: 1908, *Untersuchungen zur Grundlegung der allgemeinen Grammatik und Sprachphilosophie*. Halle, Niemeyer.

Métraux, A.: 1992, 'Kurt Lewin im philosophisch-psychologischen Rollenkonflikt,' in Schönpflug, ed., 1992a, 29–38.

Metzger, W. ed.: 1966, *Handbuch der Psychologie,* vol 1/1. *Wahrnehmung und Bewußtsein*. Hogrefe, Göttingen.

Metzger, W.: 1975, *Gesetze des Sehens*. Frankfurt, Kramer.

Metzger, W.: 1979, 'Der Einfluß von Kurt Lewin auf die Entwicklung der Sozialpsychologie,' in Heigl-Evers, ed., 1979, 7–16.

Peirce, C.S.: 1976, *The New Elements of Mathematics*, ed. by C. Eisele, The Hague, Nijhoff.

Peirce, C.S.: 1990. 'Die Grundlagen des Pragmatismus. Drei Entwürfe zu einem Aufsatz,' (translation of the manuscripts No 202–204, 1905), in *Collected Papers, Semiotische Schriften*, vol. 2 (1903-1906), ed. by C. Kloesel and H. Pape, Frankfurt, Suhrkamp, 289–391.

Petitot, J.: 1992, *Physique du Sens. De la théorie des singularités aux structures sémio-narratives*. Paris, Editions du CNRS.

Plümacher, M.: 1997, 'Gestaltpsychologie und Wahrnehmungstheorie bei Ernst Cassirer', in Rudolph and Stamatescu, eds., 1997, 171–207.

Rausch, E.: 1966, 'Das Eigenschaftsproblem in der Gestalttheorie der Wahrnehmung,' in Metzger, ed., 1966, 866–953.

Reichenbach, H.: 1924, 'Report on Lewin's Treatise, *Der Begriff der Genese in Physik, Biologie und Entwicklungsgeschichte,' Psychologische Forschungen*, **5**, 188–190.

Reichenbach, H.: 1947/1966, *Elements of Symbolic Logic*. New York, Free Press (1st edition 1947).

Reichenbach, H.: 1977, *Gesammelte Werke in sieben Bänden*, ed. by A. Kamlah and M. Reichenbach, Wiesbaden,Vieweg.

Rudolph, E. and Stamatescu, I.O. eds.: 1997, *Von der Philosophie zur Wissenschaft. Cassirers Dialog mit der Naturwissenschaft*. Hamburg, Meiner.

Sasse, H.-J.: 1993, 'Das Nomen – eine universale Kategorie?, *Sprachtypologie und Universalienforschung (STUF)*, **46**, 187–221.

Saussure, F. de: 1916/1975, *Cours de linguistique générale*. Lausanne, 1916 (new edition, 1975, Paris, Payot).

Schönpflug, W. ed.: 1992a, *Kurt Lewin – Person, Werk, Umfeld. Historische Rekonstruktionen und aktuelle Wertungen aus Anlaß seines hundersten Geburtstags*. Frankfurt, Lang.

Schönpflug, W.: 1992b, 'Kurt Lewin – eine biographische Skizze', in Schönpflug, ed., 1992a, 13–28.

Stadler, M. and Kruse, P. eds.: 1995a, *Ambiguity in Mind and Nature. Multistable Cognitive Phenomena*. Berlin, Springer.

Stadler, M. and Kruse, P.: 1995b, 'The Function of Meaning in Cognitive Order Formation', in Stadler and Kruse, eds., 1995a, 5–22.

Summer, L.W. Slater, J.G. and Wilson, F. eds.: *Pragmatism and Purpose: Essays Presented to Thomas S. Gouge*. Toronto, Toronto University Press.

Tesnière, L.: 1959/1988, *Eléments de syntaxe structurale*. Paris, Klincksieck (2nd corrected edition, 1988; first edition 1959).

Thom, R.: 1983, *Mathematical Models of Morphogenesis*. Chichester, Horwood.

Whitehead, A.N. and Russell, B.: 1986, *Principia Mathematica*, German tr. by H.Mokre, Frankfurt, Suhrkamp.

Wildgen, W.: 1982, *Catastrophe Theoretic Semantics. An Elaboration and Application of René Thom's Theory*. Amsterdam, Benjamins Publishing Company.

Wildgen, W.: 1983, 'Dialogdynamik. Erste Ansätze zu einer Erfassung der Dynamik verbaler Interaktionen am Beispiel richterlicher Anhörungen', *Papiere zur Linguistik*, **29 (2)**, 3–21.

Wildgen, W.: 1985, *Archetypensemantik. Grundlagen einer dynamischen Semantik auf der Basis der Katastrophentheorie*. Tübingen, Narr.

Wildgen, W.: 1994, *Process, Image, and Meaning. A Realistic Model of the Meanings of Sentences and Narrative Texts*. Amsterdam, Benjamins Publishing Company.

Wildgen, W.: 1998, *Das kosmische Gedächtnis. Kosmologie, Semiotik und Gedächtniskunst im Werke von Giordano Bruno* (1548-1600). Frankfurt, Lang.

Wildgen, W.: 1999, *De la grammaire au discours. Une approche morphodynamique*. Bern, Lang (series European Semiotics, no. 1).

Wildgen, W.: (forthcoming), 'The History and Future of Field Semantics. From Giordano Bruno to Dynamic Semantics', in Albertazzi, ed., forthcoming.

Wittmann, S.: 1998, *Das Frühwerk Kurt Lewins. Zu den Quellen sozialpsychologischer Ansätze im Feldkonzept und Wissenschaftstheorie*. Frankfurt, Lang.

RUGGERO PIERANTONI

CHILDREN'S DRAWINGS AS SENSIBLE PROBES INTO THE REALM OF REPRESENTATIONS

1. PROF. JAKOB BURCKHARDT AND HIS LECTURE ON THE GOLDSMITH'S ART

During his academic years in Basel, Jakob Burckhardt was accustomed to meet the intellectual 'milieu' of his city in a series of extra-university seminars and lectures. In the year 1860, when *The Civilisation of Renaissance in Italy* and *Renaissance and Architecture in Italy* were published, Burckhardt chose to entertain his sophisticated audience by discussing a quite particular and unusual subject. The topic of his lecture was three books on the goldsmith's art belonging to the Amerbach legacy. The text of Burckhardt's lecture highlights at least two main elements, both of which show his concern with the verbal representation of small, exquisitely adorned objects. He challenged the translation from the visible to linguistic world by summoning all possible syntactic and stylistic tools in order to give his descriptions the fullest and, in a certain sense, parallel dignity.

Following this strategy, numerous highly complicated drawings and decorative patterns of jewels, church ritualistic tools, silver and gold plates are 'represented in words' through an acrobatic stylistic exercise. The text reveals an endeavour to create a second-order object, this time pertaining to the world of writing and no longer to that of the goldsmith's workshop. The second point is at variance with the first. Besides providing an almost orgiastic description of an object, Burckhardt sternly reminds his audience of the crucial importance of mathematics in harnessing the production of decorative patterns. He speaks of a "merciless geometrical law" which rigorously governs the immoderate blossoming of decorative forms. Evidently, the extreme richness of decoration, its intricacy, its lace-like quality, must be controlled, brought back to reason, subjected to measurement, made more 'rational.' It is curious to note that Le Corbusier, precisely at the moment when he 'delivered' his most controversial and unexpected creation, the church of Sainte Marie des Haut at Longschamp

L. Albertazzi (ed.), The Dawn of Cognitive Science. Early European Contributors, 333–355.

writes at length about acoustics. He narrates how his building was conceived and eventually built 'according to the merciless laws of acoustics.'

2. LESSING'S PROPHETIC *DER LAOOKOON*

We may now leave Jakob Burckhardt to his lecture and make a significant step backwards in time to almost exactly one hundred years previously. We are now at the moment when Gotthold Ephraim Lessing saw his *Der Laokoon* finally published in 1766. The main goal of *Der Laocoon* was to set the rules governing the two provinces of time and space as far as the arts were concerned. In the Enlightenment and rationalistic framework of the book, the arts of space are architecture, sculpture, painting, whilst those of time are poetry, drama and music. In this apparent rationalisation of aesthetics, a significant amount of discussion is devoted to the subject who receives the messages, both in time and in space, transmitted through artistic objects. In perhaps the most interesting and prophetic passage of the book, Lessing analyses the actions related to the inner representation and evocation of the external world. According to Lessing, the operations involved in the reconstruction of three-dimensional objects in mental space are time-consuming ones. They need, in fact, some time to 'stitch' together the different views of the same object and to 'adapt' them to each other in a coherent unifying vision. Usually, we do not notice this time because it is extremely brief, and we rely on the bias that mental processes are timeless. This prejudice induces us to forget the physical nature, the 'mechanical' qualities, of thought processes. They are considered, by necessity, not to be treatable with physiological terminology and certainly not susceptible to measurement and experimentation. Another significant aspect of Lessing's treatment is his analysis of the efforts, and more literally 'fatigue,' required to follow a written or spoken description of an object, in particular if it is very detailed and complex. If we return to Burckhardt's long evocation of his plates and vases, we can understand the deliberate confusion between the acts of vision and those of understanding. Hegemonic writing is the real medium of presentation of historically filtered memories. It is not a case that the 'pure visibilistic' evocation of splendid objects ends with the magical name of Holbein and recapitulates all images under the general heading of historical biography. The main thrust of Burckhardt's aesthetic studies is a profound conviction of the fundamental importance of dramatic contrasts and intense biographical evocations, with a look, perhaps a too acute one, towards violence, wars and historical disasters. The visual evocation of the object is filtered through history, the art of writing, and the control of language. In Basel, Burckhardt may have felt more protected in his historical quest than he did in Berlin, and in fact, in 1867 he firmly refused the Ranke chair of Art History at Berlin.

3. BUT BASEL IS NOT BERLIN AND BERLIN IS NOT WIEN

The historical colouring given to the 'visibilisation' of objects and to their evocation was inherited by Heinrich Wölfflin and brought to the Florentine meetings, encounters and co-operations with Conrad Fiedler, von Marèes, von Hildebrand and Franz von Brentano, eventually concluding with B. Croce's late but significant study of 1912, *La teoria dell'arte come pura visibilità.* But a quite different route brings us to Vienna. Since the beginning of the century, the academic world of Vienna had been so dominated by Herbart's philosophy that it was called a 'stronghold of Herbartism.' The Hegelian tradition had not filtered through to Vienna, and in Herbart's version of 'practical philosophy,' the relationships between the object and the subject did not achieve dialectic dynamics, although their metaphysics were still taken for granted. The formalised approach to the relationships between object and subject rested upon empirical grounds and used taxonomic tools, the isolation of peculiar characters, and more cogently, the universal concept of 'Gestalt.'

Instead of an axiomatic geometry of dichotomic reflection between object and subject integrated into a dialectical opposition, we encounter a complex and multifarious system of reciprocal contacts which invite study by the physiologist, the physicist, and the mathematician. It is suggested, moreover, that the biology or chemistry laboratory is a more appropriate place for their study than the philosopher's room. In effect, the final step toward the scientific study of psychological processes was not taken by Herbart, who still maintained a metaphysical attitude towards the use of mathematics in psychology; it is in the many laboratories set up by Wundt that we witness the very birth of modern psychophysics.

It was thus, therefore, that mathematics and statistics were introduced into the realm of psychology and of perceptual phenomena, mainly because it became clear that a certain amount of 'measurability' was to be found in (or introduced into) peripheral sensory actions and reactions to stimuli. A dramatic development came when Hermann Ebbinghaus presented the scientific community of the time with his totally revolutionary first quantitative analysis of the 'superior processes': in particular his famous 'mathematical function of forgetting.' It may be of interest, with regard to the discussion of children's drawings which follows, that Ebbinghaus proposed (almost at the same time as Binet in France) a quantitative approach to the measurement of intelligence in children. Ebbinghaus's extension of quantitative tools to the 'superior processes' crowned the major contribution to study of the sensory modalities made by Ewald Hering, Karl Stumpf, and Hermann von Helmholtz, all of whom drew heavily on the lexicon of mathematics and physics.

The concept of the measurability of man, his actions and sensations, has been discussed – in social and technological terms as well – by Jonathan Crary in his *Techniques of the Observer.* According to Crary, the dominant force that ushered mankind into the age of 'self-measurability' was the statistical conditions under which humans lived at the end of the last century, and which, on the basis of monetary and economical evaluation of work performance, justified the equation between human activities and the corresponding payments.

The multi-disciplinary figure of Ebbinghaus again provides a link with the theory of representation. Ebbinghaus was a close friend of the senator Giovanni Morelli, alias Ivan Lermolieff, who introduced a totally new technique of stylistic analysis of the art connoiseurship and, on more practical grounds, of recognition of a specific artist technique. The basis for this new procedure was the careful attention paid to the minute details of the painted image. The act of attributing a painting to a painter, in the absence of independent documentation, was closely connected with recognition of 'non-essential,' representative and 'automatic' figurative habits which, according to Morelli, were revealing features which played a significant role in the critic's work. It is difficult to resist quoting the controversial yet seminal paper, 'Where Pelicans Kiss Seals' by Ellen Winner, which points out that scrupulous study of the *minutiae* of the scribbles drawn by a very young boy may reveal a host of figural elements, representations and forms almost invisible at first sight. Moreover, according to Ellen Winner, these signs reveal the child's thought processes and linguistic competence, but they also have a distinct representational level which reveals itself as soon as we are able to disentangle them from the 'graphical mess' in which they are immersed. A very similar problem confronted the first scholars of paleolithic art when they sought to interpret numerous superimposed images. These anticipatory notes will be recalled when I will return to this problem later. For the moment, I shall resume discussion of the use of mathematical and laboratorial tools in the study of psychology.

4. Metaphysics, Mathematics and Dialectics

Jakob Burckhardt frequently warned against the use of mathematics and geometry in analysis of decorative patterns. His invitation to regard mathematical thought as a necessary corollary to aesthetic perception has its parallel in the recent utilisation of the same techniques in perceptual studies. In the latter, however, mathematics is implemented as a factual portion of the physiological and neurological events involved in aesthetic appreciation, no longer as an external contamination between the visual structure of an object and a cultural tradition. When Ernst Mach employed the concept of the second derivative to account for a certain visual contrast phenomena, he obviously viewed the mathematical process as an essential link in a chain of biological events, not as an external metaphorical descriptive system arbitrarily applied to a given event.

Hitherto, analysis of the complex inter-relationships between the object and its beholder had followed different paths, viz.:

1. Dialectic and formalised with metaphysical overtones. Note that the German word for object, *Gegenstand*, contains an implicit sense of opposition, of mirroring, of alienness. If we approach apperception and the inner representation of an external, physical object in this manner, the technical means at our disposal with which to disentangle the problem are axiomatic and abstract formal thought and language. Mathematics may play a significant role in this procedure, but only as an external, formalised operator whose intrinsic structure does not reflect the dynamics of the relationship.
2. Empirical analysis. This approach to perception and inner evocation rests on dynamic analysis of the perceptual processes and the 'superior activities' of the central nervous system. Laboratory procedures, mathematical methods, and the statistical treatment of the experimental results provide the framework within which interpretation and analysis can be conducted of the beholder's reactions to the external object, which are here reduced to a system of proximal and distal sources of energy.
3. The historical approach. In this case, study of representational processes involves historical analysis and, perhaps, the aesthetic categories. The instruments employed are mainly classificatory tools, the taxonomic or systematic grouping of objects, the identification of assonances, analogies and descriptive operators. The inter-relationships between the object and the beholder or the imaginative mind are modulated in the first instance by the historical context, knowledge of the temporal sequencing followed by the images, and evocation of their symbolic and even magical powers.

There are therefore three different approaches: Burckhardt's historical-biographic reading of historical processes; the formalised technique with an attenuated abstract attitude that Herbart used in his work; the laboratory approach made explicit by Ebbinghaus's and Wundt's experimental techniques. But it is again Burchkardt's approach that is the most intriguing and fecund, because its historical *cotè* involves the most interesting procedures subsequently used in the analysis of children's drawings. I shall introduce these procedures with discussion of Burckhardt's *Erinnerungen aus Rubens*.

Burckhardt's studies of Rubens, the *Erinnerungen*, were published in 1893, and they are strikingly original from many points of view. Even if we take the great popularity of Rubens and his school for granted, he does not belong to the closed realm of classical beauty. His creations, the bodies he paints, the situations he evokes on immense canvases, all elude strict Winkelmannian aesthetics. The impulse to study Rubens derived from his extremely complex and fascinating life which made him a significant witness to numerous important events in European history. Apart from this biographically-based interest, there are the intrinsic suggestions that stem from 'Gestalten' at variance with the dominant conceptions of beauty, classical order, consonance and poise. Rubens was a sort of Trojan horse left 'inadvertently' in the well-defended citadel of classical aesthetics. The complex inter-relationships between the *objet d'art* and its beholder were not explored by classical aesthetics in substantially innovative ways. Instead, much use was made of a well-tested system of preconceived analytical paths which culminated in the aesthetics of Winkelmann, who produced a complete system of evaluation and a historical frame of reference. As a consequence, the apperception of an object and its internal evocation was straightforward, because the dramatic dynamics of the perceptual processes, the evocations of memory, the cognitive structures involved in these mental procedures were largely pre-conceived and pre-determined by the formal reasoning constructed around objects themselves.

5. *DAMOISELLES D'AVIGNON* AND THE LIKE

The invasion of art history by strange new forms, unexpected patterns, previously unheard-of musical tones and timbres, and enigmatic architectural structures wrought a profound revolution in the interpretation of perceptual and cognitive processes. It very soon became clear that none of the mental categories moulded around Greek archaeological remains were able to cope with the new images emerging from paleolithic caves, or from the objects brought into Europe from Africa, Asia and Oceania. In particular, they were entirely useless for those who endeavoured to interpret children's drawings.

In a certain sense, it was more difficult to recognise the importance of aesthetic values and complex cultural messages with regard to the classical arts and civilisations in their less studied, and less understood, periods. While Burckhardt had indicated how to escape the classical canons by concentrating on certain neglected 'minor arts' or the flamboyant, but unruly, art of Rubens in 1898, Wickoff, with his *Wiener Genesis* of 1895, explored a province of late antiquity which was clearly 'taboo' to the academic tradition and established significant links between the illuminated Wiener edition of the *Genesis* and the sculpture and rules of representation of the Constantine, and later periods. Wickoff, again, proposed a general theory of the Eastern origin of the arts in the 'dark periods.' This revolutionary account of the representational strategies of the newly discovered and excavated civilisations – Ur, Early Mesopotamia, Egypt, Maya, the peoples of the Pacific, American Indians, Inuits – imported an immense quantity of images which were generally quite incomprehensible to eyes accustomed to Roman-Greek forms. The consequence was a profound re-thinking of apperception, inner representation and memory.

One of the most notable of these cultural adjustments to the newly discovered forms was the almost traumatic experience of Carterwood, the illustrator of Maja ruins and *stelae* in Honduras, Nicaragua and Belize. A highly professional and accomplished illustrator, Carterwood was appointed to draw and engrave a series of plates for the future publications of the English archaeological mission to those countries. His first weeks of work were utterly frustrating and brought him to the verge of resigning his commission. He was completely unable 'to see' the stèle of Copan and consequently unable to draw it. But after two months had passed, he suddenly found he was able to draw again with his usual precision and highly detailed technique, producing the famous plates of Mayan sculptures and architecture that we know so well. The experience of a biologist who looks into a microscope for the first time is rather similar. For long hours, and sometimes weeks, he or she finds it impossible to see anything. Every object in the field appears totally illogical, incomprehensible, and inseparable from the background. But then everything falls in place and even very small details are easily extracted from the previous visual confusion.

It is of interest to compare Carterwood's dramatic, and painful learning process with the ease and full control of pictorial devices displayed by another extremely able illustrator, David Roberts. His carnet of drawings from the Near East, in particular from Petra and Egypt, shows full control and total understanding of the architectures, sculptures and landscapes represented. Roberts, in fact, still sees, and therefore represents, objects through a classically focused eye, and he is able to produce delightful and faithful images because he can rely on a sound academic training which enables him to bring the represented object into his classically structured representational field.

6. KUNSTWOLLEN

A critical mass of new images entered European visual imagery in consequence of the discoveries of the painted caves in Spain and France, and they posed immediate problems: were they really so ancient? Could they be defined works of arts? What did they mean? Did they have religious, social, psychological and symbolic significance? In this case, interpreters could not count on language, history, written documents, architectural remains, and reading. The apperception and inner understanding and representations of these images remained confined to a-critical astonishment or to confused discourse on the 'eternity of art,' the 'necessity of beauty,' the instinctive drive to produce well-balanced, harmonious 'Gestalten.' In effect, study of this immense quantity of images was characterised for many decades by the continuous repetition of standard observations which produced nothing of note until recently, when apperception of these images was connected to human cognitive processes, anatomy and physiology, mainly finger-hand-eye co-ordination and the study of generative grammars.

The main problem in understanding the paleolithic system of representation was the absence of the historical trend of events – the sequential series of facts which, for Western culture, plays an essential role in achieving understanding of any cultural manifestation. This absence of a historical orientation was by-passed in study of other cultures and representational codes and helped in the inner representation of their artefacts. In contrast to the almost unsolvable cases of the Paleolithic and Neolithic periods, those of the Egyptians, of Late Antiquity, and of the complex interrelationships among different figural code systems in the Mediterranean, offered numerous occasions for theoretical analysis, mainly because a historical design was available, even if it was fragmented and discontinuous. In 1893, Jakob Burckhardt left his teaching position in Basel and went into retirement, shortly before his dramatic death during an asthma attack. In the same year Alois Riegl published his *Stilfragen,* in which he followed the elusive 'palmetto theme' throughout the Mediterranean and utilised it as a sensible probe for his celebrated idea of '*Kunstwollen.*': complex word suggesting a strong willigness at creating manufactured objects according to the laws of order and beauty. Riegl introduced this concept in order to counter the followers of Semper, who had departed radically from the ideas of their movement's founder and his extremely prudent conclusions concerning the necessary path from textile technologies to the decorative arts. Whereas the Semperians saw a quasi-mechanical and necessary sequence of events from the primeval body ornaments and first geometries imposed by the loom technique of weaving to the ornamental strategies of vases, architectures and sculptures, Riegl discerned a new force at work. His basic idea was that it is impossible to reconstruct a continuous series of events leading from the textile industries to the ornamental arts, and that men had always worked on surfaces at a level which could not be defined simply as a consequence of mechanical forces and the imitation of manual actions. Riegl expresses this opinion repeatedly in *Stilfragen*, and also in *Spätromisches Kunstindustrie*, but he states it most clearly in the final lines of the chapter devoted to the Proto-Geometric Style :

> [...] there is something in humans which makes us rejoice over formal beauty [...] This something freely and independently determined the geometric combinations of lines without the help of any possible connecting step.[1]

This 'something' is a real, a pure 'will to produce art,' an impulse which is entirely independent of mechanical necessities, technical solutions, the local utilisation of specific materials.

Riegl also observes that 'primeval societies' cannot elaborate in the third dimension; they cannot achieve the abstraction necessary to recognise objects and persons in images (photographic images), and they cannot go much further than very simple symmetrical bi-dimensional patterns. The progressive generation of ever more complex and intricate forms originates from a complex linguistic system which is the true engine of the evolution in decorative shapes and forms. Again, mathematics plays a leading role in these developments. One notes that the very lengthy and detailed descriptions of decorative patterns in Riegl's book, although apparently similar to Burchkardt's outburst of stylistic ability, do not belong to the same province of thought. Riegl's analysis of small details has a very specific goal which closely resembles the philological analysis of a text, where every item of written matter is important. In Burchkard, writing was viewed as a device to cope with the visual world. An isolated reference to children's drawings occurs when Riegl, during his apologia for a mathematically necessary framework, points out that 'an unruly infant's scribble lacks any artistic interest.'

A quite different approach to the analysis of children's drawings is set out in another seminal book on representation: Heinrich Schafer's *Von Ägyptischer Kunst*, published in 1919 and quite recently analysed with great care and detail by John Baines in his paper "Theories and universals of representation." Since its publication, Schafer's book had occupied a leading place in Egyptology because of its extremely original treatment of the representational strategies adopted by ancient Egyptian artists. One characteristic of the book is its frequent quotation and reproduction of children's drawings – often Egyptian – in order to furnish examples of the graphical rendering of objects, body parts, furniture and architectural details. A number of items of information from Schafer's book are of relevance to this topic of this present study.

Schafer intelligently points out the simple but extremely important fact that practically all the drawings have been executed from memory and not using a model. This implies a system of presentations (*Vorstellungen)*, by which complex term is meant both the internal image and the conceptual construction that gives rise to it. According to Schafer, the Egyptians preferred a form of 'frontal representation' (*Geradvorstellung*) which underpins their peculiar and strictly formal representational system. These internal representations have the structure of a primitive form, in the sense that they are apparently independent of direct visual perception, given that they do not follow the laws of perspective and are object-centred more than observer-centred. As regards children's drawings, Schafer cogently observes that children have no problems at all when working on the depiction of inner *Vorstellungen*. The difficulties arise when they seek to give a representational or narrative content to their drawings: in short, when they want to draw what they see, and how they see it. Schafer suggested a sequence of stages in the general development of drawing abilities which can be summarised as follows:

1) Schema. The stage in which the object is depicted according to what will later be defined as canonical. The representation has more to do with 'primitives' than with the actual act of vision.
2) Realism. At this stage the child deliberately introduces elements into his or her work which depend heavily on visual experience and visual knowledge.
3) Obliqueness and foreshortening. This critical step involves conscious use of the oblique line, so that it no longer graphically represents an oblique object but instead a part of an object which 'appears' to be oblique. This is the primitive step towards the axonometric system of representation. It will be discussed later in the case of a particularly gifted child.
4) Perspective proper. The fundamental problem in perspective is elimination, during graphical elaboration, of the parallelism between parts of an object that are parallel to each other.

From this short section on '*Kunstwollen*' it is easy to note how Alois Riegl dissected the use of specific idiosyncratic elements which convey the will to represent and produce, through their singularity, the very personality and significance of a work of art. In short, artistic activities or, more prudently, representation processes are specific, individual, selective and unique. This is evidenced by historically oriented cultural processes in which the characterisation of a structure, its individuation, and its definition follow a distinct temporal sequence. A case in point is Bernhard Schweitzer's study of the development of the Greek portrait, in which he examines the parthenonic *methopae* and the psychological studies of the faces of the centhauri. Schweitzer notes how the centhauri display an intense inner feeling of violence, brutality and pain, but their adversaries show no facial sign whatever of distress, rage or terror. The centhauri are quite different from men in that their two natures, the human and the ferine, fight against each other, and the inner tensions surface in their faces. According to Schweitzer, the centhauri of the parthenonic *methopae* are the forerunners of the Greek portrait with their first precise psychological depiction of character, will and self-consciousness. This extremely seminal view on the birth of the Greek portraiture can be found in Bernhard Schweitzer's "Studien zur Entstehung des Portrats bei den Griechen."

I conclude this part of the study with a curious detail from Irenaus Eibl-Eibesfeldt's classic study, *Die Biologie des menschlichen Verhaltens: Grundiß of Humanaethologie*. In Chapter Nine, which deals with the contribution of aethology to aesthetics, Eibl-Eibesfeldt presents two drawings executed by the same little girl and both representing her mother. The first is a classic 'children's drawing': the mother's face is quite conventional, unpersonalised and with few or no facial features. The second portrait contains a large number of details, wrinkles, the nares of the nose, eye rings, eyebrows and a peculiar hair style. This second portrait was executed by the same girl, with the same subject, and at the same time: the only difference between the two was the suggestion by a teacher that the girl should draw the mother's face after a close inspection of her facial features. This mild verbal stimulus suddenly changed the whole representational procedure and the two graphical works are strikingly different. The intense characterisation of the mother's face was brought about by a distinct request and it related, ultimately, to history – in particular with the personal history of that child.

7. Children's drawings, at last. Children lost in History

One of the many events that modulated the concept of representation and its extension to other and uncommon spheres was the discovery of paleolithic cultures. These radically challenged the interpretative abilities and criticisms of art historians, who were confronted by an almost unsolvable dilemma: 'How could it be that humans completely unable to write, or to communicate with each other in a well-defined language, were able to draw such complex scenes in such a splendid manner? The most interesting paleolithic discoveries were promptly published in the *Reliquiae Aquitanicae*, and the issues of the year 1893 may be remembered as quite significant for the importance and quantity of the published images concerning with the 'cave art works.' Almost at the same time, in 1886, a curious book by Corrado Ricci, *L'arte dei bambini*, was published in Italy. Ricci's book was the first entirely devoted to children's drawings, and it contains observations, images and analyses still worth studying today. Significantly, Ricci was an archaeologist specialised in the Byzantine period: his home was in Ravenna. Shortly after the appearance of Ricci's work, Stanford University published its first book on education, edited by E. Barnes. Volume 1 contained an interesting paper by A. B. Clark which addressed a very specific issue, "The child's attitude towards perspective problems." This was 1897, the same year in which Jakob Burckhardt died in Basel.

The issue of perspective understanding and graphical 'rendering' is today a crucial one. But at the end of the last century, studies of perspective were too crude to account for all the subtleties of culture and representation. Only after 1940 did the full complexity of the ideologies behind the graphical procedures adopted in perspective begin to be appreciated. This relative lack of comprehension reflected back on the ideas about perspective related to children's drawings. But the fact that children's drawings were 'a-perspective' was immediately taken to be indicative of some 'misunderstanding,' or the 'inability' of the child to produce a 'correct' representation. At the same time, the absence of a graphical code in perspective drawing was correctly understood to be a cognitive rather than a graphical problem. Natural vision of tri-dimensional objects is, now, considered as the physiological counterpart of the inverted optics. This relatively new branch of physics establishes formal relationships between the 'inherent' properties of an object: size, state of motion, roughness of its surface, etc., and its 'image.' Whereas the generic word 'image' is a bi-dimensional distribution of luminance upon a surface. In natural vision parallel lines may not appear parallel; objects with the same dimensions vary enormously in size; angles of any value constantly change according to the vantage point; symmetrical relationships produce strongly asymmetrical percepts; and so on. On the other hand, 'knowledge' of objects acquired from touch and physical construction produces a completely different set of information, almost always at variance with visual appearance. The graphical action by which a child (and not only a child) seeks to represent three-dimensional objects must strike a compromise between cognitive and visual data. Educational processes may provide a set of 'solutions' which have been historically filtered but in substance not produced by the child him/herself, who can only accept or reject them. Perspective rendering is such a procedure, and its historical development is one of the most complex in pattern.

We are confronted with a somewhat dramatic situation: children's drawings have no history. We have no idea how Egyptian, Greek, Roman, Renaissance, Baroque children, the French children of the Revolution times, American children contemporary to George Washington, actually drew. Were their drawings all the same and therefore historically independent, or did their graphical creations reflect, in some way, the *Zeitgeist*? In short, were or were not their representations dependent on a historical process?

This complete absence of information has been partially, but significantly, overcome by a recent study: Laurie Rubin's "First Draft Artistry: Children's Drawings in the Sixteenth and Seventeenth Centuries. Children of Mercury," which presents two different sets of literary and pictorial documents. The literary ones yield very remote information about children's drawings – or at least their appreciation by adults starting from 1350. Particularly worth quoting from this study are two quite different opinions on these manufacts: the first by Boccaccio, and the second by Vasari. In the sixth day of the *Decameron*, Boccaccio describes a popular family in his Florence of 1350: the Baronci. The members of this family were famous for their almost monstrous physical gracelessness. At the same time, they claimed to be the most ancient family in Florence; their origin, according to them, being so remote that no memory of it remained. Boccaccio cruelly writes of their faces that they "were just like the faces that little children make when they first learn to draw." Their physical features were asymmetrical, irregular, devoid of any beauty. Boccaccio's phrase is interesting for two reasons: first, a man's face as represented by a child is irregular, asymmetrical and ugly; second, a child "learns how to draw."

Both aspects are worth considering because they are the two sides to the same coin: to do it is necessary to learn, and a child needs to be instructed by adults how to 'do drawing.' According to Boccaccio, a child is not naturally equipped with the knowledge necessary to produce 'decent representations.' This must be provided by adults. In Boccaccio's Florence there were numerous children employed in the *botteghe*, where they were instructed on how to mill pigment, to prepare a canvas, to sharpen the sanguigna, the lead styluses, and the charcoal. Only after this quite strenuous period of apprenticeship were they allowed to trace simple patterns or to fill unimportant sections of the painting with colours. Two centuries later, Caravaggio had to work for about twenty years before to be presented by his teacher as 'pittore.'

A quite different view on the spontaneous ability to draw 'from Nature' is advanced by Vasari, who writes at the beginning of his *Vite*:

> [...] in our own times it has been seen (as I trust I shall be able to demonstrate a little later with many examples) that simple children roughly reared in the woods, with their only model being the beautiful pictures and sculptures of Nature, and by the vivacity of their wit, have begun by themselves to make designs.

Here we are confronted by the ideology of the spontaneous gift of Nature, which is bestowed upon humans before and in the absence of education, school, or apprenticeships. In short, in the absence of history. Vasari evidently follows a philosophical trend curiously at variance with the classical tradition. It should be borne in mind that Vasari's intellectual setting was no longer that of the Renaissance, where the *antichi* were the only admissible guide to beauty, order and symmetry. *Manierismo*, of which Vasari was a leading proponent and interpreter, was the forerunner of *Verismo* and *Naturalismo*. The Zuccaris, the Carraccis, and Caravaggio would have moved in this new province of representation. Vasari's words have a curious 'Rousseau-esque' tinge to them, with their implication that a child immersed in Nature is more than equipped to represent it in a spontaneous and 'witty' way.

The historical compromise between Boccaccio's and Vasari's opinions is provided by the ancient tale of the child who was engaged in drawing 'from Nature,' but certainly not according to the laws of representation, and was observed by a painter. The great man noted the spontaneous but unruly talent of the child and took him into his *bottega*. In a few years the young man had surpassed his teacher. This very ancient tale – which Fritz Saxl traced to an Asiatic, Sassanid tradition of the Middle Ages – took material form in the life of Giotto. As Giotto was drawing his sheep from nature, Cimabue came by, looked at his marvellous drawings and took him to Florence.

In a sense, the three tales, the Cimabue-Giotto story, that of the Baronci family, and Vasari's 'witty' child, represent three accounts of children's drawings and representations. These three lines of thought can be summarised as follows:

1. The child is completely unable to represent things as they are, or as they appear. He needs guidance from the outset, models must be proposed to him, strong discipline and hard work are necessary if the child has to learn how to draw.
2. The child is intrinsically gifted and learns from himself and from Nature. Scholastic control, a sequence of educational steps, teachers and models may destroy his natural patrimony of beauty, representational strategies, skilful control of colours, forms and compositional laws.
3. The child is strongly and inherently driven by a will to represent. This intense and uncontrollable urge to draw, to hit, to carve, to 'destroy,' in a certain sense, is a necessary perquisite. Without it teaching is useless. But the child must not be left alone: he must be presented at appropriate stages in his development, with representational 'tricks,' time-proven solutions, a code of meaningful signs.

After this cursory analysis of the development dynamics of drawing as it was seen and understood in the past, we must return to graphical documents. These documents are few in number, and I begin with an indirect one: Giovanni Francesco Caroto's painting, *Boy with a Drawing*. Irving Lavin, the scholar of Bernini's life and works, in his study "Bernini as a Social satirist"[1] describes a very interesting painting. The author, Giovanni Francesco Caroto, depicts a young man, very jovial and smiling, who is pointing to a drawing of a man presumably drawn by a child. The subject is represented according the canonical representational laws used by a child of about 7-8 years old: frontal view, rigid positioning of the limbs, quite detailed dress, absence of dynamic connotations, as well as other 'classic features.' The drawing is so perfect that it seems a quite competent copy or even a formal example of 'what a child's drawing should look like.' A brief biographical note on Caroto may help to delineate his career and character. Giovanni Francesco Caroto was born in Verona in around 1480 and died in Verona in 1555. At the beginning of his career he entered the studio of Liberale and then worked steadily for Mantegna in Mantua. His peculiar ability to copy the style of Mantegna convinced the great painter to sell some of Caroto's works as his originals. We have an autograph of his dated MDXXXI, and in some circumstances he signed his works with the curious signature: F. K. ROTO. The script of the signature was intentionally infantile.[2]

Evidently this drawing was not executed by a child but was probably copied very faithfully by the adult and 'witty' painter. The man depicted is not distinguishable from the drawings collected at the beginning of this century in London in a low middle class milieu, or from those executed in Los Angeles in 1920 by very poor Italian children of the first immigrant wave. In particular, Caroto's man is almost identical to the men represented by the children who drew for Corrado Ricci's book of 1886. We may now move to a very precious graphical historical document. This is the important autographed and certified series of drawings executed during a period of about two years by the future king of France, Louis XIII, when he was aged between 3 and 5 years old. The drawings were carefully collected, annotated and saved from August 1605 to September 1607 by his teacher, Docteur Jean Héoard (the drawings are kept in the Bibliothèque Nationale in Paris).

The earliest drawing is dated 'Août 1605' and with great precision of detail represents 'two birds pecking one another.' The two birds are embedded in a complex system of scribbles. When looking at this drawing, one is immediately reminded of the 'Pelican Kissing a Seal' collected by Ellen Winner and previously commented on. In the modern drawing we are able to 'recognise' each animal because we can rely on a television recording of the drawing, which enables us to 'see' an animal in an altogether 'meaningless' mess of lines. The problem of sorting out the single graphical creatures is that we lack a very significant part of the document: the Dauphin's voice. At four years of age, in August 1606, the future king made a portrait of a child of his own age, 'Mr. Mistaudin,' son to Mr. de Liauncourt, the Dauphin's teacher of equitation. The represented boy is very precisely drawn, but the portrait is not particularly different from the numerous other drawings that we see every day. A very rich page, dated 'Septembre 1607,' again shows birds, as well as writing exercises in letters and numbers (the Dauphin was then five years old) and, more interestingly, two men riding horses. There is a drawing, collected by Ricci and included in his book, of a man riding what is perhaps a donkey and who cannot be easily distinguished from this 'royal portrait' of about three hundred years before. In particular, what makes the two drawings almost identical is the treatment of the graphic interaction between the legs of the man and the body of the horse. The horse is made 'transparent' and we can 'see through it.'

This close similarity between two drawings by children far apart in time and in social position suggests that children's drawings had not significantly 'evolved' over time. And, if this is true, we need not be particularly bitter about the loss of documentation. A short comment is required on the so-called 'X-ray drawing' technique as shown by the two drawings, which reveal a significant identity of graphical procedure. Norman H. Freeman, in his *Strategies of Representation in Young Children: Analysis of Spatial Skills and Drawing Processes*, introduced the useful concept of the 'hidden line elimination technique' upon which we will return at the end of this study. Freeman's analysis concentrates on the graphical procedure which eliminates the hidden line due to the visual interposition of an opaque obstacle between the observer and the object to be depicted. This graphical operator is not present in the child before 7–8 years of age. In general, the child tends to represent the totality of the hidden object despite its optical invisibility. Schafer, in his study of Egyptian representational strategies, had argued in the same way. In very many instances, hidden objects like a man under a blanket, a piece of bread in a vase, the head of a priest inside a ritual mask, and so on, are represented as if the container was completely transparent. Numerous Neolithic drawings of animals, as well as others produced by different civilisations like the North Pacific Indians, the Sinai population, the French classical cave painters, the Aborigines in the Hyers Rock classical settlements, show with extreme clarity the interior of animals: kangaroos, mammoths, deer, salmon, etc. One discerns here a close correspondence between children's graphical techniques, which develop in a very short span of time, and the immensely longer millennia of the duration of whole civilisations.

8. History lost in Children: the Nadia's case

In September 1973 a young girl of about four years was brought to the Development Research Unit of the University of Nottingham, England. The child, Nadia, showed all the classic signs of autism, but she also displayed another, extremely interesting behavioural trait. Nadia was able to draw extremely complex drawings with immense 'bravura.' Among the most notable of them is, for example, a rider on a horse represented with perfect fore-shortening, and this at the age of four! The classic works on Nadia have been written by the psychologist who had her as a patient for almost thirty years and who described her case history. Lorna Selfe's books are a fundamental source of information about this world-famous case.[3] But, other scholars have studied the case with subtle and intelligent insight, for instance Pariser.[4]

The basic problem, independently of the tragic human circumstance, can be summed up in a simple question. How could it happen that a child, completely bereft of a language system and officially defined as almost totally mentally retarded, could draw highly complex scenes, objects and human bodies with such precision and perfection, at an age when a normal child is unable to produce anything that comes even remotely close to these representations? Many cases of children affected with autistic syndromes at different levels of gravity and simultaneously 'gifted' with unusual graphical competence have shown a bewildering ability in representing large objects, city streets, engines, trains, architectural details in perspective. If we have to take in consideration the resistance encountered by the teachers all over the world as shown by their pupils when they come to terms with perspective teaching this strange ability of autistic children at representing objets in perspective results even more puzzling. In the next discussion on perspective graphical procedures this data may be remembered.

The particular and dramatic case of Nadia has been debated for almost forty year and its complexity defies any superficial analysis. A new corollary question, among the many unanswered ones, rises now: "Which kind of law controls the very act of drawing?"

The enigmatic case of the autistic child, graphically well gifted and extremely 'competent,' confronts us with an almost unsolvable problem. Graphical skill suggests a form of conscious control, a perfect match between the visual input and the graphical output, a system of accurately-followed representational strategies utilised in the most appropriate way. At the same time, we must acknowledge the autist's almost total linguistic incompetence, his inability to follow symbolic and social play and interactions, a total or almost complete absence of emotional response, as well as other dramatic social and psychological impairments.

We have on the one hand Nadia's mysterious and splendid horses in full gallop, frozen with perfect foreshortening, and executed with superhuman fluency and precision at the age of four. On the other, we have the rigid, puppet-like, extremely schematic and childish horses that Nadia drew for her doctor when she was thirty years old. Terrible drawings which show a terrifying inability and a graphical clumsiness absolutely not present in the normal child. We may now leave this study with these two different horses in mind, hoping that they will, one day, transform themselves into Trojan horses and lead us inside the human mind exactly as it happened with the Ebbinghaus 'experimental' studies on memory dynamics which turned out to play the key role in cognitive studies.

9. Setting the Stage for Representation: Inner and Outer Spaces

It has been noted, in the paragraph devoted to the scant historical data concerning with the children's drawings, how the critical studies upon the perspective had began only around the 1940 to grasp the full meaning and the very complex developments which have characterised its devious historical course. A quite seminal study upon perspective and its meaning was the immense bibliographical collection of studies on perspective published by Luigi Vagnetti under the typically erudite title, which may deserve full quotation: "De naturali et artificiali perspectiva. Bibliografia ragionata delle fonti teoriche e delle ricerche di storia della prospettiva; contributo alla formazione della conoscenza di un'idea razionale, nei suoi sviluppi da Euclide a Gaspard Monge, in *Studi e documenti di Architettura.* The complete title, after the Latin quotation "On the artificial and natural perspective" continues in archaic Italian as follows "A reasoned bibliography upon the theoretical sources and the inquiries on the history of Perspective: a contribution to the shaping of the knowledge of a rational idea along its development from Euclid to Monge". In the title the expression 'idea razionale,' opposes itself, with obvious intentions, to the "symbolic form" of Panofskian origin. A subsequent series of fundamental books and definitive studies on perspective had shown how this graphical procedure, or, in more modern terms 'representational algorithm,' had its deepest roots in cognitive processes, perceptual 'prejudices,' linguistic landscapes and social conventions more than in the bare graphical procedure *per se.*

We may take this representational problem, the perspective rendering of space, as the *via regia* to the understanding of the pictorial codes and procedures adopted by the child. The graphical algorithm giving rise to the 'realistic' representation of the three-dimensional world in a bi-dimensional space had reached its perfection at the Renaissance and the procedure was well known to all the artists who possessed the skills of the days. The basic assumption and the primary act was to make all the lines, edges, borders parallel to the direction of the gaze converge on a single point: the vanishing point. Additional detailing and elementary geometrical procedures were used to work on the transverse lines and to obtain their convincing perspective accumulation. The result of such simple operations was rewarding indeed. On the wall, the canvas, the sheet of paper a new world sprang to life. Distant cities were rationally connected to the nearby hill, deep and precipitous porticoes ran suggestively towards the end of the roads, trees dispersed in the highly geometrised countryside set their heights correctly and obediently to the underlying, albeit invisible, checkerboard embedded in the soil, rivers wound sinuously from one's feet to the distant horizon.

However, they were some major drawbacks and they did not escape the acute eyes of Leonardo. It was far too obvious that the whole operation was highly artificial, too many absurd assumptions had to be taken, too many logic operations had to be contradicted. In a building, the parallel edges were not parallel to each other, symmetry was shattered, the angles were never the original ones, the ratios between different parts were never respected. Leon Battista Alberti invited the architects not to use perspective for their drawings, Leonardo never did so and, Masaccio applied graphical perspective at its maximum perfection in his Cappella in Santa Maria Novella in Florence although he applied perspective to the architecture and not to the human bodies, Paolo Uccello introduced too many 'mistakes' in his *discorso*, Michelangelo never paid attention to perspective. The main reason for these difficulties lay in an unsolvable set of contradiction between the 'things' we know about objects and the things we see of them. These intellectual difficulties made the masters who gave perspective its laws uncertain and introduced them particular solutions, devices and elegant but deceiving procedures.

Thus we have a child unaware of the cultural history of space struggling to draw a cube or a box. She knows that her box has right angles, the six faces are identical , all the edges meet with the same spatial order and distribution. Why should she represent only three faces of her box when she knows so well that they are six of them? Why should she draw six or seven different angles between the short segments when there is only one angle? Why should she draw all the edges differently, when she knows for certain that they are all equal? Where are the obliques in her cube? These questions and other ones, irksome and perhaps tormenting, lie behind all the millions boxes drawn by millions of children (and not only children but also thoughtful adults) and determine their strange aspects. When Luquet summed up his ideas on children's drawings at the moment of their "intellectual realism" he wrote that "the child draws what he knows but not what he sees," thereby neglecting the fact that the majority of the "things we know" come to us through the visual channels.[5] But the most important question overlooked by Luquet has the peculiar quality of the 'representation.' In his analysis Luquet considers the children's drawings to be 'representations' in the pictorial sense of the word because he thinks what he is in presence of conscious representational acts not very dissimilar from the classical pictorial tradition. But recent results from elaborate experiments on the spatial rendering in children's drawings have shown a quite complex series of problems, and their related solutions, which have little to do with the graphical rendering of something we have seen. Some examples follow all of which converging on a broader concept of representation where the mere graphical 'depiction' of an object does not play the major role.

10. The Hidden Lines and Their Mysteries

It was the merit, among many others, of N. H. Freeman to have brought attention to a peculiar characteristic of children's drawings: the occluding line. Consider two opaque objects located in such a way as that they do not cover each other completely if looked at along a given direction of gaze. This spatial arrangement yields to an uncompleted view of the distant object which should lead to its uncompleted representation if our graphical rendering seeks to be faithful to the 'seen.' Freeman's innovative study which plays a significant role in his fundamental work, deals with the graphical depiction, or its absence, of the hidden line in children's drawings.[6] The hidden line is the invisible line which should be absent from a realistic drawing. But in general children manage to evade this difficulty by tracing the distant object totally completed but located aside of the near one. During the act of the graphical representation we see the consequence of the act of the mental representation which, probably, preceded it and determined its graphical structure. The official explanation of the phenomenon of "Hidden Line Elimination" as it has been christened by N. Freeman is that the child does not 'take the responsibility' to draw an object uncompleted, and decides to represent it in its integrity modifying its spatial relationships with the partially occluding one located in front. Behind and 'under' this idea is again the Gestalt psychology which states the primary and superior levels of the *Gestalten* which have to be respected and considered as basic and unavoidable steps in inner representation.

However, more recent discoveries and observations on the Hidden Line have shown that the mental processes at the basis of its elimination may simply be a graphical procedure which does not involve any superior and highly cognitive process. These more recent results point in another direction, towards the influences of small details in the spatial locations of objects, their reciprocal orientation, their side or frontal presentation, their reciprocal position, the vertical arrangements, their relative dimensions and similar 'lesser' parameters. It has been found, for example, that a cup presented to a child with its handle on one side or with the handle hidden behind the cup is represented in different ways, implying that knowledge of the 'cup-Gestalt' in itself does not play any determinant role. Another interesting finding concerns the 'displaced representation' of an object presented to a child as contained in a transparent box. In this case there is no Gestalt to be protected because the included object is totally visible. A large amount of other data reveal a very complex interplay between direct visual perception, graphical competence and 'spatial prejudices.'[7]

11. LEAVING THE REALM OF REPRESENTATION

It seems difficult to decide, in the children's drawings interpretation, between a formalized, axiomatic, symbolic and highly cognitive approach and an empirical, intuitive, qualitative and occasional one. The former view pays very little attention to graphical devices, executive competencies, learning experiences, act of copying, environment and the social forces around the drawing child. The second approach seems partially unaware of the innate hardware and software of our brains, the formalized linguistic primary structures, the relatively homogeneous layout of our sensory systems and perceptual processes. A third, very intricate approach takes account, and indeed considers to be the primary driving force, the family life, the emotional state of the child, his or her symbolic experiences, sexual and social maturation, response to the different living conditions, the unconscious level of private life. According to this latter attitude a drawing is a forest of symbols, of metaphors, of heraldic and allegorical devices.

As soon as the children's drawings begin to have a real history, a significant temporal span upon which to cast the a critical eye and an interdisciplinary approach linking it with its 'time and space,' it appears necessary to conjure around them all the traditional cultural and disciplinary tools we use in the reading of a 'work of art.' Exactly as it appears now to us that a 'work of art' cannot be studied without taking into consideration biographic data, social landscapes, technical backgrounds, literary, philosophical and religious backgrounds we have to follow the same total strategy in our study on children's drawings. And we may discover the simple fact that also the drawing child engrossed in her study and depiction of her cubic box is following the same steps of Leonardo during his unfathomable drawings of eddies and whirlpools, of the sequential monumental plastic decisions of Michelangelo in his quest for the ultimate form or of the painful, yet deliriously happy, struggle of Vincent Van Gogh with light.

Ruggero Pierantoni
Consiglio Nazionale della Ricerca
Istituto di Cibernetica e Biofisica
Via dei Marini, 4
I-16145 Genova

NOTES

1 Quoted from Almgren 1971.
2 See also Benezit 1941.
3 Selfe 1977, 1983, and 1995.
4 In Pariser 1981.
5 Luquet 1927.
6 Feeeman 1980.
7 Ibidem. For these interesting problems which shed a new light on the general problem of representation in general, and on the children's drawing in particular, see Light and MacIntosh 1980; Light and Foot 1986; Teske, Walts, and Shenk 1992. The more complex interplay between drawing and linguistic and conceptual procedures have been studied in detail and a rich bibliography is available. The few following titles may be used as a primitive guide line into this rather recent development: Claire Golomb's 1996 definitive study; Willats 1977; Willats and Butterworth 1977.

REFERENCES

Almgren, A.: 1971, *Die umgekehrte Perspektive und die Fluchachsenperspektive*, Uppsala, Almquist & Wiksell.

Baines, J.: 1985, 'Theories and universal of representation: Heinrich Schafer and Egyptian art', *Art History*, **8** (**1**), 1-39.

Benezit, E.: 1941, *Dictionnaire des Peintres, Sculpteurs, Dessinnateurs et Graveurs*. France, Librairie Grund.

Clark, A. B. and Barnes, E. eds.: 1897, *The Child's Attitude towards Perspective Problems*, Vol. 1. Stanford, Stanford University Press.

Crary, J.: 1992, *Techniques of the Observer*. Cambridge, Mass., MIT Press.

Croce, B.: 1912a, *Scritti vari di erudizione e di critica in onore di Rodolfo Reiner*. Torino.

Croce, B.: 1912b, 'La teoria dell'arte come pura visibilità, Von Marèes, Fiedler, Hildebrand', in Croce 1912a, 259-270.

Freeman, N. H.: 1980, 'Strategies of Representation', in *Young Children: Analysis of Spatial Skills and Drawing Processes*. New York, Academic Press.

Golomb, C.: 1996, *The Child's Creation of a Pictorial World*. Los Angeles, University of California Press.

Lavin, I.: 1971, 'Bernini as a Social Satirist,' quoted in Almgren 1971.

Light, P. H. and Foot, T.: 1986, 'Partial Occlusion in Young Children's Drawings,' *Journal of Experimental Child Psychology*, **41**, 38–48.

Light, P. H. and MacIntosh E.: 1980, 'Depth Relationships in Young Children's Drawings,' *Journal of Experimental Child Psychology*, **30**, 79–87.

Luquet, G.: 1927, *Le dessin enfantin*. Paris, Alcan.

Pariser, D.: 1981, 'Nadia's Drawings: Theorizing about an Autistic Child's Phenomenal Ability', *Studies in Art Education*, **22** (**2**), 20-32.

Ricci, C.: 1886, *L'arte dei bambini*. Ravenna. Repr. 1996, Florence, Giunti-Barbera.

Rubin, L.: 1984, 'First Draft Artistry: Children's Drawings in the Sixteenth and Seventeenth Centuries. Children of Mercury,' *Cognition*, **14**, 183-194.

Schweitzer, B.: 1939, 'Studien zur Entstehung des Portrats bei den Griechen,' *Berichte der Philosophisch-historischen Klasse*, **91**, Heft 4, 1-168.

Selfe, L.: 1977, *Nadia. A Case of Extraordinary Drawing Ability in an Autistic Child.* New York, Academic Press.

Selfe, L.: 1983, *Normal and Anomalous Representational Drawing Ability in Children.* New York, Academic Press.

Selfe, L.: 1995, 'Nadia Reconsidered,' in Golomb, ed., 1996, 196–236.

Teske, J. A., Walts, C. R. and Shenk, C. C.: 1992, 'The Effects of Object Alignment on the Representation of Depth in Young Children's Drawings,' *Journal of Experimental Child Psychology*, **54**, 19–36.

Vagnetti, L.: 1979, 'De naturali et artificiali perspectiva. Bibliografia ragionata delle fonti teoriche e delle ricerche di storia della prospettiva; contributo alla formazione della conoscenza di un'idea razionale, nei suoi sviluppi da Euclide a Gaspard Monge,' *Studi e Documenti di Architettura*, **9–10**, 1-463.

Willats, J.: 1977, 'What do the Marks in the Pictures Stand for?,' *Quarterly Journal of Experimental Psychology*, **29**, 367–82.

Willats, J. and Butterworth, G. eds.: 1977a, *The Child's Representation of the World.* New York, Plenum Press.

Willats, J. and Butterworth, G.: 1977b, 'How children learn to represent three dimensional space in drawings', in Willats and Butterworth, eds., 1997a, 122-136.

Winner, E.: 1986, 'Where Pelicans Kiss Seals,' *Psychology Today*, (**6**), 25–35.

SYNTHESE LIBRARY

1. J. M. Bochénski, *A Precis of Mathematical Logic*. Translated from French and German by O. Bird. 1959 ISBN 90-277-0073-7
2. P. Guiraud, *Problèmes et méthodes de la statistique linguistique*. 1959 ISBN 90-277-0025-7
3. H. Freudenthal (ed.), *The Concept and the Role of the Model in Mathematics and Natural and Social Sciences*. 1961 ISBN 90-277-0017-6
4. E. W. Beth, *Formal Methods*. An Introduction to Symbolic Logic and to the Study of Effective Operations in Arithmetic and Logic. 1962 ISBN 90-277-0069-9
5. B. H. Kazemier and D. Vuysje (eds.), *Logic and Language*. Studies dedicated to Professor Rudolf Carnap on the Occasion of His 70th Birthday. 1962 ISBN 90-277-0019-2
6. M. W. Wartofsky (ed.), *Proceedings of the Boston Colloquium for the Philosophy of Science, 1961–1962*. [Boston Studies in the Philosophy of Science, Vol. I] 1963 ISBN 90-277-0021-4
7. A. A. Zinov'ev, *Philosophical Problems of Many-valued Logic*. A revised edition, edited and translated (from Russian) by G. Küng and D.D. Comey. 1963 ISBN 90-277-0091-5
8. G. Gurvitch, *The Spectrum of Social Time*. Translated from French and edited by M. Korenbaum and P. Bosserman. 1964 ISBN 90-277-0006-0
9. P. Lorenzen, *Formal Logic*. Translated from German by F.J. Crosson. 1965 ISBN 90-277-0080-X
10. R. S. Cohen and M. W. Wartofsky (eds.), *Proceedings of the Boston Colloquium for the Philosophy of Science, 1962–1964*. In Honor of Philipp Frank. [Boston Studies in the Philosophy of Science, Vol. II] 1965 ISBN 90-277-9004-0
11. E. W. Beth, *Mathematical Thought*. An Introduction to the Philosophy of Mathematics. 1965 ISBN 90-277-0070-2
12. E. W. Beth and J. Piaget, *Mathematical Epistemology and Psychology*. Translated from French by W. Mays. 1966 ISBN 90-277-0071-0
13. G. Küng, *Ontology and the Logistic Analysis of Language*. An Enquiry into the Contemporary Views on Universals. Revised ed., translated from German. 1967 ISBN 90-277-0028-1
14. R. S. Cohen and M. W. Wartofsky (eds.), *Proceedings of the Boston Colloquium for the Philosophy of Sciences, 1964–1966*. In Memory of Norwood Russell Hanson. [Boston Studies in the Philosophy of Science, Vol. III] 1967 ISBN 90-277-0013-3
15. C. D. Broad, *Induction, Probability, and Causation*. Selected Papers. 1968 ISBN 90-277-0012-5
16. G. Patzig, *Aristotle's Theory of the Syllogism*. A Logical-philosophical Study of *Book A* of the *Prior Analytics*. Translated from German by J. Barnes. 1968 ISBN 90-277-0030-3
17. N. Rescher, *Topics in Philosophical Logic*. 1968 ISBN 90-277-0084-2
18. R. S. Cohen and M. W. Wartofsky (eds.), *Proceedings of the Boston Colloquium for the Philosophy of Science, 1966–1968, Part I*. [Boston Studies in the Philosophy of Science, Vol. IV] 1969 ISBN 90-277-0014-1
19. R. S. Cohen and M. W. Wartofsky (eds.), *Proceedings of the Boston Colloquium for the Philosophy of Science, 1966–1968, Part II*. [Boston Studies in the Philosophy of Science, Vol. V] 1969 ISBN 90-277-0015-X
20. J. W. Davis, D. J. Hockney and W. K. Wilson (eds.), *Philosophical Logic*. 1969 ISBN 90-277-0075-3
21. D. Davidson and J. Hintikka (eds.), *Words and Objections*. Essays on the Work of W. V. Quine. 1969, rev. ed. 1975 ISBN 90-277-0074-5; Pb 90-277-0602-6
22. P. Suppes, *Studies in the Methodology and Foundations of Science. Selected Papers from 1951 to 1969*. 1969 ISBN 90-277-0020-6
23. J. Hintikka, *Models for Modalities*. Selected Essays. 1969 ISBN 90-277-0078-8; Pb 90-277-0598-4

24. N. Rescher *et al.* (eds.), *Essays in Honor of Carl G. Hempel.* A Tribute on the Occasion of His 65th Birthday. 1969 ISBN 90-277-0085-0
25. P. V. Tavanec (ed.), *Problems of the Logic of Scientific Knowledge.* Translated from Russian. 1970 ISBN 90-277-0087-7
26. M. Swain (ed.), *Induction, Acceptance, and Rational Belief.* 1970 ISBN 90-277-0086-9
27. R. S. Cohen and R. J. Seeger (eds.), *Ernst Mach: Physicist and Philosopher.* [Boston Studies in the Philosophy of Science, Vol. VI]. 1970 ISBN 90-277-0016-8
28. J. Hintikka and P. Suppes, *Information and Inference.* 1970 ISBN 90-277-0155-5
29. K. Lambert, *Philosophical Problems in Logic.* Some Recent Developments. 1970 ISBN 90-277-0079-6
30. R. A. Eberle, *Nominalistic Systems.* 1970 ISBN 90-277-0161-X
31. P. Weingartner and G. Zecha (eds.), *Induction, Physics, and Ethics.* 1970 ISBN 90-277-0158-X
32. E. W. Beth, *Aspects of Modern Logic.* Translated from Dutch. 1970 ISBN 90-277-0173-3
33. R. Hilpinen (ed.), *Deontic Logic.* Introductory and Systematic Readings. 1971 *See also* No. 152. ISBN Pb (1981 rev.) 90-277-1302-2
34. J.-L. Krivine, *Introduction to Axiomatic Set Theory.* Translated from French. 1971 ISBN 90-277-0169-5; Pb 90-277-0411-2
35. J. D. Sneed, *The Logical Structure of Mathematical Physics.* 2nd rev. ed., 1979 ISBN 90-277-1056-2; Pb 90-277-1059-7
36. C. R. Kordig, *The Justification of Scientific Change.* 1971 ISBN 90-277-0181-4; Pb 90-277-0475-9
37. M. Čapek, *Bergson and Modern Physics.* A Reinterpretation and Re-evaluation. [Boston Studies in the Philosophy of Science, Vol. VII] 1971 ISBN 90-277-0186-5
38. N. R. Hanson, *What I Do Not Believe, and Other Essays.* Ed. by S. Toulmin and H. Woolf. 1971 ISBN 90-277-0191-1
39. R. C. Buck and R. S. Cohen (eds.), *PSA 1970.* Proceedings of the Second Biennial Meeting of the Philosophy of Science Association, Boston, Fall 1970. In Memory of Rudolf Carnap. [Boston Studies in the Philosophy of Science, Vol. VIII] 1971 ISBN 90-277-0187-3; Pb 90-277-0309-4
40. D. Davidson and G. Harman (eds.), *Semantics of Natural Language.* 1972 ISBN 90-277-0304-3; Pb 90-277-0310-8
41. Y. Bar-Hillel (ed.), *Pragmatics of Natural Languages.* 1971 ISBN 90-277-0194-6; Pb 90-277-0599-2
42. S. Stenlund, *Combinators, γ Terms and Proof Theory.* 1972 ISBN 90-277-0305-1
43. M. Strauss, *Modern Physics and Its Philosophy.* Selected Paper in the Logic, History, and Philosophy of Science. 1972 ISBN 90-277-0230-6
44. M. Bunge, *Method, Model and Matter.* 1973 ISBN 90-277-0252-7
45. M. Bunge, *Philosophy of Physics.* 1973 ISBN 90-277-0253-5
46. A. A. Zinov'ev, *Foundations of the Logical Theory of Scientific Knowledge (Complex Logic).* Revised and enlarged English edition with an appendix by G. A. Smirnov, E. A. Sidorenka, A. M. Fedina and L. A. Bobrova. [Boston Studies in the Philosophy of Science, Vol. IX] 1973 ISBN 90-277-0193-8; Pb 90-277-0324-8
47. L. Tondl, *Scientific Procedures.* A Contribution concerning the Methodological Problems of Scientific Concepts and Scientific Explanation. Translated from Czech by D. Short. Edited by R.S. Cohen and M.W. Wartofsky. [Boston Studies in the Philosophy of Science, Vol. X] 1973 ISBN 90-277-0147-4; Pb 90-277-0323-X
48. N. R. Hanson, *Constellations and Conjectures.* 1973 ISBN 90-277-0192-X

49. K. J. J. Hintikka, J. M. E. Moravcsik and P. Suppes (eds.), *Approaches to Natural Language.* 1973 ISBN 90-277-0220-9; Pb 90-277-0233-0
50. M. Bunge (ed.), *Exact Philosophy.* Problems, Tools and Goals. 1973 ISBN 90-277-0251-9
51. R. J. Bogdan and I. Niiniluoto (eds.), *Logic, Language and Probability.* 1973 ISBN 90-277-0312-4
52. G. Pearce and P. Maynard (eds.), *Conceptual Change.* 1973 ISBN 90-277-0287-X; Pb 90-277-0339-6
53. I. Niiniluoto and R. Tuomela, *Theoretical Concepts and Hypothetico-inductive Inference.* 1973 ISBN 90-277-0343-4
54. R. Fraissé, *Course of Mathematical Logic* – Volume 1: *Relation and Logical Formula.* Translated from French. 1973 ISBN 90-277-0268-3; Pb 90-277-0403-1
(For *Volume 2* see under No. 69).
55. A. Grünbaum, *Philosophical Problems of Space and Time.* Edited by R.S. Cohen and M.W. Wartofsky. 2nd enlarged ed. [Boston Studies in the Philosophy of Science, Vol. XII] 1973 ISBN 90-277-0357-4; Pb 90-277-0358-2
56. P. Suppes (ed.), *Space, Time and Geometry.* 1973 ISBN 90-277-0386-8; Pb 90-277-0442-2
57. H. Kelsen, *Essays in Legal and Moral Philosophy.* Selected and introduced by O. Weinberger. Translated from German by P. Heath. 1973 ISBN 90-277-0388-4
58. R. J. Seeger and R. S. Cohen (eds.), *Philosophical Foundations of Science.* [Boston Studies in the Philosophy of Science, Vol. XI] 1974 ISBN 90-277-0390-6; Pb 90-277-0376-0
59. R. S. Cohen and M. W. Wartofsky (eds.), *Logical and Epistemological Studies in Contemporary Physics.* [Boston Studies in the Philosophy of Science, Vol. XIII] 1973 ISBN 90-277-0391-4; Pb 90-277-0377-9
60. R. S. Cohen and M. W. Wartofsky (eds.), *Methodological and Historical Essays in the Natural and Social Sciences. Proceedings of the Boston Colloquium for the Philosophy of Science, 1969–1972.* [Boston Studies in the Philosophy of Science, Vol. XIV] 1974 ISBN 90-277-0392-2; Pb 90-277-0378-7
61. R. S. Cohen, J. J. Stachel and M. W. Wartofsky (eds.), *For Dirk Struik. Scientific, Historical and Political Essays.* [Boston Studies in the Philosophy of Science, Vol. XV] 1974 ISBN 90-277-0393-0; Pb 90-277-0379-5
62. K. Ajdukiewicz, *Pragmatic Logic.* Translated from Polish by O. Wojtasiewicz. 1974 ISBN 90-277-0326-4
63. S. Stenlund (ed.), *Logical Theory and Semantic Analysis.* Essays dedicated to Stig Kanger on His 50th Birthday. 1974 ISBN 90-277-0438-4
64. K. F. Schaffner and R. S. Cohen (eds.), *PSA 1972. Proceedings of the Third Biennial Meeting of the Philosophy of Science Association.* [Boston Studies in the Philosophy of Science, Vol. XX] 1974 ISBN 90-277-0408-2; Pb 90-277-0409-0
65. H. E. Kyburg, Jr., *The Logical Foundations of Statistical Inference.* 1974 ISBN 90-277-0330-2; Pb 90-277-0430-9
66. M. Grene, *The Understanding of Nature.* Essays in the Philosophy of Biology. [Boston Studies in the Philosophy of Science, Vol. XXIII] 1974 ISBN 90-277-0462-7; Pb 90-277-0463-5
67. J. M. Broekman, *Structuralism: Moscow, Prague, Paris.* Translated from German. 1974 ISBN 90-277-0478-3
68. N. Geschwind, *Selected Papers on Language and the Brain.* [Boston Studies in the Philosophy of Science, Vol. XVI] 1974 ISBN 90-277-0262-4; Pb 90-277-0263-2
69. R. Fraissé, *Course of Mathematical Logic* – Volume 2: *Model Theory.* Translated from French. 1974 ISBN 90-277-0269-1; Pb 90-277-0510-0
(For *Volume 1* see under No. 54)

70. A. Grzegorczyk, *An Outline of Mathematical Logic*. Fundamental Results and Notions explained with all Details. Translated from Polish. 1974 ISBN 90-277-0359-0; Pb 90-277-0447-3
71. F. von Kutschera, *Philosophy of Language*. 1975 ISBN 90-277-0591-7
72. J. Manninen and R. Tuomela (eds.), *Essays on Explanation and Understanding*. Studies in the Foundations of Humanities and Social Sciences. 1976 ISBN 90-277-0592-5
73. J. Hintikka (ed.), *Rudolf Carnap, Logical Empiricist*. Materials and Perspectives. 1975 ISBN 90-277-0583-6
74. M. Čapek (ed.), *The Concepts of Space and Time*. Their Structure and Their Development. [Boston Studies in the Philosophy of Science, Vol. XXII] 1976 ISBN 90-277-0355-8; Pb 90-277-0375-2
75. J. Hintikka and U. Remes, *The Method of Analysis*. Its Geometrical Origin and Its General Significance. [Boston Studies in the Philosophy of Science, Vol. XXV] 1974 ISBN 90-277-0532-1; Pb 90-277-0543-7
76. J. E. Murdoch and E. D. Sylla (eds.), *The Cultural Context of Medieval Learning*. [Boston Studies in the Philosophy of Science, Vol. XXVI] 1975 ISBN 90-277-0560-7; Pb 90-277-0587-9
77. S. Amsterdamski, *Between Experience and Metaphysics*. Philosophical Problems of the Evolution of Science. [Boston Studies in the Philosophy of Science, Vol. XXXV] 1975 ISBN 90-277-0568-2; Pb 90-277-0580-1
78. P. Suppes (ed.), *Logic and Probability in Quantum Mechanics*. 1976 ISBN 90-277-0570-4; Pb 90-277-1200-X
79. H. von Helmholtz: *Epistemological Writings. The Paul Hertz / Moritz Schlick Centenary Edition of 1921 with Notes and Commentary by the Editors*. Newly translated from German by M. F. Lowe. Edited, with an Introduction and Bibliography, by R. S. Cohen and Y. Elkana. [Boston Studies in the Philosophy of Science, Vol. XXXVII] 1975 ISBN 90-277-0290-X; Pb 90-277-0582-8
80. J. Agassi, *Science in Flux*. [Boston Studies in the Philosophy of Science, Vol. XXVIII] 1975 ISBN 90-277-0584-4; Pb 90-277-0612-2
81. S. G. Harding (ed.), *Can Theories Be Refuted?* Essays on the Duhem-Quine Thesis. 1976 ISBN 90-277-0629-8; Pb 90-277-0630-1
82. S. Nowak, *Methodology of Sociological Research*. General Problems. 1977 ISBN 90-277-0486-4
83. J. Piaget, J.-B. Grize, A. Szeminśska and V. Bang, *Epistemology and Psychology of Functions*. Translated from French. 1977 ISBN 90-277-0804-5
84. M. Grene and E. Mendelsohn (eds.), *Topics in the Philosophy of Biology*. [Boston Studies in the Philosophy of Science, Vol. XXVII] 1976 ISBN 90-277-0595-X; Pb 90-277-0596-8
85. E. Fischbein, *The Intuitive Sources of Probabilistic Thinking in Children*. 1975 ISBN 90-277-0626-3; Pb 90-277-1190-9
86. E. W. Adams, *The Logic of Conditionals*. An Application of Probability to Deductive Logic. 1975 ISBN 90-277-0631-X
87. M. Przełęcki and R. Wójcicki (eds.), *Twenty-Five Years of Logical Methodology in Poland*. Translated from Polish. 1976 ISBN 90-277-0601-8
88. J. Topolski, *The Methodology of History*. Translated from Polish by O. Wojtasiewicz. 1976 ISBN 90-277-0550-X
89. A. Kasher (ed.), *Language in Focus: Foundations, Methods and Systems*. Essays dedicated to Yehoshua Bar-Hillel. [Boston Studies in the Philosophy of Science, Vol. XLIII] 1976 ISBN 90-277-0644-1; Pb 90-277-0645-X

90. J. Hintikka, *The Intentions of Intentionality and Other New Models for Modalities.* 1975 ISBN 90-277-0633-6; Pb 90-277-0634-4
91. W. Stegmüller, *Collected Papers on Epistemology, Philosophy of Science and History of Philosophy.* 2 Volumes. 1977 Set ISBN 90-277-0767-7
92. D. M. Gabbay, *Investigations in Modal and Tense Logics with Applications to Problems in Philosophy and Linguistics.* 1976 ISBN 90-277-0656-5
93. R. J. Bogdan, *Local Induction.* 1976 ISBN 90-277-0649-2
94. S. Nowak, *Understanding and Prediction.* Essays in the Methodology of Social and Behavioral Theories. 1976 ISBN 90-277-0558-5; Pb 90-277-1199-2
95. P. Mittelstaedt, *Philosophical Problems of Modern Physics.* [Boston Studies in the Philosophy of Science, Vol. XVIII] 1976 ISBN 90-277-0285-3; Pb 90-277-0506-2
96. G. Holton and W. A. Blanpied (eds.), *Science and Its Public: The Changing Relationship.* [Boston Studies in the Philosophy of Science, Vol. XXXIII] 1976 ISBN 90-277-0657-3; Pb 90-277-0658-1
97. M. Brand and D. Walton (eds.), *Action Theory.* 1976 ISBN 90-277-0671-9
98. P. Gochet, *Outline of a Nominalist Theory of Propositions.* An Essay in the Theory of Meaning and in the Philosophy of Logic. 1980 ISBN 90-277-1031-7
99. R. S. Cohen, P. K. Feyerabend, and M. W. Wartofsky (eds.), *Essays in Memory of Imre Lakatos.* [Boston Studies in the Philosophy of Science, Vol. XXXIX] 1976 ISBN 90-277-0654-9; Pb 90-277-0655-7
100. R. S. Cohen and J. J. Stachel (eds.), *Selected Papers of Léon Rosenfield.* [Boston Studies in the Philosophy of Science, Vol. XXI] 1979 ISBN 90-277-0651-4; Pb 90-277-0652-2
101. R. S. Cohen, C. A. Hooker, A. C. Michalos and J. W. van Evra (eds.), *PSA 1974. Proceedings of the 1974 Biennial Meeting of the Philosophy of Science Association.* [Boston Studies in the Philosophy of Science, Vol. XXXII] 1976 ISBN 90-277-0647-6; Pb 90-277-0648-4
102. Y. Fried and J. Agassi, *Paranoia.* A Study in Diagnosis. [Boston Studies in the Philosophy of Science, Vol. L] 1976 ISBN 90-277-0704-9; Pb 90-277-0705-7
103. M. Przełęcki, K. Szaniawski and R. Wójcicki (eds.), *Formal Methods in the Methodology of Empirical Sciences.* 1976 ISBN 90-277-0698-0
104. J. M. Vickers, *Belief and Probability.* 1976 ISBN 90-277-0744-8
105. K. H. Wolff, *Surrender and Catch.* Experience and Inquiry Today. [Boston Studies in the Philosophy of Science, Vol. LI] 1976 ISBN 90-277-0758-8; Pb 90-277-0765-0
106. K. Kosík, *Dialectics of the Concrete.* A Study on Problems of Man and World. [Boston Studies in the Philosophy of Science, Vol. LII] 1976 ISBN 90-277-0761-8; Pb 90-277-0764-2
107. N. Goodman, *The Structure of Appearance.* 3rd ed. with an Introduction by G. Hellman. [Boston Studies in the Philosophy of Science, Vol. LIII] 1977 ISBN 90-277-0773-1; Pb 90-277-0774-X
108. K. Ajdukiewicz, *The Scientific World-Perspective and Other Essays, 1931-1963.* Translated from Polish. Edited and with an Introduction by J. Giedymin. 1978 ISBN 90-277-0527-5
109. R. L. Causey, *Unity of Science.* 1977 ISBN 90-277-0779-0
110. R. E. Grandy, *Advanced Logic for Applications.* 1977 ISBN 90-277-0781-2
111. R. P. McArthur, *Tense Logic.* 1976 ISBN 90-277-0697-2
112. L. Lindahl, *Position and Change.* A Study in Law and Logic. Translated from Swedish by P. Needham. 1977 ISBN 90-277-0787-1
113. R. Tuomela, *Dispositions.* 1978 ISBN 90-277-0810-X
114. H. A. Simon, *Models of Discovery and Other Topics in the Methods of Science.* [Boston Studies in the Philosophy of Science, Vol. LIV] 1977 ISBN 90-277-0812-6; Pb 90-277-0858-4

115. R. D. Rosenkrantz, *Inference, Method and Decision.* Towards a Bayesian Philosophy of Science. 1977 ISBN 90-277-0817-7; Pb 90-277-0818-5
116. R. Tuomela, *Human Action and Its Explanation.* A Study on the Philosophical Foundations of Psychology. 1977 ISBN 90-277-0824-X
117. M. Lazerowitz, *The Language of Philosophy.* Freud and Wittgenstein. [Boston Studies in the Philosophy of Science, Vol. LV] 1977 ISBN 90-277-0826-6; Pb 90-277-0862-2
118. Not published 119. J. Pelc (ed.), *Semiotics in Poland, 1894–1969.* Translated from Polish. 1979 ISBN 90-277-0811-8
120. I. Pörn, *Action Theory and Social Science.* Some Formal Models. 1977 ISBN 90-277-0846-0
121. J. Margolis, *Persons and Mind.* The Prospects of Nonreductive Materialism. [Boston Studies in the Philosophy of Science, Vol. LVII] 1977 ISBN 90-277-0854-1; Pb 90-277-0863-0
122. J. Hintikka, I. Niiniluoto, and E. Saarinen (eds.), *Essays on Mathematical and Philosophical Logic.* 1979 ISBN 90-277-0879-7
123. T. A. F. Kuipers, *Studies in Inductive Probability and Rational Expectation.* 1978 ISBN 90-277-0882-7
124. E. Saarinen, R. Hilpinen, I. Niiniluoto and M. P. Hintikka (eds.), *Essays in Honour of Jaakko Hintikka on the Occasion of His 50th Birthday.* 1979 ISBN 90-277-0916-5
125. G. Radnitzky and G. Andersson (eds.), *Progress and Rationality in Science.* [Boston Studies in the Philosophy of Science, Vol. LVIII] 1978 ISBN 90-277-0921-1; Pb 90-277-0922-X
126. P. Mittelstaedt, *Quantum Logic.* 1978 ISBN 90-277-0925-4
127. K. A. Bowen, *Model Theory for Modal Logic.* Kripke Models for Modal Predicate Calculi. 1979 ISBN 90-277-0929-7
128. H. A. Bursen, *Dismantling the Memory Machine.* A Philosophical Investigation of Machine Theories of Memory. 1978 ISBN 90-277-0933-5
129. M. W. Wartofsky, *Models.* Representation and the Scientific Understanding. [Boston Studies in the Philosophy of Science, Vol. XLVIII] 1979 ISBN 90-277-0736-7; Pb 90-277-0947-5
130. D. Ihde, *Technics and Praxis.* A Philosophy of Technology. [Boston Studies in the Philosophy of Science, Vol. XXIV] 1979 ISBN 90-277-0953-X; Pb 90-277-0954-8
131. J. J. Wiatr (ed.), *Polish Essays in the Methodology of the Social Sciences.* [Boston Studies in the Philosophy of Science, Vol. XXIX] 1979 ISBN 90-277-0723-5; Pb 90-277-0956-4
132. W. C. Salmon (ed.), *Hans Reichenbach: Logical Empiricist.* 1979 ISBN 90-277-0958-0
133. P. Bieri, R.-P. Horstmann and L. Krüger (eds.), *Transcendental Arguments in Science.* Essays in Epistemology. 1979 ISBN 90-277-0963-7; Pb 90-277-0964-5
134. M. Marković and G. Petrović (eds.), *Praxis.* Yugoslav Essays in the Philosophy and Methodology of the Social Sciences. [Boston Studies in the Philosophy of Science, Vol. XXXVI] 1979 ISBN 90-277-0727-8; Pb 90-277-0968-8
135. R. Wójcicki, *Topics in the Formal Methodology of Empirical Sciences.* Translated from Polish. 1979 ISBN 90-277-1004-X
136. G. Radnitzky and G. Andersson (eds.), *The Structure and Development of Science.* [Boston Studies in the Philosophy of Science, Vol. LIX] 1979 ISBN 90-277-0994-7; Pb 90-277-0995-5
137. J. C. Webb, *Mechanism, Mentalism and Metamathematics.* An Essay on Finitism. 1980 ISBN 90-277-1046-5
138. D. F. Gustafson and B. L. Tapscott (eds.), *Body, Mind and Method.* Essays in Honor of Virgil C. Aldrich. 1979 ISBN 90-277-1013-9
139. L. Nowak, *The Structure of Idealization.* Towards a Systematic Interpretation of the Marxian Idea of Science. 1980 ISBN 90-277-1014-7

140. C. Perelman, *The New Rhetoric and the Humanities.* Essays on Rhetoric and Its Applications. Translated from French and German. With an Introduction by H. Zyskind. 1979 ISBN 90-277-1018-X; Pb 90-277-1019-8
141. W. Rabinowicz, *Universalizability.* A Study in Morals and Metaphysics. 1979 ISBN 90-277-1020-2
142. C. Perelman, *Justice, Law and Argument.* Essays on Moral and Legal Reasoning. Translated from French and German. With an Introduction by H.J. Berman. 1980 ISBN 90-277-1089-9; Pb 90-277-1090-2
143. S. Kanger and S. Öhman (eds.), *Philosophy and Grammar.* Papers on the Occasion of the Quincentennial of Uppsala University. 1981 ISBN 90-277-1091-0
144. T. Pawlowski, *Concept Formation in the Humanities and the Social Sciences.* 1980 ISBN 90-277-1096-1
145. J. Hintikka, D. Gruender and E. Agazzi (eds.), *Theory Change, Ancient Axiomatics and Galileo's Methodology.* Proceedings of the 1978 Pisa Conference on the History and Philosophy of Science, Volume I. 1981 ISBN 90-277-1126-7
146. J. Hintikka, D. Gruender and E. Agazzi (eds.), *Probabilistic Thinking, Thermodynamics, and the Interaction of the History and Philosophy of Science.* Proceedings of the 1978 Pisa Conference on the History and Philosophy of Science, Volume II. 1981 ISBN 90-277-1127-5
147. U. Mönnich (ed.), *Aspects of Philosophical Logic.* Some Logical Forays into Central Notions of Linguistics and Philosophy. 1981 ISBN 90-277-1201-8
148. D. M. Gabbay, *Semantical Investigations in Heyting's Intuitionistic Logic.* 1981 ISBN 90-277-1202-6
149. E. Agazzi (ed.), *Modern Logic – A Survey.* Historical, Philosophical, and Mathematical Aspects of Modern Logic and Its Applications. 1981 ISBN 90-277-1137-2
150. A. F. Parker-Rhodes, *The Theory of Indistinguishables.* A Search for Explanatory Principles below the Level of Physics. 1981 ISBN 90-277-1214-X
151. J. C. Pitt, *Pictures, Images, and Conceptual Change.* An Analysis of Wilfrid Sellars' Philosophy of Science. 1981 ISBN 90-277-1276-X; Pb 90-277-1277-8
152. R. Hilpinen (ed.), *New Studies in Deontic Logic.* Norms, Actions, and the Foundations of Ethics. 1981 ISBN 90-277-1278-6; Pb 90-277-1346-4
153. C. Dilworth, *Scientific Progress.* A Study Concerning the Nature of the Relation between Successive Scientific Theories. 3rd rev. ed., 1994 ISBN 0-7923-2487-0; Pb 0-7923-2488-9
154. D. Woodruff Smith and R. McIntyre, *Husserl and Intentionality.* A Study of Mind, Meaning, and Language. 1982 ISBN 90-277-1392-8; Pb 90-277-1730-3
155. R. J. Nelson, *The Logic of Mind.* 2nd. ed., 1989 ISBN 90-277-2819-4; Pb 90-277-2822-4
156. J. F. A. K. van Benthem, *The Logic of Time.* A Model-Theoretic Investigation into the Varieties of Temporal Ontology, and Temporal Discourse. 1983; 2nd ed., 1991 ISBN 0-7923-1081-0
157. R. Swinburne (ed.), *Space, Time and Causality.* 1983 ISBN 90-277-1437-1
158. E. T. Jaynes, *Papers on Probability, Statistics and Statistical Physics.* Ed. by R. D. Rozenkrantz. 1983 ISBN 90-277-1448-7; Pb (1989) 0-7923-0213-3
159. T. Chapman, *Time: A Philosophical Analysis.* 1982 ISBN 90-277-1465-7
160. E. N. Zalta, *Abstract Objects.* An Introduction to Axiomatic Metaphysics. 1983 ISBN 90-277-1474-6
161. S. Harding and M. B. Hintikka (eds.), *Discovering Reality.* Feminist Perspectives on Epistemology, Metaphysics, Methodology, and Philosophy of Science. 1983 ISBN 90-277-1496-7; Pb 90-277-1538-6
162. M. A. Stewart (ed.), *Law, Morality and Rights.* 1983 ISBN 90-277-1519-X

163. D. Mayr and G. Süssmann (eds.), *Space, Time, and Mechanics.* Basic Structures of a Physical Theory. 1983 ISBN 90-277-1525-4
164. D. Gabbay and F. Guenthner (eds.), *Handbook of Philosophical Logic.* Vol. I: Elements of Classical Logic. 1983 ISBN 90-277-1542-4
165. D. Gabbay and F. Guenthner (eds.), *Handbook of Philosophical Logic.* Vol. II: Extensions of Classical Logic. 1984 ISBN 90-277-1604-8
166. D. Gabbay and F. Guenthner (eds.), *Handbook of Philosophical Logic.* Vol. III: Alternative to Classical Logic. 1986 ISBN 90-277-1605-6
167. D. Gabbay and F. Guenthner (eds.), *Handbook of Philosophical Logic.* Vol. IV: Topics in the Philosophy of Language. 1989 ISBN 90-277-1606-4
168. A. J. I. Jones, *Communication and Meaning.* An Essay in Applied Modal Logic. 1983 ISBN 90-277-1543-2
169. M. Fitting, *Proof Methods for Modal and Intuitionistic Logics.* 1983 ISBN 90-277-1573-4
170. J. Margolis, *Culture and Cultural Entities.* Toward a New Unity of Science. 1984 ISBN 90-277-1574-2
171. R. Tuomela, *A Theory of Social Action.* 1984 ISBN 90-277-1703-6
172. J. J. E. Gracia, E. Rabossi, E. Villanueva and M. Dascal (eds.), *Philosophical Analysis in Latin America.* 1984 ISBN 90-277-1749-4
173. P. Ziff, *Epistemic Analysis.* A Coherence Theory of Knowledge. 1984 ISBN 90-277-1751-7
174. P. Ziff, *Antiaesthetics.* An Appreciation of the Cow with the Subtile Nose. 1984 ISBN 90-277-1773-7
175. W. Balzer, D. A. Pearce, and H.-J. Schmidt (eds.), *Reduction in Science.* Structure, Examples, Philosophical Problems. 1984 ISBN 90-277-1811-3
176. A. Peczenik, L. Lindahl and B. van Roermund (eds.), *Theory of Legal Science.* Proceedings of the Conference on Legal Theory and Philosophy of Science (Lund, Sweden, December 1983). 1984 ISBN 90-277-1834-2
177. I. Niiniluoto, *Is Science Progressive?* 1984 ISBN 90-277-1835-0
178. B. K. Matilal and J. L. Shaw (eds.), *Analytical Philosophy in Comparative Perspective.* Exploratory Essays in Current Theories and Classical Indian Theories of Meaning and Reference. 1985 ISBN 90-277-1870-9
179. P. Kroes, *Time: Its Structure and Role in Physical Theories.* 1985 ISBN 90-277-1894-6
180. J. H. Fetzer, *Sociobiology and Epistemology.* 1985 ISBN 90-277-2005-3; Pb 90-277-2006-1
181. L. Haaparanta and J. Hintikka (eds.), *Frege Synthesized.* Essays on the Philosophical and Foundational Work of Gottlob Frege. 1986 ISBN 90-277-2126-2
182. M. Detlefsen, *Hilbert's Program.* An Essay on Mathematical Instrumentalism. 1986 ISBN 90-277-2151-3
183. J. L. Golden and J. J. Pilotta (eds.), *Practical Reasoning in Human Affairs.* Studies in Honor of Chaim Perelman. 1986 ISBN 90-277-2255-2
184. H. Zandvoort, *Models of Scientific Development and the Case of Nuclear Magnetic Resonance.* 1986 ISBN 90-277-2351-6
185. I. Niiniluoto, *Truthlikeness.* 1987 ISBN 90-277-2354-0
186. W. Balzer, C. U. Moulines and J. D. Sneed, *An Architectonic for Science.* The Structuralist Program. 1987 ISBN 90-277-2403-2
187. D. Pearce, *Roads to Commensurability.* 1987 ISBN 90-277-2414-8
188. L. M. Vaina (ed.), *Matters of Intelligence.* Conceptual Structures in Cognitive Neuroscience. 1987 ISBN 90-277-2460-1

189. H. Siegel, *Relativism Refuted.* A Critique of Contemporary Epistemological Relativism. 1987 ISBN 90-277-2469-5
190. W. Callebaut and R. Pinxten, *Evolutionary Epistemology.* A Multiparadigm Program, with a Complete Evolutionary Epistemology Bibliograph. 1987 ISBN 90-277-2582-9
191. J. Kmita, *Problems in Historical Epistemology.* 1988 ISBN 90-277-2199-8
192. J. H. Fetzer (ed.), *Probability and Causality.* Essays in Honor of Wesley C. Salmon, with an Annotated Bibliography. 1988 ISBN 90-277-2607-8; Pb 1-5560-8052-2
193. A. Donovan, L. Laudan and R. Laudan (eds.), *Scrutinizing Science.* Empirical Studies of Scientific Change. 1988 ISBN 90-277-2608-6
194. H.R. Otto and J.A. Tuedio (eds.), *Perspectives on Mind.* 1988 ISBN 90-277-2640-X
195. D. Batens and J.P. van Bendegem (eds.), *Theory and Experiment.* Recent Insights and New Perspectives on Their Relation. 1988 ISBN 90-277-2645-0
196. J. Österberg, *Self and Others.* A Study of Ethical Egoism. 1988 ISBN 90-277-2648-5
197. D.H. Helman (ed.), *Analogical Reasoning.* Perspectives of Artificial Intelligence, Cognitive Science, and Philosophy. 1988 ISBN 90-277-2711-2
198. J. Woleński, *Logic and Philosophy in the Lvov-Warsaw School.* 1989 ISBN 90-277-2749-X
199. R. Wójcicki, *Theory of Logical Calculi.* Basic Theory of Consequence Operations. 1988 ISBN 90-277-2785-6
200. J. Hintikka and M.B. Hintikka, *The Logic of Epistemology and the Epistemology of Logic.* Selected Essays. 1989 ISBN 0-7923-0040-8; Pb 0-7923-0041-6
201. E. Agazzi (ed.), *Probability in the Sciences.* 1988 ISBN 90-277-2808-9
202. M. Meyer (ed.), *From Metaphysics to Rhetoric.* 1989 ISBN 90-277-2814-3
203. R.L. Tieszen, *Mathematical Intuition.* Phenomenology and Mathematical Knowledge. 1989 ISBN 0-7923-0131-5
204. A. Melnick, *Space, Time, and Thought in Kant.* 1989 ISBN 0-7923-0135-8
205. D.W. Smith, *The Circle of Acquaintance.* Perception, Consciousness, and Empathy. 1989 ISBN 0-7923-0252-4
206. M.H. Salmon (ed.), *The Philosophy of Logical Mechanism.* Essays in Honor of Arthur W. Burks. With his Responses, and with a Bibliography of Burk's Work. 1990 ISBN 0-7923-0325-3
207. M. Kusch, *Language as Calculus vs. Language as Universal Medium.* A Study in Husserl, Heidegger, and Gadamer. 1989 ISBN 0-7923-0333-4
208. T.C. Meyering, *Historical Roots of Cognitive Science.* The Rise of a Cognitive Theory of Perception from Antiquity to the Nineteenth Century. 1989 ISBN 0-7923-0349-0
209. P. Kosso, *Observability and Observation in Physical Science.* 1989 ISBN 0-7923-0389-X
210. J. Kmita, *Essays on the Theory of Scientific Cognition.* 1990 ISBN 0-7923-0441-1
211. W. Sieg (ed.), *Acting and Reflecting.* The Interdisciplinary Turn in Philosophy. 1990 ISBN 0-7923-0512-4
212. J. Karpiński, *Causality in Sociological Research.* 1990 ISBN 0-7923-0546-9
213. H.A. Lewis (ed.), *Peter Geach: Philosophical Encounters.* 1991 ISBN 0-7923-0823-9
214. M. Ter Hark, *Beyond the Inner and the Outer.* Wittgenstein's Philosophy of Psychology. 1990 ISBN 0-7923-0850-6
215. M. Gosselin, *Nominalism and Contemporary Nominalism.* Ontological and Epistemological Implications of the Work of W.V.O. Quine and of N. Goodman. 1990 ISBN 0-7923-0904-9
216. J.H. Fetzer, D. Shatz and G. Schlesinger (eds.), *Definitions and Definability.* Philosophical Perspectives. 1991 ISBN 0-7923-1046-2
217. E. Agazzi and A. Cordero (eds.), *Philosophy and the Origin and Evolution of the Universe.* 1991 ISBN 0-7923-1322-4

218. M. Kusch, *Foucault's Strata and Fields*. An Investigation into Archaeological and Genealogical Science Studies. 1991 ISBN 0-7923-1462-X
219. C.J. Posy, *Kant's Philosophy of Mathematics*. Modern Essays. 1992 ISBN 0-7923-1495-6
220. G. Van de Vijver, *New Perspectives on Cybernetics*. Self-Organization, Autonomy and Connectionism. 1992 ISBN 0-7923-1519-7
221. J.C. Nyíri, *Tradition and Individuality*. Essays. 1992 ISBN 0-7923-1566-9
222. R. Howell, *Kant's Transcendental Deduction*. An Analysis of Main Themes in His Critical Philosophy. 1992 ISBN 0-7923-1571-5
223. A. García de la Sienra, *The Logical Foundations of the Marxian Theory of Value*. 1992 ISBN 0-7923-1778-5
224. D.S. Shwayder, *Statement and Referent*. An Inquiry into the Foundations of Our Conceptual Order. 1992 ISBN 0-7923-1803-X
225. M. Rosen, *Problems of the Hegelian Dialectic*. Dialectic Reconstructed as a Logic of Human Reality. 1993 ISBN 0-7923-2047-6
226. P. Suppes, *Models and Methods in the Philosophy of Science: Selected Essays*. 1993 ISBN 0-7923-2211-8
227. R. M. Dancy (ed.), *Kant and Critique: New Essays in Honor of W. H. Werkmeister*. 1993 ISBN 0-7923-2244-4
228. J. Woleński (ed.), *Philosophical Logic in Poland*. 1993 ISBN 0-7923-2293-2
229. M. De Rijke (ed.), *Diamonds and Defaults*. Studies in Pure and Applied Intensional Logic. 1993 ISBN 0-7923-2342-4
230. B.K. Matilal and A. Chakrabarti (eds.), *Knowing from Words*. Western and Indian Philosophical Analysis of Understanding and Testimony. 1994 ISBN 0-7923-2345-9
231. S.A. Kleiner, *The Logic of Discovery*. A Theory of the Rationality of Scientific Research. 1993 ISBN 0-7923-2371-8
232. R. Festa, *Optimum Inductive Methods*. A Study in Inductive Probability, Bayesian Statistics, and Verisimilitude. 1993 ISBN 0-7923-2460-9
233. P. Humphreys (ed.), *Patrick Suppes: Scientific Philosopher*. Vol. 1: Probability and Probabilistic Causality. 1994 ISBN 0-7923-2552-4
234. P. Humphreys (ed.), *Patrick Suppes: Scientific Philosopher*. Vol. 2: Philosophy of Physics, Theory Structure, and Measurement Theory. 1994 ISBN 0-7923-2553-2
235. P. Humphreys (ed.), *Patrick Suppes: Scientific Philosopher*. Vol. 3: Language, Logic, and Psychology. 1994 ISBN 0-7923-2862-0
     Set ISBN (Vols 233–235) 0-7923-2554-0
236. D. Prawitz and D. Westerståhl (eds.), *Logic and Philosophy of Science in Uppsala*. Papers from the 9th International Congress of Logic, Methodology, and Philosophy of Science. 1994 ISBN 0-7923-2702-0
237. L. Haaparanta (ed.), *Mind, Meaning and Mathematics*. Essays on the Philosophical Views of Husserl and Frege. 1994 ISBN 0-7923-2703-9
238. J. Hintikka (ed.), *Aspects of Metaphor*. 1994 ISBN 0-7923-2786-1
239. B. McGuinness and G. Oliveri (eds.), *The Philosophy of Michael Dummett*. With Replies from Michael Dummett. 1994 ISBN 0-7923-2804-3
240. D. Jamieson (ed.), *Language, Mind, and Art*. Essays in Appreciation and Analysis, In Honor of Paul Ziff. 1994 ISBN 0-7923-2810-8
241. G. Preyer, F. Siebelt and A. Ulfig (eds.), *Language, Mind and Epistemology*. On Donald Davidson's Philosophy. 1994 ISBN 0-7923-2811-6
242. P. Ehrlich (ed.), *Real Numbers, Generalizations of the Reals, and Theories of Continua*. 1994 ISBN 0-7923-2689-X

243. G. Debrock and M. Hulswit (eds.), *Living Doubt.* Essays concerning the epistemology of Charles Sanders Peirce. 1994 ISBN 0-7923-2898-1
244. J. Srzednicki, *To Know or Not to Know.* Beyond Realism and Anti-Realism. 1994 ISBN 0-7923-2909-0
245. R. Egidi (ed.), *Wittgenstein: Mind and Language.* 1995 ISBN 0-7923-3171-0
246. A. Hyslop, *Other Minds.* 1995 ISBN 0-7923-3245-8
247. L. Pólos and M. Masuch (eds.), *Applied Logic: How, What and Why.* Logical Approaches to Natural Language. 1995 ISBN 0-7923-3432-9
248. M. Krynicki, M. Mostowski and L.M. Szczerba (eds.), *Quantifiers: Logics, Models and Computation.* Volume One: Surveys. 1995 ISBN 0-7923-3448-5
249. M. Krynicki, M. Mostowski and L.M. Szczerba (eds.), *Quantifiers: Logics, Models and Computation.* Volume Two: Contributions. 1995 ISBN 0-7923-3449-3
Set ISBN (Vols 248 + 249) 0-7923-3450-7
250. R.A. Watson, *Representational Ideas from Plato to Patricia Churchland.* 1995 ISBN 0-7923-3453-1
251. J. Hintikka (ed.), *From Dedekind to Gödel.* Essays on the Development of the Foundations of Mathematics. 1995 ISBN 0-7923-3484-1
252. A. Wiśniewski, *The Posing of Questions.* Logical Foundations of Erotetic Inferences. 1995 ISBN 0-7923-3637-2
253. J. Peregrin, *Doing Worlds with Words.* Formal Semantics without Formal Metaphysics. 1995 ISBN 0-7923-3742-5
254. I.A. Kieseppä, *Truthlikeness for Multidimensional, Quantitative Cognitive Problems.* 1996 ISBN 0-7923-4005-1
255. P. Hugly and C. Sayward: *Intensionality and Truth.* An Essay on the Philosophy of A.N. Prior. 1996 ISBN 0-7923-4119-8
256. L. Hankinson Nelson and J. Nelson (eds.): *Feminism, Science, and the Philosophy of Science.* 1997 ISBN 0-7923-4162-7
257. P.I. Bystrov and V.N. Sadovsky (eds.): *Philosophical Logic and Logical Philosophy.* Essays in Honour of Vladimir A. Smirnov. 1996 ISBN 0-7923-4270-4
258. Å.E. Andersson and N-E. Sahlin (eds.): *The Complexity of Creativity.* 1996 ISBN 0-7923-4346-8
259. M.L. Dalla Chiara, K. Doets, D. Mundici and J. van Benthem (eds.): *Logic and Scientific Methods.* Volume One of the Tenth International Congress of Logic, Methodology and Philosophy of Science, Florence, August 1995. 1997 ISBN 0-7923-4383-2
260. M.L. Dalla Chiara, K. Doets, D. Mundici and J. van Benthem (eds.): *Structures and Norms in Science.* Volume Two of the Tenth International Congress of Logic, Methodology and Philosophy of Science, Florence, August 1995. 1997 ISBN 0-7923-4384-0
Set ISBN (Vols 259 + 260) 0-7923-4385-9
261. A. Chakrabarti: *Denying Existence.* The Logic, Epistemology and Pragmatics of Negative Existentials and Fictional Discourse. 1997 ISBN 0-7923-4388-3
262. A. Biletzki: *Talking Wolves.* Thomas Hobbes on the Language of Politics and the Politics of Language. 1997 ISBN 0-7923-4425-1
263. D. Nute (ed.): *Defeasible Deontic Logic.* 1997 ISBN 0-7923-4630-0
264. U. Meixner: *Axiomatic Formal Ontology.* 1997 ISBN 0-7923-4747-X
265. I. Brinck: *The Indexical 'I'.* The First Person in Thought and Language. 1997 ISBN 0-7923-4741-2
266. G. Hölmström-Hintikka and R. Tuomela (eds.): *Contemporary Action Theory.* Volume 1: Individual Action. 1997 ISBN 0-7923-4753-6; Set: 0-7923-4754-4

267. G. Hölmström-Hintikka and R. Tuomela (eds.): *Contemporary Action Theory.* Volume 2: Social Action. 1997 ISBN 0-7923-4752-8; Set: 0-7923-4754-4
268. B.-C. Park: *Phenomenological Aspects of Wittgenstein's Philosophy.* 1998 ISBN 0-7923-4813-3
269. J. Paśniczek: *The Logic of Intentional Objects.* A Meinongian Version of Classical Logic. 1998 Hb ISBN 0-7923-4880-X; Pb ISBN 0-7923-5578-4
270. P.W. Humphreys and J.H. Fetzer (eds.): *The New Theory of Reference.* Kripke, Marcus, and Its Origins. 1998 ISBN 0-7923-4898-2
271. K. Szaniawski, A. Chmielewski and J. Woleński (eds.): *On Science, Inference, Information and Decision Making.* Selected Essays in the Philosophy of Science. 1998 ISBN 0-7923-4922-9
272. G.H. von Wright: *In the Shadow of Descartes.* Essays in the Philosophy of Mind. 1998 ISBN 0-7923-4992-X
273. K. Kijania-Placek and J. Woleński (eds.): *The Lvov–Warsaw School and Contemporary Philosophy.* 1998 ISBN 0-7923-5105-3
274. D. Dedrick: *Naming the Rainbow.* Colour Language, Colour Science, and Culture. 1998 ISBN 0-7923-5239-4
275. L. Albertazzi (ed.): *Shapes of Forms.* From Gestalt Psychology and Phenomenology to Ontology and Mathematics. 1999 ISBN 0-7923-5246-7
276. P. Fletcher: *Truth, Proof and Infinity.* A Theory of Constructions and Constructive Reasoning. 1998 ISBN 0-7923-5262-9
277. M. Fitting and R.L. Mendelsohn (eds.): *First-Order Modal Logic.* 1998 Hb ISBN 0-7923-5334-X; Pb ISBN 0-7923-5335-8
278. J.N. Mohanty: *Logic, Truth and the Modalities from a Phenomenological Perspective.* 1999 ISBN 0-7923-5550-4
279. T. Placek: *Mathematical Intiutionism and Intersubjectivity.* A Critical Exposition of Arguments for Intuitionism. 1999 ISBN 0-7923-5630-6
280. A. Cantini, E. Casari and P. Minari (eds.): *Logic and Foundations of Mathematics.* 1999 ISBN 0-7923-5659-4 set ISBN 0-7923-5867-8
281. M.L. Dalla Chiara, R. Giuntini and F. Laudisa (eds.): *Language, Quantum, Music.* 1999 ISBN 0-7923-5727-2; set ISBN 0-7923-5867-8
282. R. Egidi (ed.): *In Search of a New Humanism.* The Philosophy of Georg Hendrik von Wright. 1999 ISBN 0-7923-5810-4
283. F. Vollmer: *Agent Causality.* 1999 ISBN 0-7923-5848-1
284. J. Peregrin (ed.): *Truth and Its Nature (if Any).* 1999 ISBN 0-7923-5865-1
285. M. De Caro (ed.): *Interpretations and Causes.* New Perspectives on Donald Davidson's Philosophy. 1999 ISBN 0-7923-5869-4
286. R. Murawski: *Recursive Functions and Metamathematics.* Problems of Completeness and Decidability, Gödel's Theorems. 1999 ISBN 0-7923-5904-6
287. T.A.F. Kuipers: *From Instrumentalism to Constructive Realism.* On Some Relations between Confirmation, Empirical Progress, and Truth Approximation. 2000 ISBN 0-7923-6086-9
288. G. Holmström-Hintikka (ed.): *Medieval Philosophy and Modern Times.* 2000 ISBN 0-7923-6102-4
289. E. Grosholz and H. Breger (eds.): *The Growth of Mathematical Knowledge.* 2000 ISBN 0-7923-6151-2

290. G. Sommaruga: *History and Philosophy of Constructive Type Theory*. 2000 ISBN 0-7923-6180-6
291. J. Gasser (ed.): *A Boole Anthology*. Recent and Classical Studies in the Logic of George Boole. 2000 ISBN 0-7923-6380-9
292. V.F. Hendricks, S.A. Pedersen and K.F. Jørgensen (eds.): *Proof Theory*. History and Philosophical Significance. 2000 ISBN 0-7923-6544-5
293. W.L. Craig: *The Tensed Theory of Time*. A Critical Examination. 2000 ISBN 0-7923-6634-4
294. W.L. Craig: *The Tenseless Theory of Time*. A Critical Examination. 2000 ISBN 0-7923-6635-2
295. L. Albertazzi (ed.): *The Dawn of Cognitive Science*. Early European Contributors. 2001 ISBN 0-7923-6799-5

Previous volumes are still available.

KLUWER ACADEMIC PUBLISHERS – DORDRECHT / BOSTON / LONDON

GPSR Compliance
The European Union's (EU) General Product Safety Regulation (GPSR) is a set of rules that requires consumer products to be safe and our obligations to ensure this.

If you have any concerns about our products, you can contact us on

ProductSafety@springernature.com

In case Publisher is established outside the EU, the EU authorized representative is:

Springer Nature Customer Service Center GmbH
Europaplatz 3
69115 Heidelberg, Germany

www.ingramcontent.com/pod-product-compliance
Lightning Source LLC
LaVergne TN
LVHW011907200926
848484LV00002B/554
* 9 7 8 0 7 9 2 3 6 7 9 9 4 *